The Joshua Tree Stargazing Guide
Fall 2026–Summer 2027

The Joshua Tree Stargazing Guide

Fall 2026–Summer 2027

DAGMAR MIURA
LOS ANGELES

The Joshua Tree Stargazing Guide Fall 2026–Summer 2027

First published 2026

ISBN: 979-8-89195-086-3

Contents

Ancestral Puebloan rock art at Chaco Canyon, New Mexico. The Chacoan people were keen astronomers, and their connections to the night sky are visible in the orientation of their architecture and in the art they made. Here the artist might have been depicting the supernova of 1054 CE, so bright that it was visible in daytime. The position of the spiked object near the crescent is the view the artist would have had of the supernova and the moon from the canyon on July 5, 1054.

Introduction

Joshua Tree, in the Mojave Desert, is an ideal place to view the dark sky. Most nights are free of clouds, and once your eyes adjust to the dark, you'll see the full breathtaking majesty of the Milky Way and its countless stars.

It takes about ten minutes for your eyes to start to adjust to the dark, and in twenty minutes you'll see even more stars. Looking at a phone screen or any light except red light will instantly disrupt your night vision, so leave the phone inside.

TELESCOPE VERSUS BINOCULAR ASTRONOMY

A telescope can provide close-up views of dark sky objects, but using a telescope requires time spent aiming it and adjusting the view. If you're comfortable using one, it can bring a lot of enjoyment, and has greater magnification than binoculars. A more casual way to look at night sky objects is with astronomy binoculars. You can explore the sky without much work to aim at your targets, and you can quickly pan around the sky.

The disadvantage to binocular astronomy is that it's impossible to hold binoculars perfectly still, so objects in the sky move around. But your eyes get used to it, and you'll soon be able to track objects with ease.

With a telescope or with a pair of astronomy binoculars, you'll easily be able to see craters on the moon, the Andromeda Galaxy, the pinpoints of Jupiter's moons, and objects like globular star clusters and nebulae.

QUALITY OF SEEING

Except on the rare nights when it's overcast at Joshua Tree, the quality of the darkness, what astronomy buffs call "seeing," is superb. Compared to urban areas and anywhere east of the Rockies, the desert air has much less moisture and dust in it to diffuse the light of night-sky objects.

Using This Guide

The daily guide pages that follow, starting on page 21, show the timing of sunset and twilight for each day, as well as what to expect from the moon and what to look for in the dark sky.

TWILIGHT

Once the sun has set, **civil twilight** lasts for 20 or 30 minutes. It is technically defined as lasting from sunset until the sun is 6 degrees below the horizon. The stars aren't yet visible, but the sky at Joshua Tree is dramatic, with the warm colors of sunset in the west over Big Bear, and the pink sky in the east fading as the indigo shadow of the earth rises from the eastern horizon.

Nautical twilight is when the first stars begin to appear in the sky. It is technically defined as when the sun is 6 degrees to 12 degrees below the horizon. Watch as the first bright stars appear, gradually followed by the dimmer stars. This is also the time to start watching for meteors.

The next phase is **astronomical twilight,** which lasts until the sun is 18 degrees below the horizon. After that, the sky is as dark as it is going to be for the night. Most nights the Milky Way will become visible once your eyes have adjusted to the dark.

In the pages that follow, the times listed for twilight are for nautical twilight, when the stars begin to appear in the evening, and in the morning when daylight starts to obscure the stars.

FINDING OBJECTS IN THE NIGHT SKY

Constellations are named groupings of stars. Not every constellation and planet is visible every night. Five planets are visible at different times, and constellations appear seasonally. The ones that are in the far northern sky, like the Great Bear (Big Dipper) and Cassiopeia, are visible most of the year. The ones near the ecliptic—the celestial equator, above the earth's equator—are visible in specific months of the year.

The twelve astrological constellations or signs are arranged along the ecliptic, where the sun passes through them at different times of year. The sun is always near the ecliptic, so constellations that are far from that line, like the Great Bear, are never visited by the path of the sun. When the sun is in Gemini in June, you're not going to see Gemini's stars, as they are invisible next to the sun. But six months later, in November, Gemini is opposite the sun, so its stars Castor and Pollux are visible for most of the night.

The planets follow regular patterns, but they aren't connected to the seasons. The planets are always near the ecliptic, the same as the sun, so you'll never see a planet in the far northern sky. When Mercury and Venus are visible, they are always near the sun, so they are seen just after sunset or before sunrise. Venus is often easy to see, but Mercury is a much harder target. When Mars is visible, it looks distinctly rusty red, even without a telescope or binoculars. Jupiter is usually brighter than most stars, and Saturn is about the same brightness as a bright star. If you're not sure if you're looking at a planet or a star, remember that stars are emitting light, so they twinkle and shimmer. Planets are reflecting sunlight, the same way the moon does, so their light is steady, flat, and doesn't twinkle.

This guide has a few examples of objects to look for in the dark sky. There are a variety of websites that can tell you a lot more about what's up there, what you're seeing, and what to look for. If an object is visible, you only need to know two things to locate it: which direction it is in, and how high it is in the sky.

Finding North and Other Directions

The only star in the sky that doesn't move as the earth revolves is **Polaris.** That's because it's directly over the earth's north pole. Polaris isn't a bright star, but it's alone in its part of the sky, and it's extremely useful for direction finding because it's always due north. For how to find Polaris in the night sky, see "The Great Bear" on page 10.

Altitude of Dark Sky Objects

The way to describe how high an object is in the sky, which astronomers call altitude or elevation, is with degrees. Zero degrees is the horizon, and straight overhead is 90 degrees. It's a logical system: if you point to the horizon, and with your other hand point straight up, your arms form a 90-degree right angle. From Joshua Tree, Polaris is 34 degrees above the horizon. Because of the way the lines of latitude on the earth

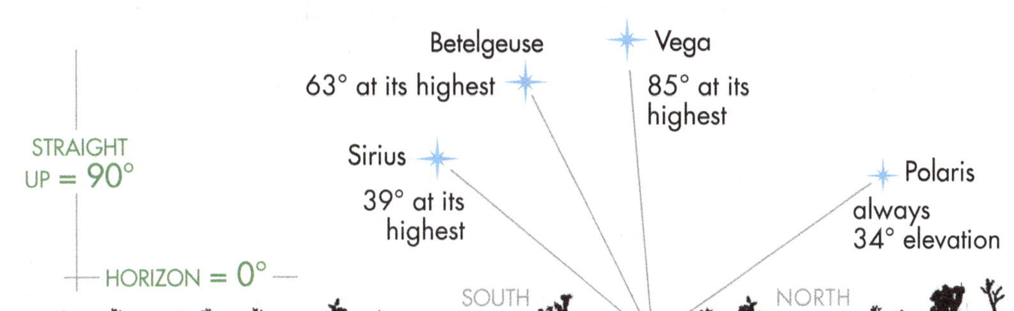

are set up, the altitude of Polaris is always the same as the latitude of where you're standing. So Joshua Tree is at 34 degrees latitude.

Things That Move in the Sky

METEORS

Meteors appear as fleeting bright streaks in the sky. Mostly they are particles of dust and grit from the asteroid belt, but the meteor showers that happen through the year are the dust trails of comets.

Sometimes the streaks of light look lumpy and seem to change direction slightly. That's an optical illusion; they're moving fast and travel in straight lines when they hit the atmosphere. Meteors can look white, yellow, even reddish or greenish. However they look, meteors happen fast, and you won't mistake them for anything else.

A rare kind of meteor is called a fireball, so named because it burns brighter than any star in the sky. Even rarer is a meteor procession, where a meteor breaks up into several distinct streaks in the sky.

More about the meteor showers that happen through the year is on page 6.

AIRCRAFT

Airplanes are easy to identify because their lights usually flicker red or white. Hundreds of commercial jets headed for LAX and other Los Angeles–area airports ply the flight paths north and south of Joshua Tree.

Because the Twentynine Palms Marine base is only a few miles away, sometimes you'll see military aircraft in the sky. They are usually much noisier than civilian aircraft. The Marines use Ospreys, which look like helicopters with wings and two rotors. They're loud and usually travel in groups of two or more. Triangular aircraft like the F117 and the B2 are spotted less frequently. The triangular layout of the lights is eerie, and sometimes the military flies with no lights at all, but you'll know they're aircraft because you'll hear the jet engines.

SATELLITES

When you're watching the whole sky without a telescope or binoculars, especially soon after evening twilight or before dawn, you might see satellites in earth orbit. They are usually in the northern sky and often look about as bright as a star. They move at a graceful constant speed. If you watch them long enough, sometimes they fade out when they enter the earth's shadow, no longer reflecting sunlight.

Some moving objects seem to brighten and dim over the course of a few seconds as they travel. This is usually **space junk,** rotating or tumbling in space and reflecting more and less sunlight in a regular cycle.

Occasionally a satellite will catch the sun in a way that makes it brighter than all the stars for a few seconds. These are called **satellite flares,** and there are several websites that predict them, including Heavens Above (www.heavens-above.com).

Another dramatic moving object sometimes seen in the sky is a **train of satellites.** A dozen or more points of light cross the sky in a line. This happens soon after Starlink or other satellite fleets are launched in a group and begin to separate.

They get dimmer and disappear a few days later as they move to a higher orbit. Astronomers feel vexed as thousands of these communications satellites in low earth orbit move through their telescope views and leak radio telescope–interfering radio waves.

Equally rare but mind-blowing when it's visible is a **rocket launch** from Vandenberg Space Force Base, on the Pacific coast north of Los Angeles. It's 240 miles away, but if the skies are clear and the rocket is headed to space, it might be visible from Joshua Tree. The most dramatic launches are right after sunset, when the sky is dark but the rocket's exhaust plume is illuminated by sunlight. A website with a schedule of upcoming launches is https://spaceflightnow.com/launch-schedule; use the QR code to check on your phone.

evening rocket launch from Vandenberg as seen from Joshua Tree

UNIDENTIFIED AERIAL PHENOMENA

I'm not going to say it's aliens, but ... objects that don't have flashing lights and change direction, objects that speed up and slow down and make no noise—those are either high-tech drones or unidentified aerial phenomena, or UAPs, another term for UFOs.

The Morongo Basin around Joshua Tree and the nearby Colorado Desert have been the site of many encounters with flying saucers and aliens. George Adamski met the blond humanoid Orthon from planet Venus near Desert Center in 1952. George Van Tassel talked to the alien Ashtar at Giant Rock, near Landers, and later boarded Ashtar's saucer, also starting in 1952. In the 1950s Van Tassel started holding UFO conventions at Giant Rock that attracted thousands of people. Those ended in the 1970s, but you can see Van Tassel's architectural work at the Integratron in Landers, and feel the spacey residual vibes at Giant Rock.

UFO conventions were held at Giant Rock, starting in the 1950s.

Selected Objects in the Night Sky

METEOR SHOWERS

The Perseids are the summer blockbuster that attracts crowds to the dark skies of Joshua Tree in August, but meteors can be seen any night of the year in lower concentrations. The best way to watch for meteors is without a telescope or binoculars. Get comfy reclining in a chair and take in a view of as much of the sky as you can.

Watching for meteors takes patience. Many of the fireballs will appear in your peripheral view rather than right where you're looking. Even during a strong shower, you might only see one or two meteors in half an hour. Usually more meteors are visible in the hours after midnight. The exception is the Draconids in early October, when viewing is better early in the evening.

The dates listed below are peak nights for each of the major showers, but you'll still see meteors for several nights before and after these nights:

Shower	Nights	Moon Interference
Draconids	October 7	excellent viewing conditions with dark skies all night
Orionids	October 20–22	moonlight interferes until the moon sets, around 2:30 a.m.
Taurids	November 4	poor conditions after the moon rises, around 2:30 a.m.
Leonids	November 16–18	good viewing conditions after the moon sets, around 11 p.m.
Geminids	December 12–14	good viewing conditions after the moon sets, around 8 p.m.
Ursids	December 21	poor conditions as moonlight washes out the shower

Quadrantids	January 2–4	excellent viewing conditions with dark skies all night
Lyrids	April 21–23	poor conditions as moonlight washes out the shower
Eta Aquarids	May 5–7	excellent viewing conditions with dark skies all night
Delta Aquarids	July 27–29	poor conditions after the moon rises, around 12:30 a.m.
Perseids	August 11–13	poor conditions as moonlight washes out the shower

THE MILKY WAY

The night sky is much darker when the moon isn't out, and the Milky Way is visible on any dark night throughout the year. It appears as a hazy band that stretches across the entire sky. At first it might look like vague clouds, but as your eyes adjust to the darkness, you'll see it as a river of light. Look at the band with binoculars and you'll see the density of the stars generating all that light.

The Milky Way is our galaxy, shaped like a spiral disk. When you see the band in the sky, you're looking at the flat part of the disk. The darker patches in the band are clouds of dust obscuring the stars that are farther away.

THE MOON

In the Mojave Desert, when the moon is full or close to it, it's so bright that you can read a book outside all night. It illuminates the landscape, and plants and structures cast clear shadows.

The moon isn't visible every night—it rises and sets at a different time every day, following a 29.5-day cycle. Starting from the **new moon,** when it's close to the sun and not illuminated, it takes about a week to reach the **first quarter,** when it's 50 percent illuminated and looks like half a pie. In another week is **full moon,** when the moon is opposite the sun and visible from sunset to sunrise. A week later is the **last quarter,** also called the third quarter, when the moon looks like half a pie again, and after the fourth week it's back to new moon.

From new moon to full moon, the moon is **waxing,** or growing, and from full moon to new moon, it is **waning.** Other terms used to describe the moon in the sky are **crescent,** meaning it appears less than half illuminated, and **gibbous,** meaning it is more than half illuminated.

The new moon as well as the crescent moons on the days near the new moon aren't usually visible because the moon rises and sets close to the sun. Around the first quarter, the moon is high in the sky at sunset and sets around midnight. Full moon rises at sunset and sets at daybreak. The third-quarter moon rises around midnight and is high in the sky by sunrise.

When the sun is up, the moon is sometimes visible in the daytime sky, especially early or late in the day. This happens more commonly in the desert because there is little haze in the air to obscure it. See the illustration for an overview of when the moon is visible in the night sky. The pages for each day of the year starting on page 21 include the moon's rise and set times as well as an image of the moon's phase.

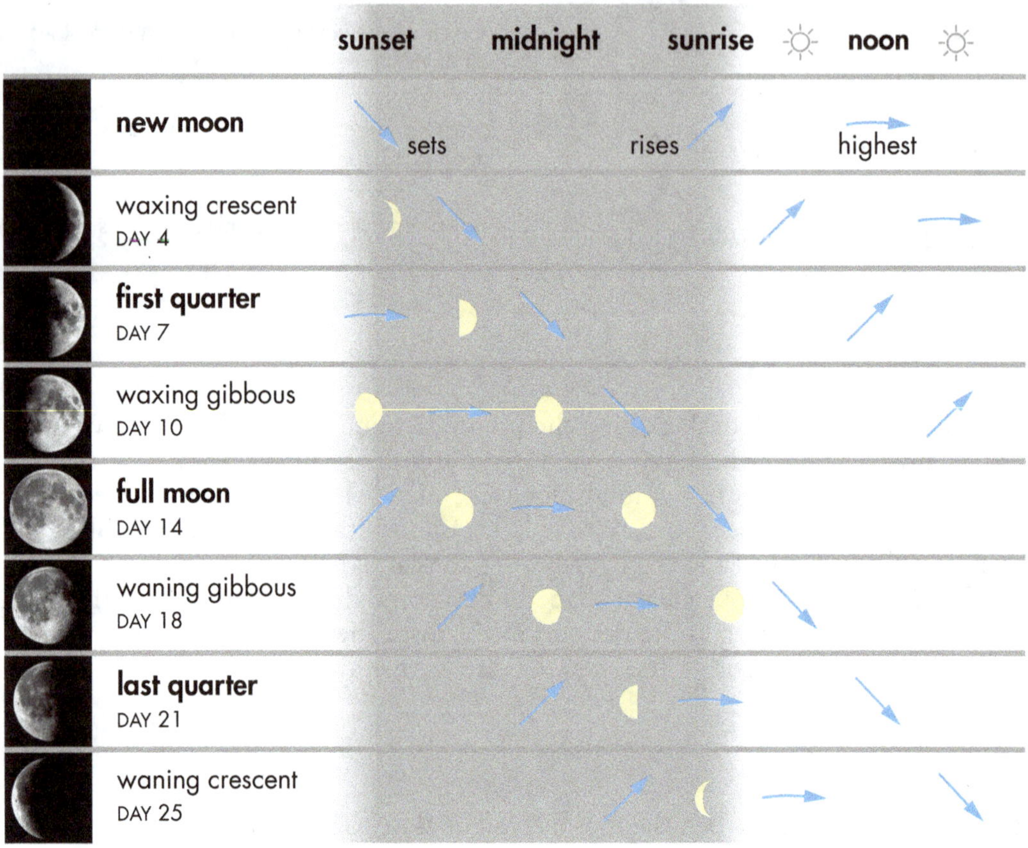

THE PLANETS

Besides the one we live on, five planets can be seen with the naked eye. The outer planets Uranus and Neptune are only visible with large telescopes. This season, all five naked-eye planets put in appearances in the night sky.

Binoculars or a telescope provide more detail on the planets, although you won't be able to see Saturn's rings or Jupiter's stripes. You'll see their colors and notice the difference between planets and stars. If you're not sure if you're looking at a star or a planet, stars are emitting light, so they twinkle and shimmer. Planets are reflecting sunlight, the same way the moon does, so their light is steady and doesn't twinkle.

Mercury: When it's visible, Mercury is always close to the sun just after sunset or just before sunrise, and it's a challenge to spot it in the dawn or dusk glare. Don't use binoculars or a telescope to look for Mercury; it's best spotted without magnification. This season's opportunities to view the planet:

October 10–14 just after sunset
November 18–22 just before sunrise
February 1–5 just after sunset
March 14–18 just before sunrise
May 26–30 just after sunset
July 12–16 just before sunrise

Venus: The planet is visible in the west after sunset until early October. It appears again in the morning sky from November to early July. It is too close to the sun to be visible for the rest of the summer.

Mars: Mars is visible every night of the season. In the fall it rises in the morning sky before the sun, and rises earlier each night until it is visible all night by February. In late February it is at its closest and brightest. By summer it sets in the west before midnight. This year Mars and Jupiter are close together in the sky from fall until early summer.

Jupiter: In the fall Jupiter rises in the east in the early morning, and rises earlier each night until it is visible all night in January and February. It is at its closest and brightest in early February. By summer the planet is in the western sky at dusk, and after late July, it is too close to the sun to observe. This season Mars and Jupiter are close together in the sky from fall until early summer, and Jupiter lingers near the constellation Leo. Read more about observing Jupiter in "The Moons of Jupiter" on page 18.

Saturn: The planet is visible most of the night in the fall. It is at its closest and brightest in late September and early October. Saturn sets earlier each night until it is visible in the west after sunset by January. By mid-March it is too close to the sun to observe, but reappears before dawn in late May, rising earlier and getting brighter each night for the rest of the year.

Retrograde Motion and Opposition

Most of the time, the planets appear to move gradually eastward against the background of the fixed stars. When the earth passes a planet at a different speed, the planet temporarily appears to move in the opposite direction, westward. Astronomers call this apparent retrograde motion. "Apparent" means it's an illusion, like when you pass a car on the freeway and it appears to approach you and then recede. Like the slower car on the freeway, the planets are still moving at the same speed and in the same direction.

When the visible outer planets—Mars, Jupiter, and Saturn—reach the midpoint of their retrograde journey, they are at their brightest and closest to the earth, in astronomy called the point of opposition. For the inner planets, Venus and Mercury, retrograde motion is linked to their proximity to the sun, when they disappear in its glare.

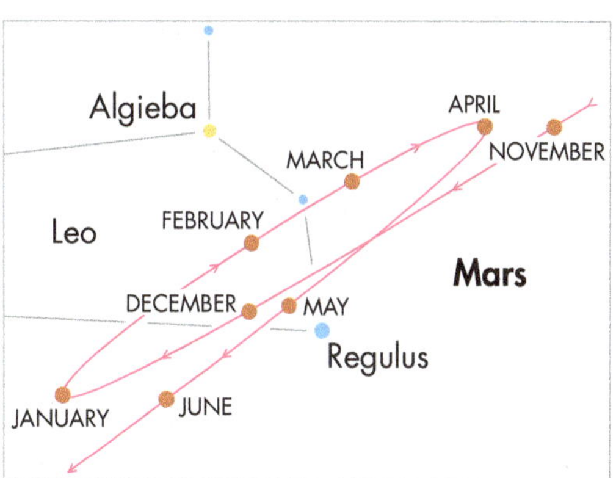

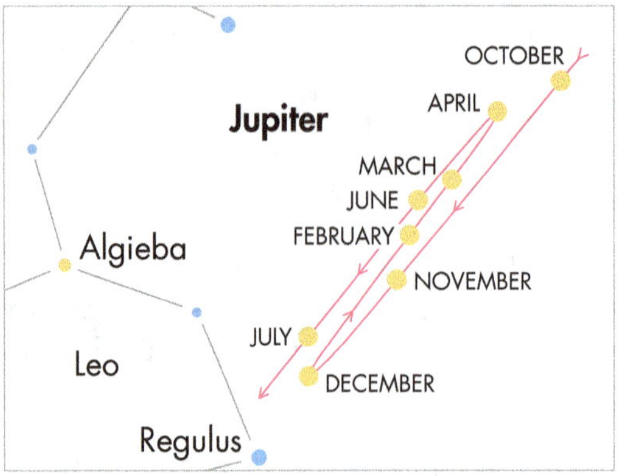

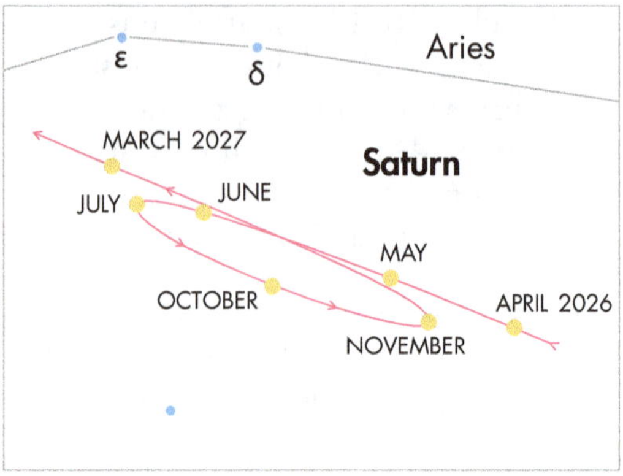

This winter, Mars goes retrograde from January to April, and is at opposition, its closest and brightest, around February 19. Jupiter goes retrograde from December to April and is at opposition around February 10. Saturn is moving retrograde until the end of November, and is at opposition around October 4.

THE GREAT BEAR

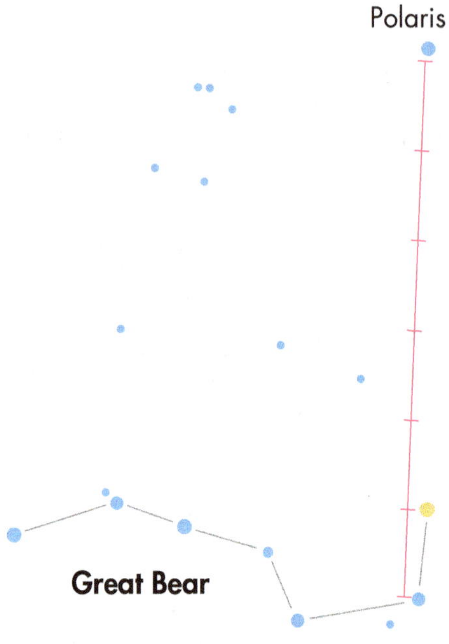

The seven main stars of the constellation Ursa Major, the Great Bear, are distinctive and have been described since prehistory in different cultures as a plow, a wagon, a ladle, three hunters chasing a bear, or three sets of footprints following a gazelle. Also called the Big Dipper, it is one of the most prominent constellations in the sky, and it's visible for much of the year at some time during the night.

The Great Bear can be used to find **Polaris,** the north star. Read more about the unique significance of Polaris above in "Finding North," page 3. Viewing the Great Bear as a ladle or dipper, draw an imaginary line through the two outside stars, the ones opposite the handle. Follow the line in the direction above the bowl. At about five times the distance between the two stars you'll see a medium-bright star on its own—Polaris.

Polaris remains stationary in the sky, and as the earth revolves, the rest of the stars appear to rotate around Polaris. Through the night, constellations like the Great Bear and Cassiopeia rise and set the same way the sun and moon do. In different seasons the constellations are higher or lower in the sky, always aligned with Polaris.

CASSIOPEIA

Cassiopeia is visible in the northern sky all year. Depending on the time of night and the season, it's shaped like an *M,* a *W,* a capital Sigma *(Σ),* or sometimes a *3.* If it's dark enough for the Milky Way to be visible, Cassiopeia is in the thick of it. Several star clusters around Cassiopeia make great binocular targets on moonless nights. See the chart to get oriented so you can pan from the bright stars to the clusters.

NGC 663 is a classic ball-shaped cluster, and NGC 869 and NGC 884, the **Double Cluster,** are loaded with blue-white stars and some red giants. The Double Cluster is easy to spot because both groups of stars fit in the binocular field.

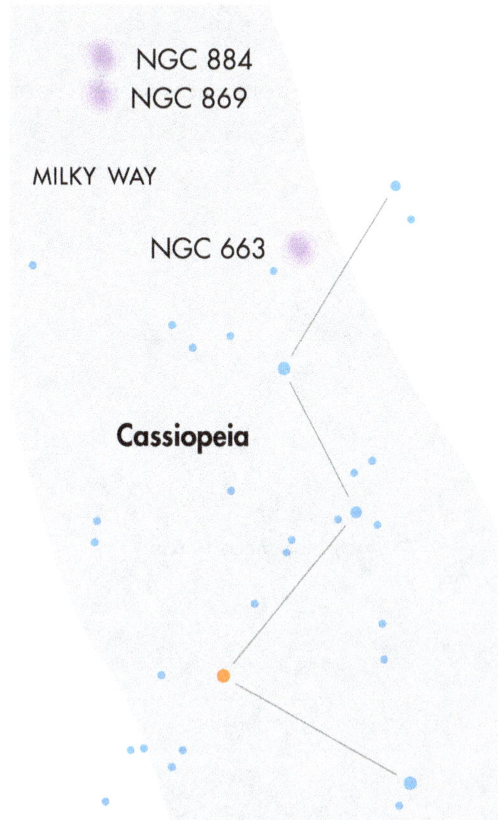

THE ANDROMEDA GALAXY

Every star you can see in the sky is in our galaxy, the Milky Way. One of the only things you can see that's outside our galaxy is the Andromeda Galaxy, visible much of the year and high in the sky from fall to February.

People with sharp vision can see it without a telescope or binoculars on a dark night with good seeing. Through a telescope or binoculars it makes a delightful target, and appears as a fuzzy or cloudy oval with a brighter center. You're looking at the light of approximately a trillion stars, much bigger than the Milky Way's 300 billion. The Andromeda Galaxy is actually moving toward us, and it might merge with the Milky Way into a larger galaxy in 3 or 4 billion years.

How to find the Andromeda Galaxy: Find **Cassiopeia** in the northern sky. From the top of the left side of the *M* shape, or left from the bottom of the *Σ*, follow an imaginary line about five times the distance of the height of the *M*. The Andromeda Galaxy is a fuzzy oval. Another way to find the Andromeda Galaxy, using the star Mirach, is described on page 14.

In the 1920s, Edwin Hubble, working at Mount Wilson Observatory, not far from Joshua Tree in the San Gabriel Mountains above Los Angeles, used his observations of the Andromeda Galaxy to figure out that the universe was much bigger than the stars we can see, and that the Milky Way Galaxy was just one of many. Before that moment, the Andromeda Galaxy had been considered a nebula, just another cloud of dust among the nearby stars. With Hubble's insight, our understanding of the size of the universe changed.

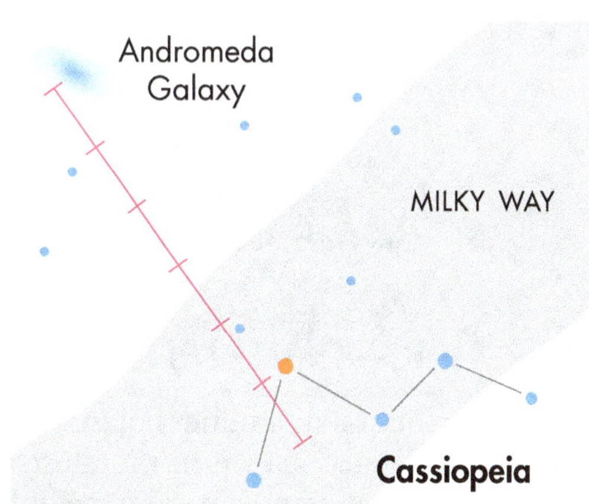

CASSIOPEIA AND THE GREAT BEAR

The two constellations shown over six months, with stationary Polaris always at the center:

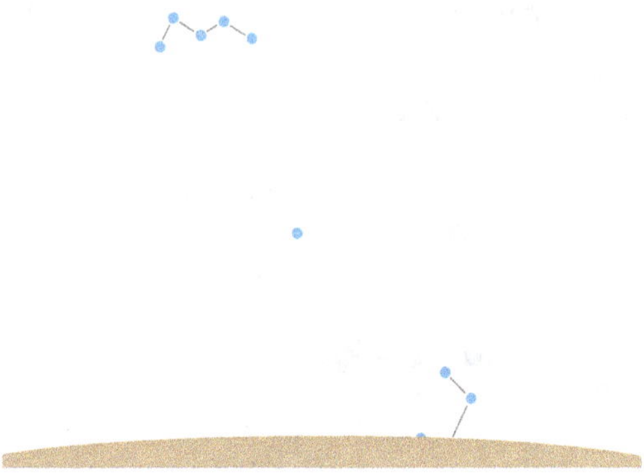

early November, 10 p.m.

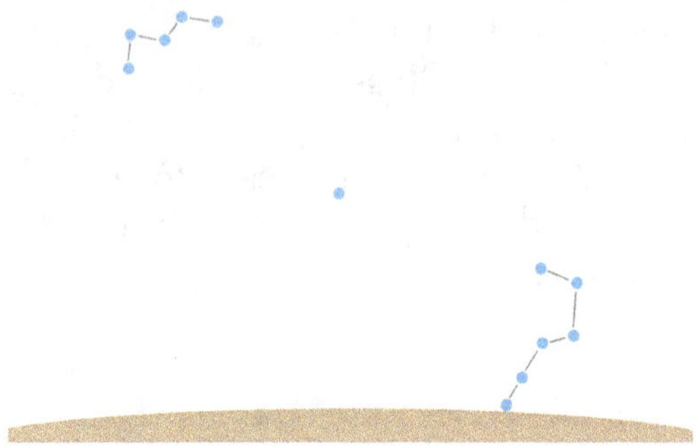

early December, 10 p.m.

early January, 10 p.m.

early February, 10 p.m.

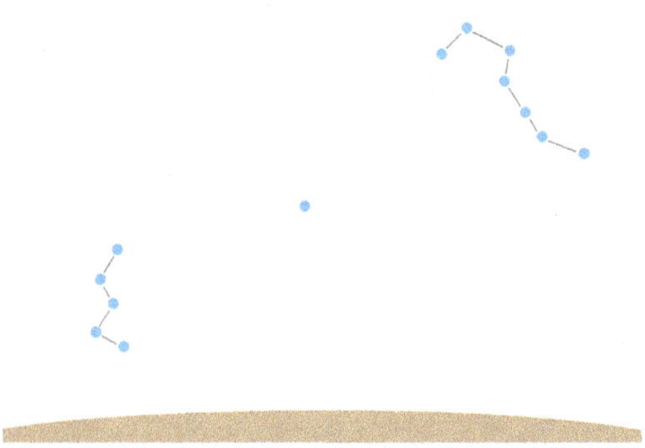

early March, 10 p.m.

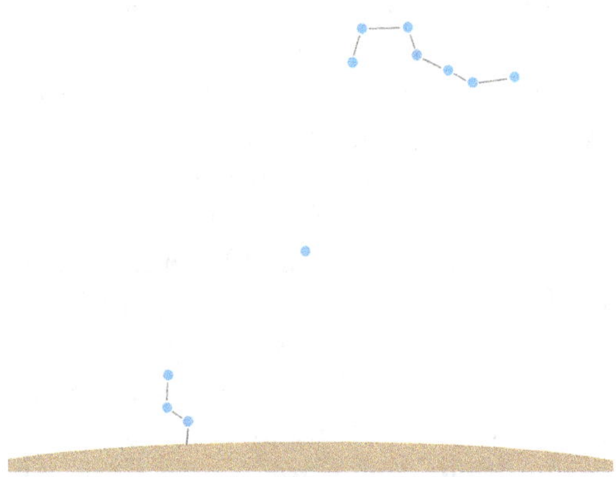

early April, 10 p.m.

NORTHERN CONSTELLATIONS

In the northern sky around Polaris are several constellations that you might be able to visualize, as shown in the illustration. Cassiopeia and the Great Bear have distinctive bright stars, but their neighbors mostly comprise dimmer stars. Once you've identified Polaris, you can spot Ursa Minor, also called the Little Dipper or the Little Bear, and then find the line of stars in Draco the dragon snaking between the Great Bear and Ursa Minor. Cepheus is on its own and not too hard to find, but Camelopardalis is a challenge even on a very dark night.

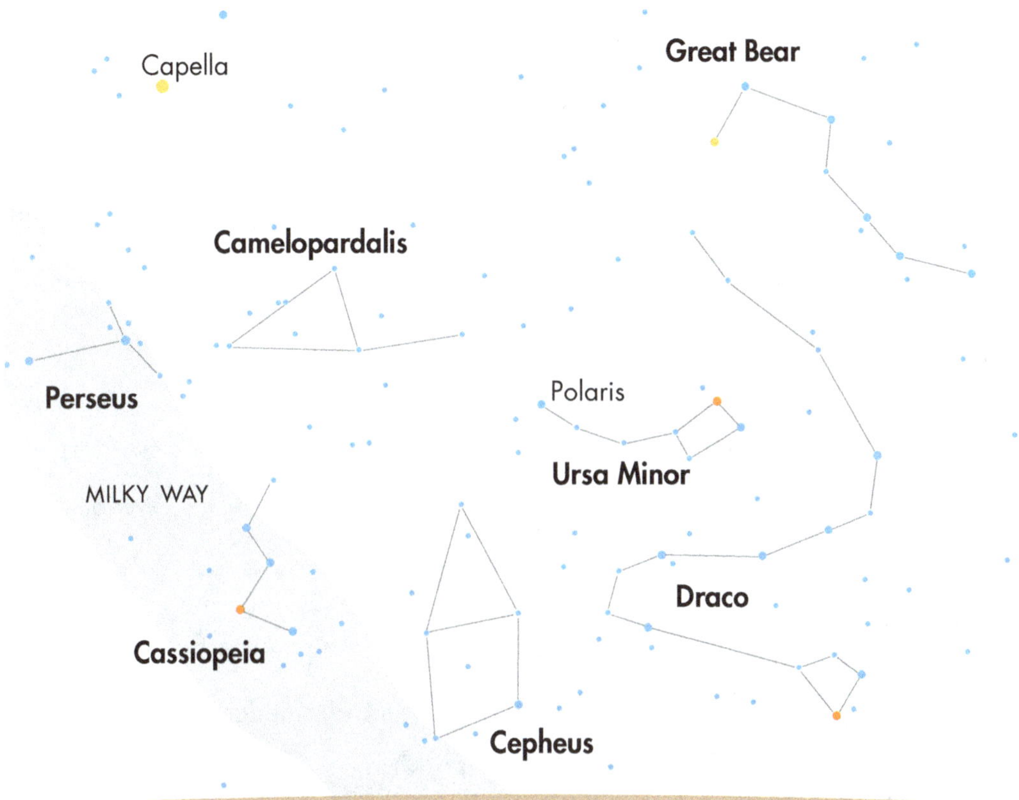

The northern sky before dawn in late November, around midnight in late January, and after sunset in late March

ARC TO ARCTURUS

One of the brightest stars in the sky, **Arcturus** is visible at some time during the night from November to September. You can use the Great Bear to find Arcturus. Viewing the Great Bear as a ladle or dipper, follow the handle of the dipper in an arc about three times the length of the handle. Arcturus is unmistakable because of its brightness, with no nearby peers. It's a red giant, and looks reddish to the naked eye or through binoculars. Following the arc farther south, about the same distance, leads to another bright star, **Spica.** Visible in the evening from March to August, Spica never gets higher than 40 degrees in the southern sky.

Arcturus is at its most prominent in the evening sky from April to July, and by September the star is descending in the west. In May and June, the similarly bright star Vega rises earlier each evening in the northeast, following Arcturus. By summer

Arcturus is fading in the west in the evening, but Vega is almost directly overhead, presiding over the night sky. More about Vega appears below in "The Summer Triangle," page 18.

PERSEUS AND ANDROMEDA

From September to April, the constellations Perseus and Andromeda are visible high in the sky. Perseus is anchored by the bright star **Mirfak,** surrounded by an array of less bright stars in a wide cluster that makes a nice binocular target. Find Mirfak from Cassiopeia by imagining a line through the bright middle star and the next star in the constellation. Viewing Cassiopeia as an *M,* use the middle star and the one to its right. At a distance of about twice the width of Cassiopeia is Mirfak; see the illustration.

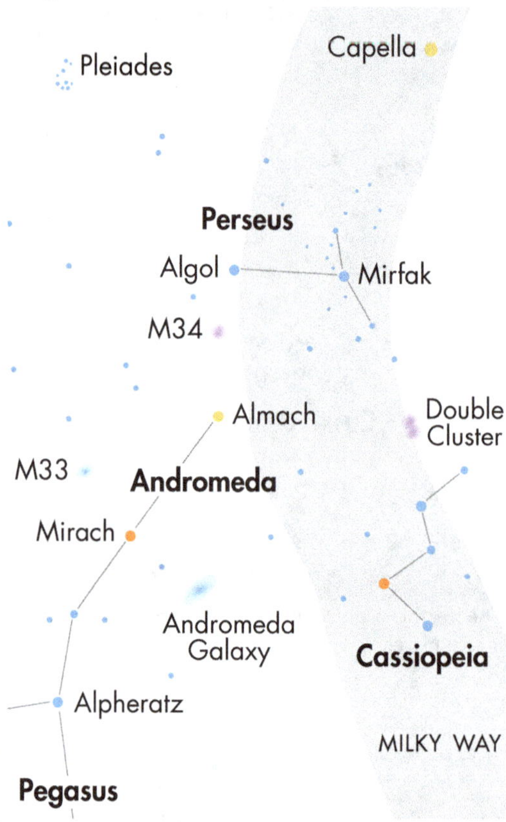

Above the top of the *M* shape of Cassiopeia, or left of its *Σ* shape, and next to Perseus are a line of bright stars that form Andromeda. Closest to Mirfak is yellowish **Almach.** It is almost in line with Mirfak and very bright Capella, with Mirfak the same distance from both. The second-brightest star in Perseus is **Algol,** and between Algol and Almach is **M34,** a dim star cluster that you can find with binoculars on a dark night. On the opposite side of Algol is the distinctive Pleiades, leading toward Orion. Read about the Pleiades on page 17.

Farther from Perseus, the next star in Andromeda beyond Almach is reddish **Mirach.** It can be used to find the **Andromeda Galaxy,** just a binocular field away, and in the opposite direction, **M33,** a much more difficult binocular target. M33 is a spiral galaxy that is much fainter than the Andromeda Galaxy, but on a dark night with no moon and good seeing, you might spot it through binoculars as a fuzzy patch. More about the Andromeda Galaxy is on page 11.

The last star in Andromeda is **Alpheratz,** one of the four corners of the Great Square of Pegasus. The square is vast and doesn't have bright stars within it, so it should jump out when you survey that part of the sky with the naked eye. Pegasus is visible until late February, and Andromeda disappears during March, but Perseus is visible until late April.

CASTOR AND POLLUX

From fall to spring, medium-bright **Castor** and **Pollux,** the twin stars in the constellation **Gemini,** are prominent, along with nearby **Capella,** one of the brightest stars in the sky. From January the Great Bear can help locate Castor and Pollux. Viewing the Great Bear as a dipper, follow the arc of the handle beyond the bowl, in the direction opposite the handle, about one and a half times the length of the Great Bear.

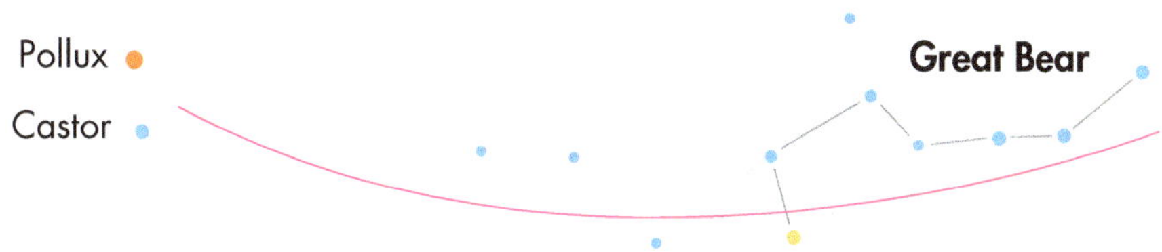

From September to April, Orion can also help locate Castor and Pollux. Imagine a line from Orion's belt through bright orange Betelgeuse toward the twin stars. Capella is the bright yellow-white star on its own to the northwest.

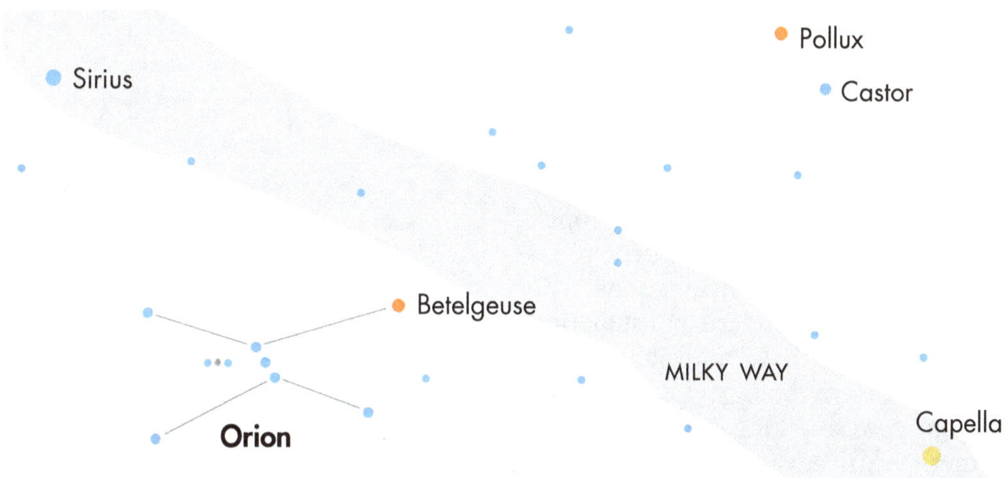

Finding Castor, Pollux, and Capella from Orion, seen in the early morning in the fall, high overhead in the winter, and in the western sky after dusk in March and April

Capella is in the band of the Milky Way, and Castor and Pollux are just outside it. Capella is almost exactly opposite Vega in the northern sky—when both are above the horizon, you can imagine a straight line between the two that passes through stationary Polaris. That means Capella is prominent in winter, while Vega is the bright center of the summer sky. Read more about Vega in "The Summer Triangle," page 18.

Castor looks like a single bluish star, but it's actually a system of six stars. Pollux is an orange giant that is known to have at least one planet orbiting it. Of the two stars, Pollux appears slightly brighter and distinctly orangey-red.

NAMING THE STARS AND CONSTELLATIONS

Prominent stars have traditional names, like Polaris and Vega and Castor and Pollux, but in the most common cataloging system, stars are named in order of brightness with lowercase Greek letters plus the adjective form of the name of the constellation they're in. The third-brightest star in the constellation Taurus is named with the third letter of the Greek alphabet, γ (Gamma), as γ Tauri. Because Vega is the brightest star in its constellation, Lyra, it's also named α Lyrae, as α (Alpha) is the first letter in the Greek alphabet. Stars farther down the alphabet often don't have names apart from their Greek letter, and even less visible stars are named with clunky catalog numbers.

With many cloudless dark nights, similar to Joshua Tree, it makes sense that stargazing would be important in the deserts of the Middle East, which explains why lots of stars have names rooted in Arabic. Many start with *Al-*, the Arabic equivalent of "the"—think of Altair, discussed on page 18, whose name means "the flying eagle," and Aldebaran, page 17, "the follower," because it appears to follow the Pleiades. More examples are in the illustration of the array of stars in Taurus and Orion that have Arabic-derived names.

Asterisms, meaning stars that form a pattern to the human eye, have been named and grouped differently in different cultures since antiquity. These days astronomers use clearly defined constellations that include all the stars in a given patch of sky, not just the historic asterisms like the Great Bear and Cassiopeia. Since 1928 the entire sky has been delineated into 88 official constellations.

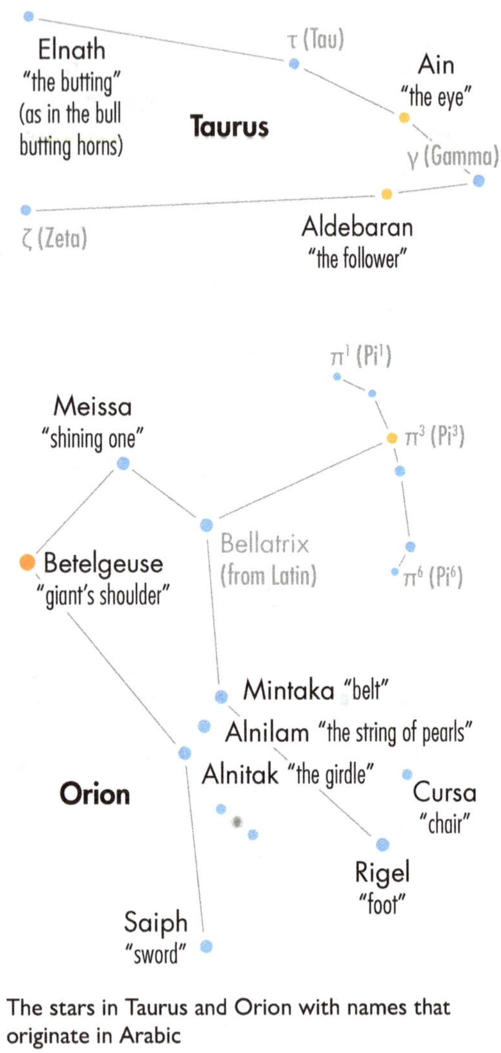

The stars in Taurus and Orion with names that originate in Arabic

Many of the modern constellations used by scientists have traditional names from Greek mythology. Cepheus and Cassiopeia were a king and queen from northern Africa, and their daughter, Andromeda, lent her name to a nearby constellation, as did her partner, the hero Perseus. Perseus and Andromeda are shown in the star chart on page 14.

In the southern hemisphere, many of the modern constellations were named during the Enlightenment, when European scientists first got a look at the southern sky. The constellation Musca is named for the housefly, Mensa was named for a mountain near an observatory in South Africa, and Reticulum was named by a French astronomer for the reticule, basically the crosshairs in the eyepiece of his telescope. These names are starkly prosaic compared to the likes of Perseus, the slayer of monsters, or Cassiopeia, who managed to anger Poseidon, god of the sea, so much that he flooded her entire kingdom.

None of these southern constellations are visible from Joshua Tree, but Reticulum might be familiar to paranormal enthusiasts: In the 1960s, Betty and Barney Hill got scooped up by aliens visiting New Hampshire from one of the planets around ζ (Zeta) Reticuli.

ORION

Orion is one of the most familiar constellations in the sky, most visible in the winter months. Orion's belt is a trio of stars, and below it hangs his sword. The **Orion Nebula,** the middle "star" in the sword, makes a great target with a telescope or with binoculars. It appears fuzzy and colorful, and in addition to the nebula you'll see a small group of stars, called the Trapezium Cluster.

Another interesting target is **Betelgeuse,** a bright orange star at the top left of the constellation, Orion's shoulder. Betelgeuse has been acting weird lately, unexpectedly brightening and dimming. Some scientists think it might go supernova soon. In astronomy, "soon" is a relative term, meaning in the next 100,000 years.

To the right of Orion, higher in the sky in the early evening, is another bright-orange star, **Aldebaran,** the eye of Taurus the bull. Around the star is the **Hyades,** a star cluster than overlaps but isn't related to Aldebaran. The Hyades make a delightful binocular target, with dozens of stars visible in a V shape near the brighter star.

Also part of Taurus and even higher in the sky than Aldebaran and the Hyades is the **Pleiades,** seven visible stars in a grouping that's relatively near the earth. They are visible to the naked eye and resolve nicely with binoculars. The group is also called the Seven Sisters, and in Japan it's called Subaru—the distinctive group of stars features in the logo of the car company with that name.

Left of Orion and rising later than the constellation is **Sirius,** unmistakable as the brightest star in the sky, twice as brilliant as the next brightest star. It rises around midnight in November, and from January rises around sunset and is visible all night. By April it is high in the sky at dusk. Some ancient observers said Sirius looked reddish, but today it's distinctly blue-white. At 8.6 light-years away, it's one of our sun's closest neighbors. Sirius is more prominent for observers in the southern hemisphere, and from Joshua Tree it never gets higher in the sky than 40 degrees.

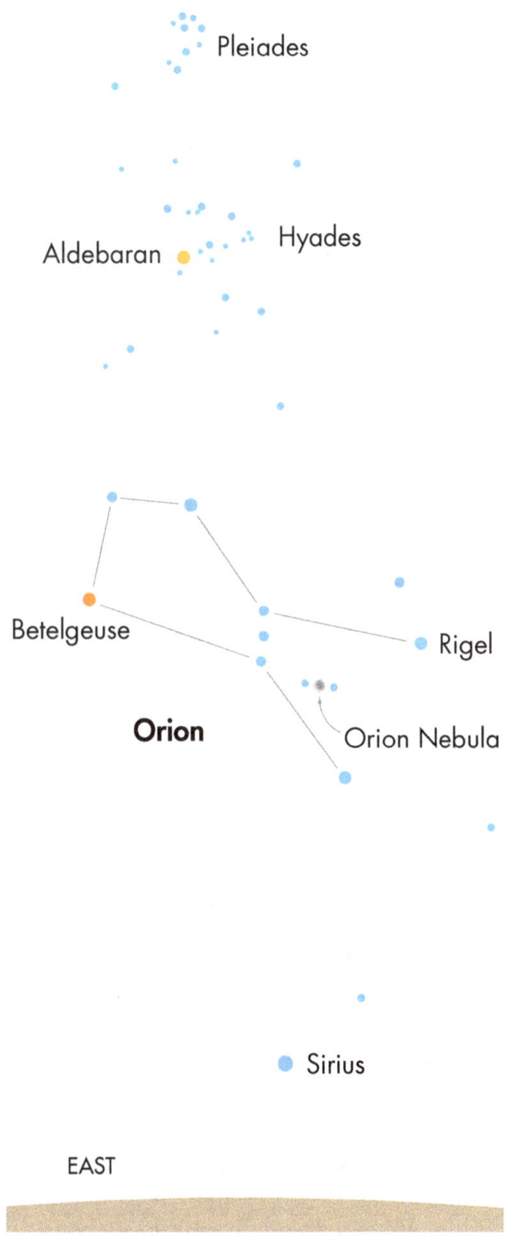

Looking east at 2 a.m. in early October, 11 p.m. in early November, 9 p.m. in early December, 7 p.m. in early January, or after dusk in early February

THE MOONS OF JUPITER

The largest and brightest planet in the night sky, Jupiter is visible most of the night until July, and is especially bright and prominent in January and February. With a telescope or with binoculars, it's possible to see the moons of Jupiter, although it might take some patience.

In 1610, Galileo Galilei was the first human to grasp what Jupiter's moons were. He thought he was looking at stars near the large planet, but then he realized they were moving in a regular pattern. This realization blew his mind and changed our concept of the cosmos.

Once you've located Jupiter, as described in the daily listings that follow, aim binoculars or a telescope and look for pinpricks of light near the planet. The giveaway is that they are lined up—from Earth we see the orbital plane of Jupiter's moons from the edge. Sometimes the moons are clustered on one side of the planet, and sometimes spread out on both sides. Two or three of the four large moons are usually visible, but they orbit the planet quickly, so the pattern is different every night.

From brightest to dimmest, the moons are Ganymede, Io, Europa, and Callisto. Scientists think there's a chance Europa is home to nonterrestrial life in the liquid ocean under its icy surface.

Galileo's sketches of the weird moving "stars" around Jupiter

View of Jupiter and its moons through a telescope or binoculars on a night with good seeing

THE SUMMER TRIANGLE

The summer triangle is anchored by **Vega,** one of the brightest stars in the sky. The other corners are Altair, a bright star on the other side of the Milky Way, and Deneb, right in the Milky Way. **Deneb** is the tail of **Cygnus the swan,** one of the constellations that actually looks like its name. Early in the summer the swan's head points south, and by September it's pointing west. It's always aligned with the Milky Way, with its wings extending out on either side of the band.

Altair can be identified by its less bright neighboring stars, reddish γ (Gamma) and β (Beta) Aquilae, one on either side, in the constellation Aquila. The line of three stars is sometimes called the shaft of Aquila.

The summer triangle isn't a constellation but rather an asterism, or a collection of stars that have a shape. Each of the three stars is in a different constellation. The Milky Way around bright Deneb and yellowish **Sadr,** the star where the swan's wings meet its body, is full of cloudy nebulae (dark gray in the illustration) and star clusters (purple). Study the area with binoculars to see these features.

Another interesting binocular target is ε (Epsilon) Lyrae, nicknamed the **Double Double.** Epsilon Lyrae is a double star that looks like a single star to the naked eye, but through binoculars it's clearly two similar stars, called ε^1 and ε^2. With the powerful magnification of a large telescope, each of these stars appears as a pair of binary stars, hence the Double Double moniker. Look for ε Lyrae in the same binocular field as Vega, in the direction of Deneb.

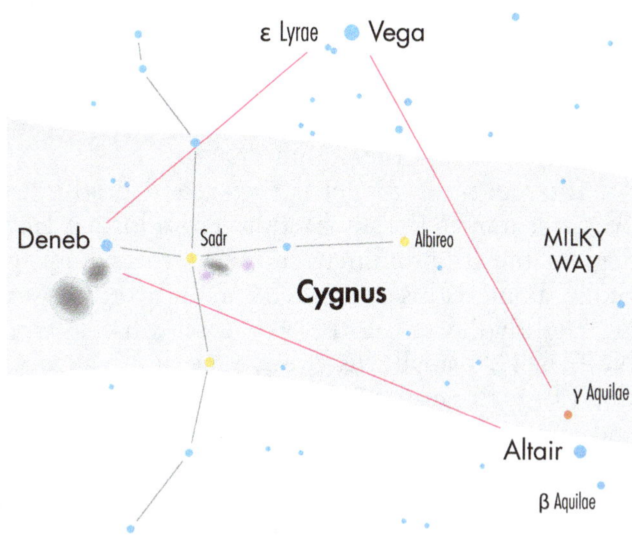

Albireo, the head of the swan, looks yellowish to the naked eye, but through binoculars or a telescope you might be able to make out two stars in very close proximity. The main star is larger and amber-yellow, and its much smaller companion is a delightful shade of blue.

SCORPIUS AND SAGITTARIUS

Scorpius is most visible in the evening in summer, and if you squint it might actually look like a scorpion. It's both the modern astronomical constellation Scorpius and the Zodiac sign Scorpio. The sun is in Scorpio in October and November, and when the earth is on the opposite side of the sun, around six months later, Scorpius is at its most visible in the night sky. From March to September it's visible at some point during the night.

The brightest star in Scorpius is **Antares,** called "the heart of the Scorpion." A red supergiant star, Antares looks reddish to the naked eye and through binoculars.

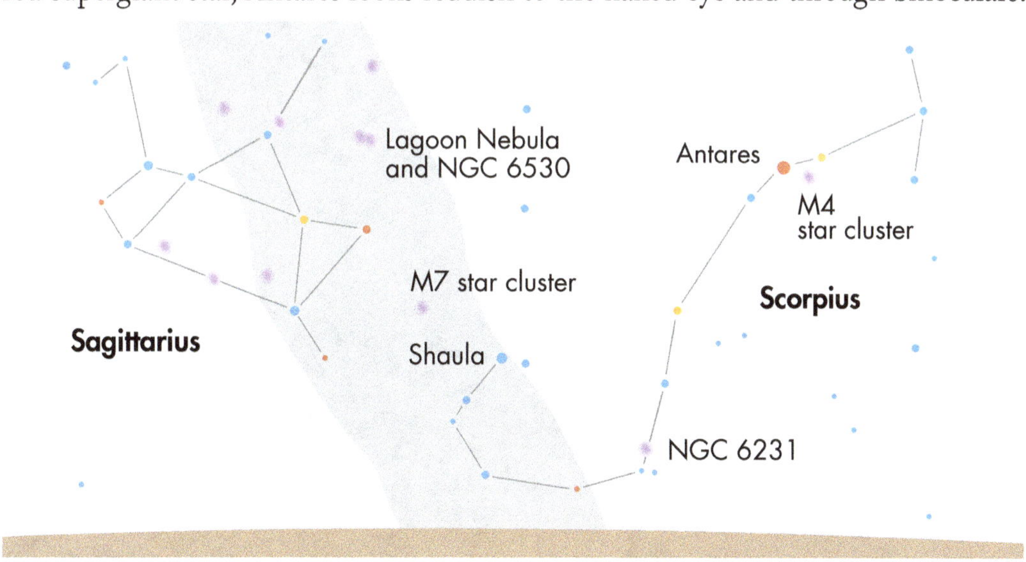

Looking south at 1:30 a.m. in early June, 11:30 p.m. in early July, or 9:30 p.m. in early August

Another interesting element of Scorpius is the **M4** star cluster. On a clear night it's visible as a fuzzy spot to the naked eye, and with binoculars it appears as a distinct ball. It's the closest globular cluster to our solar system. M4 is a good target to use **averted vision**—look slightly to the side rather than directly at it, and you'll get a better view of it in the periphery.

The scorpion's stinger is the second-brightest star in Scorpius and one of the brightest stars in the sky, **Shaula.** From Joshua Tree it is never higher than about 18 degrees, but it's prominent in the southern hemisphere and features on the flag of Brazil, along with seven other stars in Scorpius as well as Sirius and Spica.

Slightly lower in the sky than Shaula and visible through binoculars is **NGC 6231,** a small cluster of young bluish stars. One of its names is the Northern Jewel Box. Because it is low in the sky, we see it through more of the atmosphere, and it appears to sparkle.

If you have binoculars in hand, look to the east of Scorpius, at the constellation **Sagittarius.** It rises after Scorpius, so when Scorpius is in the southwest, Sagittarius is high in the southern sky. The center of our Milky Way galaxy is in Sagittarius, and that part of the sky is densely packed. Pan around with binoculars and you'll find some mesmerizing objects, including a couple of star clusters and star clouds (bright agglomerations of stars) as well as nebulas. The **M7** star cluster can be seen with the naked eye, and with binoculars the bright **Lagoon Nebula** is in the same field as the star cluster **NGC 6530.**

MORE NIGHT SKY TARGETS

An excellent website or app to point out what's in the sky is Stellarium, available at https://stellarium-web.org. It can show you what's coming up in the sky later tonight, or tomorrow night, and even includes the names of the artificial satellites you might see cruising among the stars.

Daily Dark Sky Guide

2026

Tuesday, September 1

TONIGHT

sunset: **7:11 p.m.**
twilight begins: **8:07 p.m.**

IN THE MORNING

twilight begins: **5:24 a.m.**
sunrise: **6:19 a.m.**

MOON PHASE

moonrise: **9:34 p.m.**
moonset: **10:59 a.m.**

IN THE SKY

Bright **Venus** is in the western sky at dusk and sets in the west around 8:40 p.m.

The **summer triangle**'s three bright stars are Vega, Altair, and Deneb. **Vega** is almost directly overhead around dusk and sets in the northwest around 4:45 a.m. **Altair** is at its highest, around 65 degrees, at 9:50 p.m. and sets in the west around 4:10 a.m. Altair can be identified by its less bright neighboring stars, one on either side, together called the shaft of Aquila. **Deneb** is in the band of the Milky Way. See the illustration of the summer triangle on page 19.

Saturn rises in the east around 8:55 p.m. and at its highest, around 60 degrees, at 3 a.m. Unlike bright Jupiter and Venus and ruddy Mars, Saturn is harder to distinguish from the stars. Stars twinkle and shimmer, while the planets, including Saturn, show a steady flat light.

Explore the **dark sky** after twilight and before the moon rises, around 9:15 p.m.

The **Great Bear (Big Dipper)** sets in the northwest around 10:15 p.m. See the illustration on page 10.

Bright star **Capella** rises in the northeast around 10:25 p.m. Read about Capella on page 14.

Scorpius sets in the southwest around 10:45 p.m. See the illustration on page 19.

The bright star **Arcturus** sets in the northwest around 11:05 p.m.

Just east of Scorpius, **Sagittarius** sets in the southwest around 12:15 a.m.

Rust-red and not very bright, **Mars** rises in the northeast around 2 a.m. and is in the eastern sky before dawn.

Orion rises in the east around 2:05 a.m. and is in the eastern sky before dawn. Look for the bright star **Aldebaran** and the **Pleiades** near Orion. See the illustration on page 17.

Castor and Pollux, the twin stars in the constellation Gemini, rise in the northeast around 3 a.m. and are in the eastern sky before dawn.

Cassiopeia, shaped like an *M,* is at its highest, around 65 degrees, at 3:15 a.m. See the illustration on page 11.

The brightest star in the sky, **Sirius** rises in the east around 3:45 a.m. and is in the eastern sky before dawn.

Jupiter rises in the east around 4:25 a.m. and is in the eastern sky before dawn.

2026

Wednesday, September 2

TONIGHT

sunset: **7:10 p.m.**
twilight begins: **8:05 p.m.**

IN THE MORNING

twilight begins: **5:25 a.m.**
sunrise: **6:20 a.m.**

MOON PHASE

moonrise: **10:15 p.m.**
moonset: **12:09 p.m.**

IN THE SKY

Bright **Venus** is in the western sky at dusk and sets in the west around 8:40 p.m.

The **summer triangle**'s three bright stars are Vega, Altair, and Deneb. **Vega** is almost directly overhead around dusk and sets in the northwest around 4:45 a.m. **Altair** is at its highest, around 65 degrees, at 9:50 p.m. and sets in the west around 4:10 a.m. Altair can be identified by its less bright neighboring stars, one on either side, together called the shaft of Aquila. **Deneb** is in the band of the Milky Way. See the illustration of the summer triangle on page 19.

Look for the **Andromeda Galaxy** with binoculars or a telescope. It becomes a good target around 8:40 p.m. in the northeastern sky. See the guide on page 11.

Saturn rises in the east around 8:50 p.m. and at its highest, around 60 degrees, at 2:55 a.m. Unlike bright Jupiter and Venus and ruddy Mars, Saturn is harder to distinguish from the stars. Stars twinkle and shimmer, while the planets, including Saturn, show a steady flat light.

Explore the **dark sky** after twilight and before the moon rises, around 9:55 p.m.

The **Great Bear (Big Dipper)** sets in the northwest around 10:10 p.m. See the illustration on page 10.

Bright star **Capella** rises in the northeast around 10:20 p.m. Read about Capella on page 14.

Scorpius sets in the southwest around 10:40 p.m. See the illustration on page 19.

The bright star **Arcturus** sets in the northwest around 11 p.m.

Just east of Scorpius, **Sagittarius** sets in the southwest around 12:15 a.m.

Rust-red and not very bright, **Mars** rises in the northeast around 2 a.m. and is in the eastern sky before dawn.

Orion rises in the east around 2:05 a.m. and is in the eastern sky before dawn. Look for the bright star **Aldebaran** and the **Pleiades** near Orion. See the illustration on page 17.

Castor and Pollux, the twin stars in the constellation Gemini, rise in the northeast around 3 a.m. and are in the eastern sky before dawn.

Cassiopeia, shaped like an *M,* is at its highest, around 65 degrees, at 3:10 a.m. See the illustration on page 11.

The brightest star in the sky, **Sirius** rises in the east around 3:40 a.m. and is in the eastern sky before dawn.

Jupiter rises in the east around 4:25 a.m. and is in the eastern sky before dawn.

2026

Thursday, September 3

TONIGHT

sunset: **7:08 p.m.**
twilight begins: **8:04 p.m.**

IN THE MORNING

twilight begins: **5:26 a.m.**
sunrise: **6:21 a.m.**

MOON PHASE

moonrise: **11:05 p.m.**
moonset: **1:19 p.m.**

IN THE SKY

Bright **Venus** is in the western sky at dusk and sets in the west around 8:35 p.m.

The **summer triangle**'s three bright stars are Vega, Altair, and Deneb. **Vega** is almost directly overhead around dusk and sets in the northwest around 4:40 a.m. **Altair** is at its highest, around 65 degrees, at 9:45 p.m. and sets in the west around 4:05 a.m. Altair can be identified by its less bright neighboring stars, one on either side, together called the shaft of Aquila. **Deneb** is in the band of the Milky Way. See the illustration of the summer triangle on page 19.

Saturn rises in the east around 8:45 p.m. and at its highest, around 60 degrees, at 2:50 a.m. Unlike bright Jupiter and Venus and ruddy Mars, Saturn is harder to distinguish from the stars. Stars twinkle and shimmer, while the planets, including Saturn, show a steady flat light.

The **Great Bear (Big Dipper)** sets in the northwest around 10:05 p.m. See the illustration on page 10.

Bright star **Capella** rises in the northeast around 10:15 p.m. Read about Capella on page 14.

Scorpius sets in the southwest around 10:40 p.m. See the illustration on page 19.

Explore the **dark sky** after twilight and before the moon rises, around 10:45 p.m.

The bright star **Arcturus** sets in the northwest around 10:55 p.m.

Just east of Scorpius, **Sagittarius** sets in the southwest around 12:10 a.m.

Before the moon rises, until around 10:45 p.m., look for the **Andromeda Galaxy** with binoculars or a telescope. It becomes a good target around 8:40 p.m. in the northeastern sky. See the guide on page 11.

Rust-red and not very bright, **Mars** rises in the northeast around 2 a.m. and is in the eastern sky before dawn.

Orion rises in the east around 2 a.m. and is in the eastern sky before dawn. Look for the bright star **Aldebaran** and the **Pleiades** near Orion. See the illustration on page 17.

Castor and Pollux, the twin stars in the constellation Gemini, rise in the northeast around 2:55 a.m. and are in the eastern sky before dawn.

Cassiopeia, shaped like an *M,* is at its highest, around 65 degrees, at 3:05 a.m. See the illustration on page 11.

The brightest star in the sky, **Sirius** rises in the east around 3:35 a.m. and is in the eastern sky before dawn.

Jupiter rises in the east around 4:20 a.m. and is in the eastern sky before dawn.

The **Great Bear (Big Dipper)** rises in the northeast around 5:05 a.m. and is in the northeastern sky before dawn. See the illustration on page 10.

2026
Friday, September 4

TONIGHT

sunset: **7:07 p.m.**
twilight begins: **8:02 p.m.**

IN THE MORNING

twilight begins: **5:26 a.m.**
sunrise: **6:22 a.m.**

MOON PHASE

moonrise: **12:04 a.m.**
moonset: **2:27 p.m.**

IN THE SKY

Bright **Venus** is in the western sky at dusk and sets in the west around 8:35 p.m.

The **summer triangle**'s three bright stars are Vega, Altair, and Deneb. **Vega** is almost directly overhead around dusk and sets in the northwest around 4:35 a.m. **Altair** is at its highest, around 65 degrees, at 9:40 p.m. and sets in the west around 4 a.m. Altair can be identified by its less bright neighboring stars, one on either side, together called the shaft of Aquila. **Deneb** is in the band of the Milky Way. See the illustration of the summer triangle on page 19.

Saturn rises in the east around 8:40 p.m. and at its highest, around 60 degrees, at 2:45 a.m. Unlike bright Jupiter and Venus and ruddy Mars, Saturn is harder to distinguish from the stars. Stars twinkle and shimmer, while the planets, including Saturn, show a steady flat light.

Before the moon rises, around midnight, look for the **Andromeda Galaxy** with binoculars or a telescope. It becomes a good target around 8:35 p.m. in the northeastern sky. See the guide on page 11.

The **Great Bear (Big Dipper)** is in the northwestern sky at dusk and sets in the northwest around 10 p.m. See the illustration on page 10.

Bright star **Capella** rises in the northeast around 10:10 p.m. Read about Capella on page 14.

Scorpius sets in the southwest around 10:35 p.m. See the illustration on page 19.

The bright star **Arcturus** sets in the northwest around 10:50 p.m.

Orion rises in the east around 1:55 a.m. and is in the eastern sky before dawn. Look for the bright star **Aldebaran** and the **Pleiades** near Orion. See the illustration on page 17.

Just east of Scorpius, **Sagittarius** sets in the southwest around 12:05 a.m.

Rust-red and not very bright, **Mars** rises in the northeast around 2 a.m. and is in the eastern sky before dawn.

Castor and Pollux, the twin stars in the constellation Gemini, rise in the northeast around 2:50 a.m. and are in the eastern sky before dawn.

Cassiopeia, shaped like an *M,* is at its highest, around 65 degrees, at 3 a.m. See the illustration on page 11.

The brightest star in the sky, **Sirius** rises in the east around 3:30 a.m. and is in the eastern sky before dawn.

Jupiter rises in the east around 4:20 a.m. and is in the eastern sky before dawn.

The **Great Bear (Big Dipper)** rises in the northeast around 5 a.m. and is in the northwestern sky at dusk See the illustration on page 10.

2026

Saturday, September 5

TONIGHT

sunset: **7:06 p.m.**
twilight begins: **8:01 p.m.**

MOON PHASE

IN THE MORNING

twilight begins: **5:27 a.m.**
sunrise: **6:22 a.m.**

moonrise: **1:11 a.m.**
moonset: **3:29 p.m.**

IN THE SKY

Bright **Venus** is in the western sky at dusk and sets in the west around 8:30 p.m.

The **summer triangle**'s three bright stars are Vega, Altair, and Deneb. **Vega** is almost directly overhead around dusk and sets in the northwest around 4:30 a.m. **Altair** is at its highest, around 65 degrees, at 9:35 p.m. and sets in the west around 3:55 a.m. Altair can be identified by its less bright neighboring stars, one on either side, together called the shaft of Aquila. **Deneb** is in the band of the Milky Way. See the illustration of the summer triangle on page 19.

Saturn rises in the east around 8:35 p.m. and at its highest, around 60 degrees, at 2:40 a.m. Unlike bright Jupiter and Venus and ruddy Mars, Saturn is harder to distinguish from the stars. Stars twinkle and shimmer, while the planets, including Saturn, show a steady flat light.

Before the moon rises, around 1 a.m., look for the **Andromeda Galaxy** with binoculars or a telescope. It becomes a good target around 8:35 p.m. in the northeastern sky. See the guide on page 11.

The **Great Bear (Big Dipper)** is in the northwestern sky at dusk and sets in the northwest around 9:55 p.m. See the illustration on page 10.

Scorpius sets in the southwest around 10:30 p.m. See the illustration on page 19.

Just east of Scorpius, **Sagittarius** sets in the southwest around midnight.

Bright star **Capella** rises in the northeast around 10:10 p.m. Read about Capella on page 14.

The bright star **Arcturus** sets in the northwest around 10:50 p.m.

Orion rises in the east around 1:50 a.m. and is in the eastern sky before dawn. Look for the bright star **Aldebaran** and the **Pleiades** near Orion. See the illustration on page 17.

Rust-red and not very bright, **Mars** rises in the northeast around 2 a.m. and is in the eastern sky before dawn.

Castor and Pollux, the twin stars in the constellation Gemini, rise in the northeast around 2:45 a.m. and are in the eastern sky before dawn.

Cassiopeia, shaped like an *M,* is at its highest, around 65 degrees, at 2:55 a.m. See the illustration on page 11.

The brightest star in the sky, **Sirius** rises in the east around 3:30 a.m. and is in the eastern sky before dawn.

Jupiter rises in the east around 4:15 a.m. and is in the eastern sky before dawn.

The **Great Bear (Big Dipper)** rises in the northeast around 4:55 a.m. and is in the northeastern sky before dawn. See the illustration on page 10.

2026
Sunday, September 6

TONIGHT

sunset: **7:04 p.m.**
twilight begins: **8 p.m.**

IN THE MORNING

twilight begins: **5:28 a.m.**
sunrise: **6:23 a.m.**

MOON PHASE

moonrise: **2:23 a.m.**
moonset: **4:23 p.m.**

IN THE SKY

Bright **Venus** sets in the west around 8:30 p.m.

The **summer triangle**'s three bright stars are Vega, Altair, and Deneb. **Vega** is in the western sky before dawn and sets in the northwest around 4:30 a.m. **Altair** is at its highest, around 65 degrees, at 9:30 p.m. and sets in the west around 3:55 a.m. Altair can be identified by its less bright neighboring stars, one on either side, together called the shaft of Aquila. **Deneb** is in the band of the Milky Way. See the illustration of the summer triangle on page 19.

The **Great Bear (Big Dipper)** is in the northwestern sky at dusk and sets in the northwest around 9:55 p.m. See the illustration on page 10.

With the dark sky, look for the **Andromeda Galaxy** with binoculars or a telescope. It becomes a good target around 8:30 p.m. in the northeastern sky. See the guide on page 11.

Saturn rises in the east around 8:35 p.m. and at its highest, around 60 degrees, at 2:35 a.m. Unlike bright Jupiter and Venus and ruddy Mars, Saturn is harder to distinguish from the stars. Stars twinkle and shimmer, while the planets, including Saturn, show a steady flat light.

The absence of the **moon** tonight provides excellent dark-sky viewing conditions.

Bright star **Capella** rises in the northeast around 10:05 p.m. Read about Capella on page 14.

Scorpius sets in the southwest around 10:25 p.m. See the illustration on page 19.

The bright star **Arcturus** sets in the northwest around 10:45 p.m.

Just east of Scorpius, **Sagittarius** sets in the southwest around 11:55 p.m.

Cassiopeia, shaped like an *M,* is at its highest, around 65 degrees, at 2:55 a.m. See the illustration on page 11.

Orion rises in the east around 1:45 a.m. and is in the eastern sky before dawn. Look for the bright star **Aldebaran** and the **Pleiades** near Orion. See the illustration on page 17.

Rust-red and not very bright, **Mars** rises in the northeast around 1:55 a.m. and is in the eastern sky before dawn.

Castor and Pollux, the twin stars in the constellation Gemini, rise in the northeast around 2:45 a.m. and are in the eastern sky before dawn.

The brightest star in the sky, **Sirius** rises in the east around 3:25 a.m. and is in the eastern sky before dawn.

Jupiter rises in the east around 4:10 a.m. and is in the eastern sky before dawn.

The **Great Bear (Big Dipper)** rises in the northeast around 4:50 a.m. and is in the northeastern sky before dawn. See the illustration on page 10.

2026

Monday, September 7

TONIGHT

sunset: **7:03 p.m.**
twilight begins: **7:58 p.m.**

IN THE MORNING

twilight begins: **5:29 a.m.**
sunrise: **6:24 a.m.**

MOON PHASE

moonrise: **3:35 a.m.**
moonset: **5:07 p.m.**

IN THE SKY

Bright **Venus** sets in the west around 8:25 p.m.

The **summer triangle**'s three bright stars are Vega, Altair, and Deneb. **Vega** is in the western sky before dawn and sets in the northwest around 4:25 a.m. **Altair** is at its highest, around 65 degrees, at 9:30 p.m. and sets in the west around 3:50 a.m. Altair can be identified by its less bright neighboring stars, one on either side, together called the shaft of Aquila. **Deneb** is in the band of the Milky Way. See the illustration of the summer triangle on page 19.

Saturn rises in the east around 8:30 p.m. and at its highest, around 60 degrees, at 2:35 a.m. Unlike bright Jupiter and Venus and ruddy Mars, Saturn is harder to distinguish from the stars. Stars twinkle and shimmer, while the planets, including Saturn, show a steady flat light.

With the dark sky, look for the **Andromeda Galaxy** with binoculars or a telescope. It becomes a good target around 8:25 p.m. in the northeastern sky. See the guide on page 11.

The absence of the **moon** tonight provides excellent dark-sky viewing conditions.

The **Great Bear (Big Dipper)** is in the northwestern sky at dusk and sets in the northwest around 9:50 p.m. See the illustration on page 10.

Bright star **Capella** rises in the northeast around 10 p.m. Read about Capella on page 14.

Scorpius sets in the southwest around 10:20 p.m. See the illustration on page 19.

The bright star **Arcturus** sets in the northwest around 10:40 p.m.

Just east of Scorpius, **Sagittarius** sets in the southwest around 11:55 p.m.

Orion rises in the east around 1:45 a.m. and is in the eastern sky before dawn. Look for the bright star **Aldebaran** and the **Pleiades** near Orion. See the illustration on page 17.

Rust-red and not very bright, **Mars** rises in the northeast around 1:55 a.m. and is in the eastern sky before dawn.

Castor and Pollux, the twin stars in the constellation Gemini, rise in the northeast around 2:40 a.m. and are in the eastern sky before dawn.

Cassiopeia, shaped like an *M,* is at its highest, around 65 degrees, at 2:50 a.m. See the illustration on page 11.

The brightest star in the sky, **Sirius** rises in the east around 3:20 a.m. and is in the eastern sky before dawn.

Jupiter rises in the east around 4:10 a.m. and is in the eastern sky before dawn.

The **Great Bear (Big Dipper)** rises in the northeast around 4:45 a.m. and is in the northeastern sky before dawn. See the illustration on page 10.

2026

Tuesday, September 8

MOON PHASE

TONIGHT

sunset: **7:01 p.m.**
twilight begins: **7:57 p.m.**

IN THE MORNING

twilight begins: **5:30 a.m.**
sunrise: **6:24 a.m.**

moonrise: **4:46 a.m.**
moonset: **5:44 p.m.**

IN THE SKY

Bright **Venus** sets in the west around 8:25 p.m.

The **summer triangle**'s three bright stars are Vega, Altair, and Deneb. **Vega** is in the western sky before dawn and sets in the northwest around 4:20 a.m. **Altair** is at its highest, around 65 degrees, at 9:25 p.m. and sets in the west around 3:45 a.m. Altair can be identified by its less bright neighboring stars, one on either side, together called the shaft of Aquila. **Deneb** is in the band of the Milky Way. See the illustration of the summer triangle on page 19.

The absence of the **moon** tonight provides excellent dark-sky viewing conditions.

With the dark sky, look for the **Andromeda Galaxy** with binoculars or a telescope. It becomes a good target around 8:20 p.m. in the northeastern sky. See the guide on page 11.

The **Great Bear (Big Dipper)** is in the northwestern sky at dusk and sets in the northwest around 9:45 p.m. See the illustration on page 10.

Bright star **Capella** rises in the northeast around 9:55 p.m. Read about Capella on page 14.

Scorpius sets in the southwest around 10:20 p.m. See the illustration on page 19.

The bright star **Arcturus** sets in the northwest around 10:35 p.m.

Just east of Scorpius, **Sagittarius** sets in the southwest around 11:50 p.m.

Orion rises in the east around 1:40 a.m. and is in the eastern sky before dawn. Look for the bright star **Aldebaran** and the **Pleiades** near Orion. See the illustration on page 17.

Rust-red and not very bright, **Mars** rises in the northeast around 1:55 a.m. and is in the eastern sky before dawn.

Saturn is in the eastern sky at dusk and at its highest, around 60 degrees, at 2:30 a.m. Unlike bright Jupiter and Venus and ruddy Mars, Saturn is harder to distinguish from the stars. Stars twinkle and shimmer, while the planets, including Saturn, show a steady flat light.

Castor and Pollux, the twin stars in the constellation Gemini, rise in the northeast around 2:35 a.m. and are in the eastern sky before dawn.

Cassiopeia, shaped like an *M,* is at its highest, around 65 degrees, at 2:45 a.m. See the illustration on page 11.

The brightest star in the sky, **Sirius** rises in the east around 3:15 a.m. and is in the eastern sky before dawn.

Jupiter rises in the east around 4:05 a.m. and is in the eastern sky before dawn.

With the dark sky, use binoculars or a telescope to look for the **moons of Jupiter.** See the guide on page 18.

2026

Wednesday, September 9

TONIGHT

sunset: **7 p.m.**
twilight begins: **7:55 p.m.**

IN THE MORNING

twilight begins: **5:30 a.m.**
sunrise: **6:25 a.m.**

MOON PHASE

moonrise: **5:54 a.m.**
moonset: **6:16 p.m.**

IN THE SKY

Bright **Venus** sets in the west around 8:20 p.m.

The **summer triangle**'s three bright stars are Vega, Altair, and Deneb. **Vega** is in the western sky before dawn and sets in the northwest around 4:15 a.m. **Altair** is at its highest, around 65 degrees, at 9:20 p.m. and sets in the west around 3:40 a.m. Altair can be identified by its less bright neighboring stars, one on either side, together called the shaft of Aquila. **Deneb** is in the band of the Milky Way. See the illustration of the summer triangle on page 19.

The absence of the **moon** tonight provides excellent dark-sky viewing conditions.

With the dark sky, look for the **Andromeda Galaxy** with binoculars or a telescope. It becomes a good target around 8:15 p.m. in the northeastern sky. See the guide on page 11.

The **Great Bear (Big Dipper)** is in the northwestern sky at dusk and sets in the northwest around 9:40 p.m. See the illustration on page 10.

Bright star **Capella** rises in the northeast around 9:50 p.m. Read about Capella on page 14.

Scorpius sets in the southwest around 10:15 p.m. See the illustration on page 19.

The bright star **Arcturus** sets in the northwest around 10:30 p.m.

Just east of Scorpius, **Sagittarius** sets in the southwest around 11:45 p.m.

Orion rises in the east around 1:35 a.m. and is in the eastern sky before dawn. Look for the bright star **Aldebaran** and the **Pleiades** near Orion. See the illustration on page 17.

Rust-red and not very bright, **Mars** rises in the northeast around 1:55 a.m. and is in the eastern sky before dawn.

Saturn is in the eastern sky at dusk and at its highest, around 60 degrees, at 2:25 a.m. Unlike bright Jupiter and Venus and ruddy Mars, Saturn is harder to distinguish from the stars. Stars twinkle and shimmer, while the planets, including Saturn, show a steady flat light.

Castor and Pollux, the twin stars in the constellation Gemini, rise in the northeast around 2:30 a.m. and are in the eastern sky before dawn.

Cassiopeia, shaped like an *M,* is at its highest, around 65 degrees, at 2:40 a.m. See the illustration on page 11.

The brightest star in the sky, **Sirius** rises in the east around 3:10 a.m. and is in the eastern sky before dawn.

Jupiter rises in the east around 4:05 a.m. and is in the eastern sky before dawn.

With the dark sky, use binoculars or a telescope to look for the **moons of Jupiter.** See the guide on page 18.

2026

Thursday, September 10

TONIGHT

sunset: **6:59 p.m.**
twilight begins: **7:54 p.m.**

IN THE MORNING

twilight begins: **5:31 a.m.**
sunrise: **6:26 a.m.**

MOON PHASE

moonrise: **6:59 a.m.**
moonset: **6:45 p.m.**

IN THE SKY

Bright Venus sets in the west around 8:20 p.m.

The **summer triangle**'s three bright stars are Vega, Altair, and Deneb. **Vega** is in the western sky before dawn and sets in the northwest around 4:10 a.m. **Altair** is at its highest, around 65 degrees, at 9:15 p.m. and sets in the west around 3:35 a.m. Altair can be identified by its less bright neighboring stars, one on either side, together called the shaft of Aquila. **Deneb** is in the band of the Milky Way. See the illustration of the summer triangle on page 19.

The absence of the **moon** tonight provides excellent dark-sky viewing conditions.

After the moon sets, from around 7:05 p.m., look for the **Andromeda Galaxy** with binoculars or a telescope. It is in the northeastern sky in the evening. See the guide on page 11.

The **Great Bear (Big Dipper)** is in the northwestern sky at dusk and sets in the northwest around 9:35 p.m. See the illustration on page 10.

Bright star **Capella** rises in the northeast around 9:50 p.m. Read about Capella on page 14.

Scorpius sets in the southwest around 10:10 p.m. See the illustration on page 19.

The bright star **Arcturus** sets in the northwest around 10:30 p.m.

Just east of Scorpius, **Sagittarius** sets in the southwest around 11:40 p.m.

Orion rises in the east around 1:30 a.m. and is in the eastern sky before dawn. Look for the bright star **Aldebaran** and the **Pleiades** near Orion. See the illustration on page 17.

Rust-red and not very bright, Mars rises in the northeast around 1:50 a.m. and is in the eastern sky before dawn.

Saturn is in the eastern sky at dusk and at its highest, around 60 degrees, at 2:20 a.m. Unlike bright Jupiter and Venus and ruddy Mars, Saturn is harder to distinguish from the stars. Stars twinkle and shimmer, while the planets, including Saturn, show a steady flat light.

Castor and Pollux, the twin stars in the constellation Gemini, rise in the northeast around 2:25 a.m. and are in the eastern sky before dawn.

Cassiopeia, shaped like an *M,* is at its highest, around 65 degrees, at 2:40 a.m. See the illustration on page 11.

The brightest star in the sky, **Sirius** rises in the east around 3:10 a.m. and is in the eastern sky before dawn.

Jupiter rises in the east around 4 a.m. and is in the eastern sky before dawn.

Use binoculars or a telescope to look for the **moons of Jupiter.** See the guide on page 18.

2026

Friday, September 11

TONIGHT

sunset: **6:57 p.m.**
twilight begins: **7:52 p.m.**

IN THE MORNING

twilight begins: **5:32 a.m.**
sunrise: **6:26 a.m.**

MOON PHASE

moonrise: **8:03 a.m.**
moonset: **7:13 p.m.**

IN THE SKY

Bright **Venus** sets in the west around 8:15 p.m.

The **summer triangle**'s three bright stars are Vega, Altair, and Deneb. **Vega** is in the western sky before dawn and sets in the northwest around 4:10 a.m. **Altair** is at its highest, around 65 degrees, at 9:15 p.m. and sets in the west around 3:35 a.m. Altair can be identified by its less bright neighboring stars, one on either side, together called the shaft of Aquila. **Deneb** is in the band of the Milky Way. See the illustration of the summer triangle on page 19.

The **Great Bear (Big Dipper)** is in the northwestern sky at dusk and sets in the northwest around 9:35 p.m. See the illustration on page 10.

After the moon sets, from around 7:35 p.m., look for the **Andromeda Galaxy** with binoculars or a telescope. It is in the northeastern sky in the evening. See the guide on page 11.

The absence of the **moon** tonight provides excellent dark-sky viewing conditions.

Bright star **Capella** rises in the northeast around 9:45 p.m. Read about Capella on page 14.

Scorpius sets in the southwest around 10:05 p.m. See the illustration on page 19.

The bright star **Arcturus** sets in the northwest around 10:25 p.m.

Just east of Scorpius, **Sagittarius** sets in the southwest around 11:40 p.m.

Orion rises in the east around 1:25 a.m. and is in the eastern sky before dawn. Look for the bright star **Aldebaran** and the **Pleiades** near Orion. See the illustration on page 17.

Rust-red and not very bright, **Mars** rises in the northeast around 1:50 a.m. and is in the eastern sky before dawn.

Saturn is in the eastern sky at dusk and at its highest, around 60 degrees, at 2:15 a.m. Unlike bright Jupiter and Venus and ruddy Mars, Saturn is harder to distinguish from the stars. Stars twinkle and shimmer, while the planets, including Saturn, show a steady flat light.

Cassiopeia, shaped like an *M,* is at its highest, around 65 degrees, at 2:35 a.m. See the illustration on page 11.

Castor and Pollux, the twin stars in the constellation Gemini, rise in the northeast around 2:25 a.m. and are in the eastern sky before dawn.

The brightest star in the sky, **Sirius** rises in the east around 3:05 a.m. and is in the eastern sky before dawn.

Jupiter rises in the east around 4 a.m. and is in the eastern sky before dawn.

Use binoculars or a telescope to look for the **moons of Jupiter.** See the guide on page 18.

2026

Saturday, September 12

TONIGHT

sunset: **6:56 p.m.**
twilight begins: **7:51 p.m.**

IN THE MORNING

twilight begins: **5:33 a.m.**
sunrise: **6:27 a.m.**

MOON PHASE

moonrise: **8:03 a.m.**
moonset: **7:40 p.m.**

IN THE SKY

Bright **Venus** sets in the west around 8:15 p.m.

The **summer triangle**'s three bright stars are Vega, Altair, and Deneb. **Vega** is in the western sky before dawn and sets in the northwest around 4:05 a.m. **Altair** is at its highest, around 65 degrees, at 9:10 p.m. and sets in the west around 3:30 a.m. Altair can be identified by its less bright neighboring stars, one on either side, together called the shaft of Aquila. **Deneb** is in the band of the Milky Way. See the illustration of the summer triangle on page 19.

The **Great Bear (Big Dipper)** is in the northwestern sky at dusk and sets in the northwest around 9:30 p.m. See the illustration on page 10.

The absence of the **moon** tonight provides excellent dark-sky viewing conditions.

After the moon sets, from around 8 p.m., look for the **Andromeda Galaxy** with binoculars or a telescope. It is in the northeastern sky in the evening. See the guide on page 11.

Bright star **Capella** rises in the northeast around 9:40 p.m. Read about Capella on page 14.

Scorpius sets in the southwest around 10:05 p.m. See the illustration on page 19.

The bright star **Arcturus** sets in the northwest around 10:20 p.m.

Just east of Scorpius, **Sagittarius** sets in the southwest around 11:35 p.m.

Orion rises in the east around 1:20 a.m. and is in the eastern sky before dawn. Look for the bright star **Aldebaran** and the **Pleiades** near Orion. See the illustration on page 17.

Rust-red and not very bright, **Mars** rises in the northeast around 1:50 a.m. and is in the eastern sky before dawn.

Saturn is in the eastern sky at dusk and at its highest, around 60 degrees, at 2:10 a.m. Unlike bright Jupiter and Venus and ruddy Mars, Saturn is harder to distinguish from the stars. Stars twinkle and shimmer, while the planets, including Saturn, show a steady flat light.

Castor and Pollux, the twin stars in the constellation Gemini, rise in the northeast around 2:20 a.m. and are in the eastern sky before dawn.

Cassiopeia, shaped like an *M,* is at its highest, around 65 degrees, at 2:30 a.m. See the illustration on page 11.

The brightest star in the sky, **Sirius** rises in the east around 3 a.m. and is in the eastern sky before dawn.

Jupiter rises in the east around 3:55 a.m. and is in the eastern sky before dawn.

Use binoculars or a telescope to look for the **moons of Jupiter.** See the guide on page 18.

2026

Sunday, September 13

TONIGHT

sunset: **6:55 p.m.**
twilight begins: **7:49 p.m.**

MOON PHASE

IN THE MORNING

twilight begins: **5:33 a.m.**
sunrise: **6:28 a.m.**

moonrise: **9:06 a.m.**
moonset: **8:08 p.m.**

IN THE SKY

Bright **Venus** sets in the west around 8:10 p.m.

The **summer triangle**'s three bright stars are Vega, Altair, and Deneb. **Vega** is in the western sky before dawn and sets in the northwest around 4 a.m. **Altair** is at its highest, around 65 degrees, at 9:05 p.m. and sets in the west around 3:25 a.m. Altair can be identified by its less bright neighboring stars, one on either side, together called the shaft of Aquila. **Deneb** is in the band of the Milky Way. See the illustration of the summer triangle on page 19.

The absence of the **moon** tonight provides excellent dark-sky viewing conditions. Explore the **dark sky** after the moon sets, from around 8:30 p.m.

After the moon sets, from around 8:30 p.m., look for the **Andromeda Galaxy** with binoculars or a telescope. It is in the northeastern sky in the evening. See the guide on page 11.

The **Great Bear (Big Dipper)** is in the northwestern sky at dusk and sets in the northwest around 9:25 p.m. See the illustration on page 10.

Bright star **Capella** rises in the northeast around 9:35 p.m. Read about Capella on page 14.

Scorpius sets in the southwest around 10 p.m. See the illustration on page 19.

The bright star **Arcturus** sets in the northwest around 10:15 p.m.

Just east of Scorpius, **Sagittarius** sets in the southwest around 11:30 p.m.

Orion rises in the east around 1:15 a.m. and is in the eastern sky before dawn. Look for the bright star **Aldebaran** and the **Pleiades** near Orion. See the illustration on page 17.

Rust-red and not very bright, **Mars** rises in the northeast around 1:50 a.m. and is in the eastern sky before dawn.

Saturn is in the eastern sky at dusk and at its highest, around 60 degrees, at 2:10 a.m. Unlike bright Jupiter and Venus and ruddy Mars, Saturn is harder to distinguish from the stars. Stars twinkle and shimmer, while the planets, including Saturn, show a steady flat light.

Castor and Pollux, the twin stars in the constellation Gemini, rise in the northeast around 2:15 a.m. and are in the eastern sky before dawn.

Cassiopeia, shaped like an *M,* is at its highest, around 65 degrees, at 2:25 a.m. See the illustration on page 11.

The brightest star in the sky, **Sirius** rises in the east around 2:55 a.m. and is in the eastern sky before dawn.

Jupiter rises in the east around 3:50 a.m. and is in the eastern sky before dawn.

Use binoculars or a telescope to look for the **moons of Jupiter.** See the guide on page 18.

2026
Monday, September 14

TONIGHT

sunset: **6:53 p.m.**
twilight begins: **7:48 p.m.**

IN THE MORNING

twilight begins: **5:34 a.m.**
sunrise: **6:28 a.m.**

MOON PHASE

moonrise: **10:08 a.m.**
moonset: **8:39 p.m.**

IN THE SKY

Bright **Venus** sets in the west around 8:10 p.m.

The **summer triangle**'s three bright stars are Vega, Altair, and Deneb. **Vega** is in the western sky before dawn and sets in the northwest around 3:55 a.m. **Altair** is at its highest, around 65 degrees, at 9 p.m. and sets in the west around 3:20 a.m. Altair can be identified by its less bright neighboring stars, one on either side, together called the shaft of Aquila. **Deneb** is in the band of the Milky Way. See the illustration of the summer triangle on page 19.

The absence of the **moon** tonight provides excellent dark-sky viewing conditions. Explore the **dark sky** after the moon sets, from around 9 p.m.

After the moon sets, from around 9 p.m., look for the **Andromeda Galaxy** with binoculars or a telescope. It is in the northeastern sky in the evening. See the guide on page 11.

The **Great Bear (Big Dipper)** is in the northwestern sky at dusk and sets in the northwest around 9:20 p.m. See the illustration on page 10.

Bright star **Capella** rises in the northeast around 9:35 p.m. Read about Capella on page 14.

Scorpius sets in the southwest around 9:55 p.m. See the illustration on page 19.

The bright star **Arcturus** sets in the northwest around 10:15 p.m.

Just east of Scorpius, **Sagittarius** sets in the southwest around 11:25 p.m.

Orion rises in the east around 1:10 a.m. and is in the eastern sky before dawn. Look for the bright star **Aldebaran** and the **Pleiades** near Orion. See the illustration on page 17.

Rust-red and not very bright, **Mars** rises in the northeast around 1:50 a.m. and is in the eastern sky before dawn.

Saturn is in the eastern sky at dusk and at its highest, around 60 degrees, at 2:05 a.m. Unlike bright Jupiter and Venus and ruddy Mars, Saturn is harder to distinguish from the stars. Stars twinkle and shimmer, while the planets, including Saturn, show a steady flat light.

Castor and Pollux, the twin stars in the constellation Gemini, rise in the northeast around 2:10 a.m. and are in the eastern sky before dawn.

Mars and the bright stars **Castor and Pollux** appear close together in the morning sky. Look east near the horizon around 2:30 a.m.

The brightest star in the sky, **Sirius** rises in the east around 2:55 a.m. and is in the eastern sky before dawn.

Jupiter rises in the east around 3:50 a.m. and is in the eastern sky before dawn.

Use binoculars or a telescope to look for the **moons of Jupiter.** See the guide on page 18.

2026

Tuesday, September 15

TONIGHT

sunset: **6:52 p.m.**
twilight begins: **7:46 p.m.**

IN THE MORNING

twilight begins: **5:35 a.m.**
sunrise: **6:29 a.m.**

MOON PHASE

moonrise: **11:10 a.m.**
moonset: **9:14 p.m.**

IN THE SKY

At its brightest, **Venus** sets in the west around 8:05 p.m.

The **summer triangle**'s three bright stars are Vega, Altair, and Deneb. **Vega** is in the western sky before dawn and sets in the northwest around 3:50 a.m. **Altair** is at its highest, around 65 degrees, at 8:55 p.m. and sets in the west around 3:15 a.m. Altair can be identified by its less bright neighboring stars, one on either side, together called the shaft of Aquila. **Deneb** is in the band of the Milky Way. See the illustration of the summer triangle on page 19.

The **moon** is prominent in the sky early in the evening. Using a telescope or binoculars, take a look. You should be able to see rough mountains and craters along the boundary between light and dark, in astronomy called the terminator.

The **Great Bear (Big Dipper)** is in the northwestern sky at dusk and sets in the northwest around 9:15 p.m. See the illustration on page 10.

Bright star **Capella** rises in the northeast around 9:30 p.m. Read about Capella on page 14.

Explore the **dark sky** after the moon sets, from around 9:35 p.m.

After the moon sets, from around 9:35 p.m., look for the **Andromeda Galaxy** with binoculars or a telescope. It is in the northeastern sky in the evening. See the guide on page 11.

Scorpius sets in the southwest around 9:50 p.m. See the illustration on page 19.

The bright star **Arcturus** sets in the northwest around 10:10 p.m.

Just east of Scorpius, **Sagittarius** sets in the southwest around 11:20 p.m.

Rust-red and not very bright, **Mars** rises in the northeast around 1:45 a.m. and is in the eastern sky before dawn.

Saturn is in the eastern sky at dusk and at its highest, around 60 degrees, at 2 a.m. Unlike bright Jupiter and Venus and ruddy Mars, Saturn is harder to distinguish from the stars. Stars twinkle and shimmer, while the planets, including Saturn, show a steady flat light.

Castor and Pollux, the twin stars in the constellation Gemini, rise in the northeast around 2:05 a.m. and are in the eastern sky before dawn.

Cassiopeia, shaped like an *M,* is at its highest, around 65 degrees, at 2:20 a.m. See the illustration on page 11.

Mars and the bright stars **Castor and Pollux** appear close together in the morning sky. Look east near the horizon around 2:30 a.m.

Jupiter rises in the east around 3:45 a.m. and is in the eastern sky before dawn.

Use binoculars or a telescope to look for the **moons of Jupiter.** See the guide on page 18.

2026

Wednesday, September 16

TONIGHT

sunset: **6:50 p.m.**
twilight begins: **7:45 p.m.**

IN THE MORNING

twilight begins: **5:36 a.m.**
sunrise: **6:30 a.m.**

MOON PHASE

moonrise: **12:11 p.m.**
moonset: **9:54 p.m.**

IN THE SKY

At its brightest, **Venus** sets in the west around 8 p.m.

The **moon** is prominent in the sky early in the evening. Using a telescope or binoculars, take a look. You should be able to see rough mountains and craters along the boundary between light and dark, in astronomy called the terminator.

The **Great Bear (Big Dipper)** is in the northwestern sky at dusk and sets in the northwest around 9:15 p.m. See the illustration on page 10.

Bright star **Capella** rises in the northeast around 9:25 p.m. Read about Capella on page 14.

Scorpius sets in the southwest around 9:45 p.m. See the illustration on page 19.

The bright star **Arcturus** sets in the northwest around 10:05 p.m.

Explore the **dark sky** after the moon sets, from around 10:15 p.m.

After the moon sets, from around 10:15 p.m., look for the **Andromeda Galaxy** with binoculars or a telescope. It is in the northeastern sky in the evening. See the guide on page 11.

Just east of Scorpius, **Sagittarius** sets in the southwest around 11:20 p.m.

Orion rises in the east around 1:05 a.m. and is in the eastern sky before dawn. Look for the bright star **Aldebaran** and the **Pleiades** near Orion. See the illustration on page 17.

Rust-red and not very bright, **Mars** rises in the northeast around 1:45 a.m. and is in the eastern sky before dawn.

Saturn is in the eastern sky at dusk and at its highest, around 60 degrees, at 1:55 a.m. Unlike bright Jupiter and Venus and ruddy Mars, Saturn is harder to distinguish from the stars. Stars twinkle and shimmer, while the planets, including Saturn, show a steady flat light.

Castor and Pollux, the twin stars in the constellation Gemini, rise in the northeast around 2:05 a.m. and are in the eastern sky before dawn.

Cassiopeia, shaped like an *M,* is at its highest, around 65 degrees, at 2:15 a.m. See the illustration on page 11.

The brightest star in the sky, **Sirius** rises in the east around 2:45 a.m. and is in the eastern sky before dawn.

Mars and the bright stars **Castor and Pollux** appear close together in the morning sky. Look east near the horizon around 2:30 a.m.

Jupiter rises in the east around 3:45 a.m. and is in the eastern sky before dawn.

Use binoculars or a telescope to look for the **moons of Jupiter.** See the guide on page 18.

The **Great Bear (Big Dipper)** rises in the northeast around 4:10 a.m. and is in the northeastern sky before dawn. See the illustration on page 10.

2026

Thursday, September 17

TONIGHT

sunset: **6:49 p.m.**
twilight begins: **7:43 p.m.**

IN THE MORNING

twilight begins: **5:36 a.m.**
sunrise: **6:31 a.m.**

MOON PHASE

moonrise: **1:09 p.m.**
moonset: **10:40 p.m.**

IN THE SKY

At its brightest, **Venus** sets in the west around 8 p.m.

The **moon** is prominent in the sky early in the evening. Using a telescope or binoculars, take a look. You should be able to see rough mountains and craters along the boundary between light and dark, in astronomy called the terminator.

The **summer triangle**'s three bright stars are Vega, Altair, and Deneb. **Vega** is in the western sky before dawn and sets in the northwest around 3:45 a.m. **Altair** is at its highest, around 65 degrees, at 8:50 p.m. and sets in the west around 3:10 a.m. Altair can be identified by its less bright neighboring stars, one on either side, together called the shaft of Aquila. **Deneb** is in the band of the Milky Way. See the illustration of the summer triangle on page 19.

The **Great Bear (Big Dipper)** is in the northwestern sky at dusk and sets in the northwest around 9:10 p.m. See the illustration on page 10.

Bright star **Capella** rises in the northeast around 9:20 p.m. Read about Capella on page 14.

The bright star **Arcturus** sets in the northwest around 10 p.m.

Explore the **dark sky** after the moon sets, from around 11 p.m.

After the moon sets, from around 11 p.m., look for the **Andromeda Galaxy** with binoculars or a telescope. It is in the northeastern sky in the evening. See the guide on page 11.

Just east of Scorpius, **Sagittarius** sets in the southwest around 11:15 p.m.

Rust-red and not very bright, **Mars** rises in the northeast around 1:45 a.m. and is in the eastern sky before dawn.

Saturn is visible most of the night and is in the eastern sky at dusk and at its highest, around 60 degrees, at 1:50 a.m. Unlike bright Jupiter and Venus and ruddy Mars, Saturn is harder to distinguish from the stars. Stars twinkle and shimmer, while the planets, including Saturn, show a steady flat light.

Castor and Pollux, the twin stars in the constellation Gemini, rise in the northeast around 2 a.m. and are in the eastern sky before dawn.

Cassiopeia, shaped like an *M,* is at its highest, around 65 degrees, at 2:10 a.m. See the illustration on page 11.

Mars and the bright stars **Castor and Pollux** appear close together in the morning sky. Look east near the horizon around 2:30 a.m.

The brightest star in the sky, **Sirius** rises in the east around 2:40 a.m. and is in the eastern sky before dawn.

Jupiter rises in the east around 3:40 a.m. and is in the eastern sky before dawn.

Use binoculars or a telescope to look for the **moons of Jupiter.** See the guide on page 18.

2026

Friday, September 18

TONIGHT

sunset: **6:47 p.m.**
twilight begins: **7:42 p.m.**

IN THE MORNING

twilight begins: **5:37 a.m.**
sunrise: **6:31 a.m.**

MOON PHASE

moonrise: **2:02 p.m.**
moonset: **11:31 p.m.**

IN THE SKY

At its brightest, Venus sets in the west around 7:55 p.m.

The **summer triangle**'s three bright stars are Vega, Altair, and Deneb. **Vega** is in the western sky before dawn and sets in the northwest around 3:40 a.m. **Altair** is at its highest, around 65 degrees, at 8:45 p.m. and sets in the west around 3:05 a.m. Altair can be identified by its less bright neighboring stars, one on either side, together called the shaft of Aquila. **Deneb** is in the band of the Milky Way. See the illustration of the summer triangle on page 19.

Bright star **Capella** rises in the northeast around 9:15 p.m. Read about Capella on page 14.

Scorpius is in the western sky at dusk and sets in the southwest around 9:40 p.m. See the illustration on page 19.

Explore the **dark sky** after the moon sets, from around 11:50 p.m.

After the moon sets, from around 11:50 p.m., look for the **Andromeda Galaxy** with binoculars or a telescope. It is in the northeastern sky in the evening. See the guide on page 11.

Orion rises in the east around 12:55 a.m. and is in the eastern sky before dawn. Look for the bright star **Aldebaran** and the **Pleiades** near Orion. See the illustration on page 17.

Rust-red and not very bright, Mars rises in the northeast around 1:45 a.m. and is in the eastern sky before dawn.

Saturn is visible most of the night and is in the eastern sky at dusk and at its highest, around 60 degrees, at 1:45 a.m. Unlike bright Jupiter and Venus and ruddy Mars, Saturn is harder to distinguish from the stars. Stars twinkle and shimmer, while the planets, including Saturn, show a steady flat light.

Cassiopeia, shaped like an *M,* is at its highest, around 65 degrees, at 2:05 a.m. See the illustration on page 11.

Mars and the bright stars **Castor and Pollux** appear close together in the morning sky. Look east near the horizon after 2 a.m.

The brightest star in the sky, **Sirius** rises in the east around 2:35 a.m. and is in the eastern sky before dawn.

Jupiter rises in the east around 3:35 a.m. and is in the eastern sky before dawn.

Use binoculars or a telescope to look for the **moons of Jupiter.** See the guide on page 18.

2026

Saturday, September 19

TONIGHT

sunset: **6:46 p.m.**
twilight begins: **7:40 p.m.**

IN THE MORNING

twilight begins: **5:38 a.m.**
sunrise: **6:32 a.m.**

MOON PHASE

moonrise: **2:50 p.m.**
moonset: **12:27 a.m.**

IN THE SKY

At its brightest, **Venus** sets in the southwest around 7:55 p.m.

The **summer triangle**'s three bright stars are Vega, Altair, and Deneb. **Vega** is in the western sky before dawn and sets in the northwest around 3:35 a.m. **Altair** is at its highest, around 65 degrees, at 8:40 p.m. and sets in the west around 3 a.m. Altair can be identified by its less bright neighboring stars, one on either side, together called the shaft of Aquila. **Deneb** is in the band of the Milky Way. See the illustration of the summer triangle on page 19.

The **moon** is prominent in the sky early in the evening. Using a telescope or binoculars, take a look. You should be able to see rough mountains and craters along the boundary between light and dark, in astronomy called the terminator.

The **Great Bear (Big Dipper)** is in the northwestern sky at dusk and sets in the northwest around 9 p.m. See the illustration on page 10.

Bright star **Capella** rises in the northeast around 9:15 p.m. Read about Capella on page 14.

Scorpius is in the western sky at dusk and sets in the southwest around 9:35 p.m. See the illustration on page 19.

The bright star **Arcturus** sets in the northwest around 9:55 p.m.

Just east of Scorpius, **Sagittarius** sets in the southwest around 11:05 p.m.

Orion rises in the east around 12:50 a.m. and is in the eastern sky before dawn. Look for the bright star **Aldebaran** and the **Pleiades** near Orion. See the illustration on page 17.

Saturn is visible most of the night and is in the eastern sky at dusk and at its highest, around 60 degrees, at 1:40 a.m. Unlike bright Jupiter and Venus and ruddy Mars, Saturn is harder to distinguish from the stars. Stars twinkle and shimmer, while the planets, including Saturn, show a steady flat light.

Rust-red and not very bright, **Mars** rises in the northeast around 1:45 a.m. and is in the eastern sky before dawn.

Castor and Pollux, the twin stars in the constellation Gemini, rise in the northeast around 1:50 a.m. and are in the eastern sky before dawn.

Cassiopeia, shaped like an *M,* is at its highest, around 65 degrees, at 2 a.m. See the illustration on page 11.

The brightest star in the sky, **Sirius** rises in the east around 2:35 a.m. and is in the eastern sky before dawn.

Jupiter rises in the east around 3:35 a.m. and is in the eastern sky before dawn.

Use binoculars or a telescope to look for the **moons of Jupiter.** See the guide on page 18.

2026
Sunday, September 20

TONIGHT

sunset: **6:45 p.m.**
twilight begins: **7:39 p.m.**

IN THE MORNING

twilight begins: **5:39 a.m.**
sunrise: **6:33 a.m.**

MOON PHASE

moonrise: **3:32 p.m.**
moonset: **1:26 a.m.**

IN THE SKY

At its brightest, **Venus** sets in the southwest around 7:50 p.m.

The **summer triangle**'s three bright stars are Vega, Altair, and Deneb. **Vega** is in the western sky before dawn and sets in the northwest around 3:35 a.m. **Altair** is at its highest, around 65 degrees, around dusk and sets in the west around 3 a.m. Altair can be identified by its less bright neighboring stars, one on either side, together called the shaft of Aquila. **Deneb** is in the band of the Milky Way. See the illustration of the summer triangle on page 19.

The **Great Bear (Big Dipper)** is in the northwestern sky at dusk and sets in the northwest around 9 p.m. See the illustration on page 10.

Bright star **Capella** rises in the northeast around 9:10 p.m. Read about Capella on page 14.

Scorpius is in the western sky at dusk and sets in the southwest around 9:30 p.m. See the illustration on page 19.

The bright star **Arcturus** sets in the northwest around 9:50 p.m.

Just east of Scorpius, **Sagittarius** sets in the southwest around 11 p.m.

Orion rises in the east around 12:50 a.m. and is in the eastern sky before dawn. Look for the bright star **Aldebaran** and the **Pleiades** near Orion. See the illustration on page 17.

Rust-red and not very bright, **Mars** rises in the northeast around 1:40 a.m. and is in the eastern sky before dawn.

Saturn is visible most of the night and is in the eastern sky at dusk and at its highest, around 60 degrees, at 1:40 a.m. Unlike bright Jupiter and Venus and ruddy Mars, Saturn is harder to distinguish from the stars. Stars twinkle and shimmer, while the planets, including Saturn, show a steady flat light.

Castor and Pollux, the twin stars in the constellation Gemini, rise in the northeast around 1:50 a.m. and are in the eastern sky before dawn.

Cassiopeia, shaped like an *M,* is at its highest, around 65 degrees, at 2 a.m. See the illustration on page 11.

The brightest star in the sky, **Sirius** rises in the east around 2:30 a.m. and is in the eastern sky before dawn.

Jupiter rises in the east around 3:30 a.m. and is in the eastern sky before dawn.

The **Great Bear (Big Dipper)** rises in the northeast around 3:55 a.m. and is in the northeastern sky before dawn. See the illustration on page 10.

2026

Monday, September 21

TONIGHT

sunset: **6:43 p.m.**
twilight begins: **7:38 p.m.**

IN THE MORNING

twilight begins: **5:39 a.m.**
sunrise: **6:33 a.m.**

MOON PHASE

moonrise: **4:09 p.m.**
moonset: **2:27 a.m.**

IN THE SKY

At its brightest, **Venus** sets in the southwest around 7:45 p.m.

The **summer triangle**'s three bright stars are Vega, Altair, and Deneb. **Vega** sets in the northwest around 3:30 a.m. **Altair** is at its highest, around 65 degrees, around dusk and sets in the west around 2:55 a.m. Altair can be identified by its less bright neighboring stars, one on either side, together called the shaft of Aquila. **Deneb** is in the band of the Milky Way. See the illustration of the summer triangle on page 19.

The **Great Bear (Big Dipper)** is in the northwestern sky at dusk and sets in the northwest around 8:55 p.m. See the illustration on page 10.

Bright star **Capella** rises in the northeast around 9:05 p.m. Read about Capella on page 14.

Scorpius is in the western sky at dusk and sets in the southwest around 9:25 p.m. See the illustration on page 19.

The bright star **Arcturus** sets in the northwest around 9:45 p.m.

Just east of Scorpius, **Sagittarius** sets in the southwest around 11 p.m.

Orion rises in the east around 12:45 a.m. and is in the eastern sky before dawn. Look for the bright star **Aldebaran** and the **Pleiades** near Orion. See the illustration on page 17.

Saturn is visible most of the night and is in the eastern sky at dusk and at its highest, around 60 degrees, at 1:35 a.m. Unlike bright Jupiter and Venus and ruddy Mars, Saturn is harder to distinguish from the stars. Stars twinkle and shimmer, while the planets, including Saturn, show a steady flat light.

Rust-red and not very bright, **Mars** rises in the northeast around 1:40 a.m. and is in the eastern sky before dawn.

Castor and Pollux, the twin stars in the constellation Gemini, rise in the northeast around 1:45 a.m. and are in the eastern sky before dawn.

Cassiopeia, shaped like an *M,* is at its highest, around 65 degrees, at 1:55 a.m. See the illustration on page 11.

The brightest star in the sky, **Sirius** rises in the east around 2:25 a.m. and is in the eastern sky before dawn.

Jupiter rises in the east around 3:30 a.m. and is in the eastern sky before dawn.

The **Great Bear (Big Dipper)** rises in the northeast around 3:50 a.m. and is in the northeastern sky before dawn. See the illustration on page 10.

2026

Tuesday, September 22

TONIGHT

sunset: **6:42 p.m.**
twilight begins: **7:36 p.m.**

IN THE MORNING

twilight begins: **5:40 a.m.**
sunrise: **6:34 a.m.**

MOON PHASE

moonrise: **4:41 p.m.**
moonset: **3:28 a.m.**

IN THE SKY

Bright **Venus** sets in the southwest around 7:45 p.m.

The nearly **full moon** tonight acts like a spotlight. Look around at the desert landscape at how much illumination the moon provides.

The **summer triangle**'s three bright stars are Vega, Altair, and Deneb. **Vega** sets in the northwest around 3:25 a.m. **Altair** is at its highest, around 65 degrees, around dusk and sets in the west around 2:50 a.m. Altair can be identified by its less bright neighboring stars, one on either side, together called the shaft of Aquila. **Deneb** is in the band of the Milky Way. See the illustration of the summer triangle on page 19.

The **Great Bear (Big Dipper)** is in the northwestern sky at dusk and sets in the northwest around 8:50 p.m. See the illustration on page 10.

Bright star **Capella** rises in the northeast around 9 p.m. Read about Capella on page 14.

Scorpius is in the western sky at dusk and sets in the southwest around 9:25 p.m. See the illustration on page 19.

The bright star **Arcturus** sets in the northwest around 9:40 p.m.

Just east of Scorpius, **Sagittarius** sets in the southwest around 10:55 p.m.

Orion rises in the east around 12:40 a.m. and is in the eastern sky before dawn. Look for the bright star **Aldebaran** and the **Pleiades** near Orion. See the illustration on page 17.

Saturn is visible most of the night and is in the eastern sky at dusk and at its highest, around 60 degrees, at 1:30 a.m. Unlike bright Jupiter and Venus and ruddy Mars, Saturn is harder to distinguish from the stars. Stars twinkle and shimmer, while the planets, including Saturn, show a steady flat light.

Castor and Pollux, the twin stars in the constellation Gemini, rise in the northeast around 1:35 a.m. and are in the eastern sky before dawn.

Rust-red **Mars** rises in the northeast around 1:40 a.m. and is in the eastern sky before dawn.

Cassiopeia, shaped like an *M,* is at its highest, around 65 degrees, at 1:50 a.m. See the illustration on page 11.

The brightest star in the sky, **Sirius** rises in the east around 2:20 a.m. and is in the eastern sky before dawn.

Jupiter rises in the east around 3:25 a.m. and is in the eastern sky before dawn.

The **Great Bear (Big Dipper)** rises in the northeast around 3:50 a.m. and is in the northeastern sky before dawn. See the illustration on page 10.

2026

Wednesday, September 23

TONIGHT

sunset: **6:40 p.m.**
twilight begins: **7:35 p.m.**

MOON PHASE

IN THE MORNING

twilight begins: **5:41 a.m.**
sunrise: **6:35 a.m.**

moonrise: **5:10 p.m.**
moonset: **4:30 a.m.**

IN THE SKY

Bright **Venus** sets in the southwest around 7:40 p.m.

The **summer triangle**'s three bright stars are Vega, Altair, and Deneb. **Vega** sets in the northwest around 3:20 a.m. **Altair** is at its highest, around 65 degrees, around dusk and sets in the west around 2:45 a.m. Altair can be identified by its less bright neighboring stars, one on either side, together called the shaft of Aquila. **Deneb** is in the band of the Milky Way. See the illustration of the summer triangle on page 19.

The **full moon** tonight acts like a spotlight. Look around at the desert landscape at how much illumination the moon provides.

The **Great Bear (Big Dipper)** is in the northwestern sky at dusk and sets in the northwest around 8:45 p.m. See the illustration on page 10.

Bright star **Capella** rises in the northeast around 8:55 p.m. Read about Capella on page 14.

Scorpius is in the western sky at dusk and sets in the southwest around 9:20 p.m. See the illustration on page 19.

The bright star **Arcturus** sets in the northwest around 9:35 p.m.

Just east of Scorpius, **Sagittarius** sets in the southwest around 10:50 p.m.

Orion rises in the east around 12:35 a.m. and is in the eastern sky before dawn. Look for the bright star **Aldebaran** and the **Pleiades** near Orion. See the illustration on page 17.

Saturn is visible most of the night and is in the eastern sky at dusk and at its highest, around 60 degrees, at 1:25 a.m. Unlike bright Jupiter and Venus and ruddy Mars, Saturn is harder to distinguish from the stars. Stars twinkle and shimmer, while the planets, including Saturn, show a steady flat light.

Castor and Pollux, the twin stars in the constellation Gemini, rise in the northeast around 1:30 a.m. and are in the eastern sky before dawn.

Rust-red **Mars** rises in the northeast around 1:40 a.m. and is in the eastern sky before dawn.

Cassiopeia, shaped like an *M,* is at its highest, around 65 degrees, at 1:45 a.m. See the illustration on page 11.

The brightest star in the sky, **Sirius** rises in the east around 2:15 a.m. and is in the eastern sky before dawn.

Jupiter rises in the east around 3:20 a.m. and is in the eastern sky before dawn.

The **Great Bear (Big Dipper)** rises in the northeast around 3:45 a.m. and is in the northeastern sky before dawn. See the illustration on page 10.

2026

Thursday, September 24

TONIGHT

sunset: **6:39 p.m.**
twilight begins: **7:33 p.m.**

IN THE MORNING

twilight begins: **5:42 a.m.**
sunrise: **6:36 a.m.**

MOON PHASE

moonrise: **5:38 p.m.**
moonset: **5:31 a.m.**

IN THE SKY

Bright **Venus** sets in the southwest around 7:35 p.m.

The **summer triangle**'s three bright stars are Vega, Altair, and Deneb. **Vega** sets in the northwest around 3:15 a.m. **Altair** is at its highest, around 65 degrees, around dusk and sets in the west around 2:40 a.m. Altair can be identified by its less bright neighboring stars, one on either side, together called the shaft of Aquila. **Deneb** is in the band of the Milky Way. See the illustration of the summer triangle on page 19.

The **full moon** tonight acts like a spotlight. Look around at the desert landscape at how much illumination the moon provides.

The **Great Bear (Big Dipper)** is in the northwestern sky at dusk and sets in the northwest around 8:40 p.m. See the illustration on page 10.

Bright star **Capella** rises in the northeast around 8:55 p.m. Read about Capella on page 14.

Scorpius is in the western sky at dusk and sets in the southwest around 9:15 p.m. See the illustration on page 19.

The bright star **Arcturus** sets in the northwest around 9:35 p.m.

Just east of Scorpius, **Sagittarius** sets in the southwest around 10:45 p.m.

Orion rises in the east around 12:35 a.m. and is in the eastern sky before dawn. Look for the bright star **Aldebaran** and the **Pleiades** near Orion. See the illustration on page 17.

Saturn is visible most of the night and is in the eastern sky at dusk and at its highest, around 60 degrees, at 1:20 a.m. Unlike bright Jupiter and Venus and ruddy Mars, Saturn is harder to distinguish from the stars. Stars twinkle and shimmer, while the planets, including Saturn, show a steady flat light.

Castor and Pollux, the twin stars in the constellation Gemini, rise in the northeast around 1:30 a.m. and are in the eastern sky before dawn.

Rust-red **Mars** rises in the northeast around 1:40 a.m. and is in the eastern sky before dawn.

Cassiopeia, shaped like an *M,* is at its highest, around 65 degrees, at 1:45 a.m. See the illustration on page 11.

The brightest star in the sky, **Sirius** rises in the east around 2:15 a.m. and is in the eastern sky before dawn.

Jupiter rises in the east around 3:20 a.m. and is in the eastern sky before dawn.

The **Great Bear (Big Dipper)** rises in the northeast around 3:40 a.m. and is in the northeastern sky before dawn. See the illustration on page 10.

2026

Friday, September 25

TONIGHT

sunset: **6:38 p.m.**
twilight begins: **7:32 p.m.**

IN THE MORNING

twilight begins: **5:42 a.m.**
sunrise: **6:36 a.m.**

MOON PHASE

moonrise: **6:04 p.m.**
moonset: **6:34 a.m.**

IN THE SKY

Bright **Venus** sets in the southwest around 7:30 p.m.

The **summer triangle**'s three bright stars are Vega, Altair, and Deneb. **Vega** sets in the northwest around 3:15 a.m. **Altair** is at its highest, around 65 degrees, around dusk and sets in the west around 2:40 a.m. Altair can be identified by its less bright neighboring stars, one on either side, together called the shaft of Aquila. **Deneb** is in the band of the Milky Way. See the illustration of the summer triangle on page 19.

The **full moon** tonight acts like a spotlight. Look around at the desert landscape at how much illumination the moon provides.

The **Great Bear (Big Dipper)** is in the northwestern sky at dusk and sets in the northwest around 8:40 p.m. See the illustration on page 10.

Bright star **Capella** rises in the northeast around 8:50 p.m. Read about Capella on page 14.

Scorpius is in the western sky at dusk and sets in the southwest around 9:10 p.m. See the illustration on page 19.

The bright star **Arcturus** sets in the northwest around 9:30 p.m.

Just east of Scorpius, **Sagittarius** sets in the southwest around 10:45 p.m.

Orion rises in the east around 12:30 a.m. and is in the eastern sky before dawn. Look for the bright star **Aldebaran** and the **Pleiades** near Orion. See the illustration on page 17.

Saturn is visible most of the night and is in the eastern sky at dusk and at its highest, around 60 degrees, at 1:15 a.m. Unlike bright Jupiter and Venus and ruddy Mars, Saturn is harder to distinguish from the stars. Stars twinkle and shimmer, while the planets, including Saturn, show a steady flat light.

Castor and Pollux, the twin stars in the constellation Gemini, rise in the northeast around 1:25 a.m. and are in the eastern sky before dawn.

Rust-red **Mars** rises in the northeast around 1:35 a.m. and is in the eastern sky before dawn.

Cassiopeia, shaped like an *M,* is at its highest, around 65 degrees, at 1:40 a.m. See the illustration on page 11.

The brightest star in the sky, **Sirius** rises in the east around 2:10 a.m. and is in the eastern sky before dawn.

Jupiter rises in the east around 3:15 a.m. and is in the eastern sky before dawn.

The **Great Bear (Big Dipper)** rises in the northeast around 3:35 a.m. and is in the northeastern sky before dawn. See the illustration on page 10.

2026

Saturday, September 26

TONIGHT

sunset: **6:36 p.m.**
twilight begins: **7:30 p.m.**

IN THE MORNING

twilight begins: **5:43 a.m.**
sunrise: **6:37 a.m.**

MOON PHASE

moonrise: **6:31 p.m.**
moonset: **7:39 a.m.**

IN THE SKY

Bright **Venus** sets in the southwest around 7:30 p.m.

The **summer triangle**'s three bright stars are Vega, Altair, and Deneb. **Vega** sets in the northwest around 3:10 a.m. **Altair** is at its highest, around 65 degrees, around dusk and sets in the west around 2:35 a.m. Altair can be identified by its less bright neighboring stars, one on either side, together called the shaft of Aquila. **Deneb** is in the band of the Milky Way. See the illustration of the summer triangle on page 19.

The **full moon** tonight acts like a spotlight. Look around at the desert landscape at how much illumination the moon provides.

The **Great Bear (Big Dipper)** is in the northwestern sky at dusk and sets in the northwest around 8:35 p.m. See the illustration on page 10.

Bright star **Capella** rises in the northeast around 8:45 p.m. Read about Capella on page 14.

Scorpius is in the western sky at dusk and sets in the southwest around 9:10 p.m. See the illustration on page 19.

The bright star **Arcturus** sets in the northwest around 9:25 p.m.

Just east of Scorpius, **Sagittarius** sets in the southwest around 10:40 p.m.

Orion rises in the east around 12:25 a.m. and is in the eastern sky before dawn. Look for the bright star **Aldebaran** and the **Pleiades** near Orion. See the illustration on page 17.

Saturn is visible most of the night and is in the eastern sky at dusk and at its highest, around 60 degrees, at 1:15 a.m. Unlike bright Jupiter and Venus and ruddy Mars, Saturn is harder to distinguish from the stars. Stars twinkle and shimmer, while the planets, including Saturn, show a steady flat light.

Castor and Pollux, the twin stars in the constellation Gemini, rise in the northeast around 1:20 a.m. and are in the eastern sky before dawn.

Rust-red **Mars** rises in the northeast around 1:35 a.m. and is in the eastern sky before dawn.

Cassiopeia, shaped like an *M,* is at its highest, around 65 degrees, at 1:35 a.m. See the illustration on page 11.

The brightest star in the sky, **Sirius** rises in the east around 2:05 a.m. and is in the eastern sky before dawn.

Jupiter rises in the east around 3:15 a.m. and is in the eastern sky before dawn.

The **Great Bear (Big Dipper)** rises in the northeast around 3:30 a.m. and is in the northeastern sky before dawn. See the illustration on page 10.

2026

Sunday, September 27

TONIGHT

sunset: **6:35 p.m.**
twilight begins: **7:29 p.m.**

IN THE MORNING

twilight begins: **5:44 a.m.**
sunrise: **6:38 a.m.**

MOON PHASE

moonrise: **7:01 p.m.**
moonset: **7:39 a.m.**

IN THE SKY

Bright **Venus** sets in the southwest around 7:25 p.m.

The **summer triangle**'s three bright stars are Vega, Altair, and Deneb. **Vega** sets in the northwest around 3:05 a.m. **Altair** is at its highest, around 65 degrees, around dusk and sets in the west around 2:30 a.m. Altair can be identified by its less bright neighboring stars, one on either side, together called the shaft of Aquila. **Deneb** is in the band of the Milky Way. See the illustration of the summer triangle on page 19.

The **full moon** tonight acts like a spotlight. Look around at the desert landscape at how much illumination the moon provides.

The **Great Bear (Big Dipper)** is in the northwestern sky at dusk and sets in the northwest around 8:30 p.m. See the illustration on page 10.

Bright star **Capella** rises in the northeast around 8:40 p.m. Read about Capella on page 14.

Scorpius is in the western sky at dusk and sets in the southwest around 9:05 p.m. See the illustration on page 19.

The bright star **Arcturus** sets in the northwest around 9:20 p.m.

Just east of Scorpius, **Sagittarius** sets in the southwest around 10:35 p.m.

Orion rises in the east around 12:20 a.m. and is in the eastern sky before dawn. Look for the bright star **Aldebaran** and the **Pleiades** near Orion. See the illustration on page 17.

Saturn is visible most of the night and is in the eastern sky at dusk and at its highest, around 60 degrees, at 1:10 a.m. Unlike bright Jupiter and Venus and ruddy Mars, Saturn is harder to distinguish from the stars. Stars twinkle and shimmer, while the planets, including Saturn, show a steady flat light.

Castor and Pollux, the twin stars in the constellation Gemini, rise in the northeast around 1:15 a.m. and are in the eastern sky before dawn.

Cassiopeia, shaped like an *M,* is at its highest, around 65 degrees, at 1:30 a.m. See the illustration on page 11.

Rust-red **Mars** rises in the northeast around 1:35 a.m. and is in the eastern sky before dawn.

The brightest star in the sky, **Sirius** rises in the east around 2 a.m. and is in the eastern sky before dawn.

Jupiter rises in the east around 3:10 a.m. and is in the eastern sky before dawn.

The **Great Bear (Big Dipper)** rises in the northeast around 3:30 a.m. and is in the northeastern sky before dawn. See the illustration on page 10.

2026

Monday, September 28

TONIGHT

sunset: **6:33 p.m.**
twilight begins: **7:28 p.m.**

IN THE MORNING

twilight begins: **5:45 a.m.**
sunrise: **6:38 a.m.**

MOON PHASE

moonrise: **7:35 p.m.**
moonset: **8:47 a.m.**

IN THE SKY

Bright **Venus** sets in the southwest around 7:20 p.m.

Scorpius is in the western sky at dusk and sets in the southwest around 9 p.m. See the illustration on page 19.

The **summer triangle**'s three bright stars are Vega, Altair, and Deneb. **Vega** sets in the northwest around 3 a.m. **Altair** is at its highest, around 65 degrees, around dusk and sets in the west around 2:25 a.m. Altair can be identified by its less bright neighboring stars, one on either side, together called the shaft of Aquila. **Deneb** is in the band of the Milky Way. See the illustration of the summer triangle on page 19.

The nearly **full moon** tonight acts like a spotlight. Look around at the desert landscape at how much illumination the moon provides.

The **Great Bear (Big Dipper)** is in the northwestern sky at dusk and sets in the northwest around 8:25 p.m. See the illustration on page 10.

Bright star **Capella** rises in the northeast around 8:40 p.m. Read about Capella on page 14.

The bright star **Arcturus** sets in the northwest around 9:20 p.m.

Just east of Scorpius, **Sagittarius** sets in the southwest around 10:30 p.m.

Orion rises in the east around 12:15 a.m. and is in the eastern sky before dawn. Look for the bright star **Aldebaran** and the **Pleiades** near Orion. See the illustration on page 17.

At its closest and brightest for the year, **Saturn** is visible most of the night and is in the eastern sky at dusk and at its highest, around 60 degrees, at 1:05 a.m. Unlike bright Jupiter and Venus and ruddy Mars, Saturn is harder to distinguish from the stars. Stars twinkle and shimmer, while the planets, including Saturn, show a steady flat light.

Castor and Pollux, the twin stars in the constellation Gemini, rise in the northeast around 1:10 a.m. and are in the eastern sky before dawn.

Cassiopeia, shaped like an *M,* is at its highest, around 65 degrees, at 1:25 a.m. See the illustration on page 11.

Rust-red **Mars** rises in the northeast around 1:35 a.m. and is in the eastern sky before dawn.

The brightest star in the sky, **Sirius** rises in the east around 1:55 a.m. and is in the eastern sky before dawn.

Jupiter rises in the east around 3:05 a.m. and is in the eastern sky before dawn.

The **Great Bear (Big Dipper)** rises in the northeast around 3:25 a.m. and is in the northeastern sky before dawn. See the illustration on page 10.

2026

Tuesday, September 29

TONIGHT

sunset: **6:32 p.m.**
twilight begins: **7:26 p.m.**

IN THE MORNING

twilight begins: **5:45 a.m.**
sunrise: **6:39 a.m.**

MOON PHASE

moonrise: **8:14 p.m.**
moonset: **9:58 a.m.**

IN THE SKY

Bright **Venus** sets in the southwest around 7:15 p.m.

Scorpius is in the western sky at dusk and sets in the southwest around 8:55 p.m. See the illustration on page 19.

The **summer triangle**'s three bright stars are Vega, Altair, and Deneb. **Vega** sets in the northwest around 2:55 a.m. **Altair** is at its highest, around 65 degrees, around dusk and sets in the west around 2:20 a.m. Altair can be identified by its less bright neighboring stars, one on either side, together called the shaft of Aquila. **Deneb** is in the band of the Milky Way. See the illustration of the summer triangle on page 19.

The **Great Bear (Big Dipper)** is in the northwestern sky at dusk and sets in the northwest around 8:20 p.m. See the illustration on page 10.

Bright star **Capella** rises in the northeast around 8:35 p.m. Read about Capella on page 14.

The bright star **Arcturus** is in the western sky at dusk and sets in the northwest around 9:15 p.m.

Just east of Scorpius, **Sagittarius** sets in the southwest around 10:25 p.m.

Orion rises in the east around 12:15 a.m. and is in the eastern sky before dawn. Look for the bright star **Aldebaran** and the **Pleiades** near Orion. See the illustration on page 17.

At its closest and brightest for the year, **Saturn** is visible most of the night and is in the eastern sky at dusk and at its highest, around 60 degrees, at 1 a.m. Unlike bright Jupiter and Venus and ruddy Mars, Saturn is harder to distinguish from the stars. Stars twinkle and shimmer, while the planets, including Saturn, show a steady flat light.

Castor and Pollux, the twin stars in the constellation Gemini, rise in the northeast around 1:10 a.m. and are in the eastern sky before dawn.

Cassiopeia, shaped like an *M,* is at its highest, around 65 degrees, at 1:25 a.m. See the illustration on page 11.

Rust-red **Mars** rises in the northeast around 1:30 a.m. and is in the eastern sky before dawn.

The brightest star in the sky, **Sirius** rises in the east around 1:55 a.m. and is in the eastern sky before dawn.

Jupiter rises in the east around 3:05 a.m. and is in the eastern sky before dawn.

The **Great Bear (Big Dipper)** rises in the northeast around 3:20 a.m. and is in the northeastern sky before dawn. See the illustration on page 10.

2026
Wednesday, September 30

TONIGHT

sunset: **6:31 p.m.**
twilight begins: **7:25 p.m.**

IN THE MORNING

twilight begins: **5:46 a.m.**
sunrise: **6:40 a.m.**

MOON PHASE

moonrise: **9:02 p.m.**
moonset: **11:10 a.m.**

IN THE SKY

Bright **Venus** sets in the southwest around 7:15 p.m.

The **summer triangle**'s three bright stars are Vega, Altair, and Deneb. **Vega** sets in the northwest around 2:55 a.m. **Altair** is at its highest, around 65 degrees, around dusk and sets in the west around 2:20 a.m. Altair can be identified by its less bright neighboring stars, one on either side, together called the shaft of Aquila. **Deneb** is in the band of the Milky Way. See the illustration of the summer triangle on page 19.

The **Great Bear (Big Dipper)** is in the northwestern sky at dusk and sets in the northwest around 8:20 p.m. See the illustration on page 10.

Bright star **Capella** rises in the northeast around 8:30 p.m. Read about Capella on page 14.

Before the moon rises, until around 8:40 p.m., look for the **Andromeda Galaxy** with binoculars or a telescope. It is in the northeastern sky in the evening. See the guide on page 11.

Explore the **dark sky** after twilight and before the moon rises, around 8:40 p.m.

Scorpius is in the western sky at dusk and sets in the southwest around 8:50 p.m. See the illustration on page 19.

The bright star **Arcturus** is in the western sky at dusk and sets in the northwest around 9:10 p.m.

Just east of Scorpius, **Sagittarius** sets in the southwest around 10:25 p.m.

Orion rises in the east around 12:10 a.m. and is in the eastern sky before dawn. Look for the bright star **Aldebaran** and the **Pleiades** near Orion. See the illustration on page 17.

At its closest and brightest for the year, **Saturn** is visible most of the night and is in the eastern sky at dusk and at its highest, around 60 degrees, at 12:50 a.m. Unlike bright Jupiter and Venus and ruddy Mars, Saturn is harder to distinguish from the stars. Stars twinkle and shimmer, while the planets, including Saturn, show a steady flat light.

Castor and Pollux, the twin stars in the constellation Gemini, rise in the northeast around 1:05 a.m. and are in the eastern sky before dawn.

Cassiopeia, shaped like an *M,* is at its highest, around 65 degrees, at 1:20 a.m. See the illustration on page 11.

Rust-red **Mars** rises in the northeast around 1:30 a.m. and is in the eastern sky before dawn.

The brightest star in the sky, **Sirius** rises in the east around 1:50 a.m. and is in the eastern sky before dawn.

Jupiter rises in the east around 3 a.m. and is in the eastern sky before dawn.

2026
Thursday, October 1

TONIGHT

sunset: **6:29 p.m.**
twilight begins: **7:23 p.m.**

IN THE MORNING

twilight begins: **5:47 a.m.**
sunrise: **6:41 a.m.**

MOON PHASE

moonrise: **9:58 p.m.**
moonset: **12:20 p.m.**

IN THE SKY

Bright **Venus** sets in the southwest around 7:10 p.m.

The **summer triangle**'s three bright stars are Vega, Altair, and Deneb. **Vega** sets in the northwest around 2:50 a.m. **Altair** is at its highest, around 65 degrees, around dusk and sets in the west around 2:15 a.m. Altair can be identified by its less bright neighboring stars, one on either side, together called the shaft of Aquila. **Deneb** is in the band of the Milky Way. See the illustration of the summer triangle on page 19.

The **Great Bear (Big Dipper)** is in the northwestern sky at dusk and sets in the northwest around 8:15 p.m. See the illustration on page 10.

Bright star **Capella** rises in the northeast around 8:25 p.m. and is almost directly overhead around dawn. Read about Capella on page 14.

Scorpius is in the western sky at dusk and sets in the southwest around 8:50 p.m. See the illustration on page 19.

The bright star **Arcturus** is in the western sky at dusk and sets in the northwest around 9:05 p.m.

Explore the **dark sky** after twilight and before the moon rises, around 9:40 p.m.

Before the moon rises, until around 9:40 p.m., look for the **Andromeda Galaxy** with binoculars or a telescope. It is in the northeastern sky in the evening. See the guide on page 11.

Just east of Scorpius, **Sagittarius** sets in the southwest around 10:20 p.m.

Orion rises in the east around 12:05 a.m. and is in the eastern sky before dawn. Look for the bright star **Aldebaran** and the **Pleiades** near Orion. See the illustration on page 17.

At its closest and brightest for the year, **Saturn** is visible most of the night and is at its highest, around 60 degrees, at 12:50 a.m. Unlike bright Jupiter and Venus and ruddy Mars, Saturn is harder to distinguish from the stars. Stars twinkle and shimmer, while the planets, including Saturn, show a steady flat light.

Castor and Pollux, the twin stars in the constellation Gemini, rise in the northeast around 1 a.m. and are in the eastern sky before dawn. Read how to find Castor and Pollux on page 14.

Cassiopeia, shaped like an *M,* is at its highest, around 65 degrees, at 1:15 a.m. See the illustration on page 11.

Rust-red **Mars** rises in the northeast around 1:30 a.m. and is in the eastern sky before dawn.

The brightest star in the sky, **Sirius** rises in the east around 1:45 a.m. and is in the eastern sky before dawn.

Jupiter rises in the east around 3 a.m. and is in the eastern sky before dawn.

2026

Friday, October 2

TONIGHT

sunset: **6:28 p.m.**
twilight begins: **7:22 p.m.**

IN THE MORNING

twilight begins: **5:48 a.m.**
sunrise: **6:41 a.m.**

MOON PHASE

moonrise: **11:03 p.m.**
moonset: **1:24 p.m.**

IN THE SKY

Bright **Venus** sets in the southwest around 7:05 p.m.

The **summer triangle**'s three bright stars are Vega, Altair, and Deneb. **Vega** sets in the northwest around 2:45 a.m. **Altair** is at its highest, around 65 degrees, around dusk and sets in the west around 2:10 a.m. Altair can be identified by its less bright neighboring stars, one on either side, together called the shaft of Aquila. **Deneb** is in the band of the Milky Way. See the illustration of the summer triangle on page 19.

The bright star **Arcturus** is in the western sky at dusk and sets in the northwest around 9 p.m.

Bright star **Capella** rises in the northeast around 8:20 p.m. and is almost directly overhead around dawn. Read about Capella on page 14.

The **Great Bear (Big Dipper)** is in the northwestern sky at dusk and sets in the northwest around 8:10 p.m. See the illustration on page 10.

Scorpius is in the western sky at dusk and sets in the southwest around 8:45 p.m. See the illustration on page 19.

Just east of Scorpius, **Sagittarius** sets in the southwest around 10:15 p.m.

Explore the **dark sky** after twilight and before the moon rises, around 10:45 p.m.

Before the moon rises, until around 10:45 p.m., look for the **Andromeda Galaxy** with binoculars or a telescope. It is in the northeastern sky in the evening. See the guide on page 11.

Orion rises in the east around midnight and is in the eastern sky before dawn. Look for the bright star **Aldebaran** and the **Pleiades** near Orion. See the illustration on page 17.

At its closest and brightest for the year, **Saturn** is visible most of the night and is at its highest, around 60 degrees, at 12:45 a.m. Unlike bright Jupiter and Venus and ruddy Mars, Saturn is harder to distinguish from the stars. Stars twinkle and shimmer, while the planets, including Saturn, show a steady flat light.

Castor and Pollux, the twin stars in the constellation Gemini, rise in the northeast around 12:55 a.m. and are in the eastern sky before dawn. Read how to find Castor and Pollux on page 14.

Cassiopeia, shaped like an *M,* is at its highest, around 65 degrees, at 1:10 a.m. See the illustration on page 11.

Rust-red **Mars** rises in the northeast around 1:30 a.m. and is in the eastern sky before dawn.

The brightest star in the sky, **Sirius** rises in the east around 1:40 a.m. and is in the eastern sky before dawn.

Jupiter rises in the east around 2:55 a.m. and is in the eastern sky before dawn.

2026

Saturday, October 3

TONIGHT

sunset: **6:27 p.m.**
twilight begins: **7:21 p.m.**

IN THE MORNING

twilight begins: **5:48 a.m.**
sunrise: **6:42 a.m.**

MOON PHASE

moonrise: **12:13 a.m.**
moonset: **2:19 p.m.**

IN THE SKY

Bright **Venus** sets in the southwest around 7 p.m.

The **summer triangle**'s three bright stars are Vega, Altair, and Deneb. **Vega** sets in the northwest around 2:40 a.m. **Altair** is at its highest, around 65 degrees, around dusk and sets in the west around 2:05 a.m. Altair can be identified by its less bright neighboring stars, one on either side, together called the shaft of Aquila. **Deneb** is in the band of the Milky Way. See the illustration of the summer triangle on page 19.

Before the moon rises, until about midnight, look for the **Andromeda Galaxy** with binoculars or a telescope. It is in the northeastern sky in the evening. See the guide on page 11.

The **Great Bear (Big Dipper)** is in the northwestern sky at dusk and sets in the northwest around 8:05 p.m. See the illustration on page 10.

Bright star **Capella** rises in the northeast around 8:20 p.m. and is almost directly overhead around dawn. Read about Capella on page 14.

Scorpius is in the western sky at dusk and sets in the southwest around 8:40 p.m. See the illustration on page 19.

The bright star **Arcturus** is in the western sky at dusk and sets in the northwest around 9 p.m.

Just east of Scorpius, **Sagittarius** sets in the southwest around 10:10 p.m.

Orion rises in the east around 11:55 p.m. Look for the bright star **Aldebaran** and the **Pleiades** near Orion. See the illustration on page 17.

At its closest and brightest for the year, **Saturn** is visible most of the night and is at its highest, around 60 degrees, at 12:40 a.m. Unlike bright Jupiter and Venus and ruddy Mars, Saturn is harder to distinguish from the stars. Stars twinkle and shimmer, while the planets, including Saturn, show a steady flat light.

Castor and Pollux, the twin stars in the constellation Gemini, rise in the northeast around 12:50 a.m. and are in the eastern sky before dawn. Read how to find Castor and Pollux on page 14.

Cassiopeia, shaped like an *M,* is at its highest, around 65 degrees, at 1:05 a.m. See the illustration on page 11.

Rust-red **Mars** rises in the northeast around 1:30 a.m. and is in the eastern sky before dawn.

The brightest star in the sky, **Sirius** rises in the east around 1:40 a.m. and is in the eastern sky before dawn.

Jupiter rises in the east around 2:50 a.m. and is in the eastern sky before dawn.

The **Great Bear (Big Dipper)** rises in the northeast around 3:05 a.m. and is in the northeastern sky before dawn. See the illustration on page 10.

2026
Sunday, October 4

TONIGHT

sunset: **6:25 p.m.**
twilight begins: **7:19 p.m.**

IN THE MORNING

twilight begins: **5:49 a.m.**
sunrise: **6:43 a.m.**

MOON PHASE

moonrise: **1:24 a.m.**
moonset: **3:06 p.m.**

IN THE SKY

Bright **Venus** sets in the southwest around 6:55 p.m.

Scorpius is in the western sky at dusk and sets in the southwest around 8:35 p.m. See the illustration on page 19.

The **summer triangle**'s three bright stars are Vega, Altair, and Deneb. **Vega** sets in the northwest around 2:40 a.m. **Altair** is at its highest, around 65 degrees, around dusk and sets in the west around 2:05 a.m. Altair can be identified by its less bright neighboring stars, one on either side, together called the shaft of Aquila. **Deneb** is in the band of the Milky Way. See the illustration of the summer triangle on page 19.

Bright star **Capella** rises in the northeast around 8:15 p.m. and is almost directly overhead around dawn. Read about Capella on page 14.

The bright star **Arcturus** is in the western sky at dusk and sets in the northwest around 8:55 p.m.

Just east of Scorpius, **Sagittarius** sets in the southwest around 10:05 p.m.

Before the moon rises, look for the **Andromeda Galaxy** with binoculars or a telescope. It is in the northeastern sky in the evening. See the guide on page 11.

Orion rises in the east around 11:55 p.m. Look for the bright star **Aldebaran** and the **Pleiades** near Orion. See the illustration on page 17.

At its closest and brightest for the year, **Saturn** is visible most of the night and is at its highest, around 60 degrees, at 12:35 a.m. Unlike bright Jupiter and Venus and ruddy Mars, Saturn is harder to distinguish from the stars. Stars twinkle and shimmer, while the planets, including Saturn, show a steady flat light.

Cassiopeia, shaped like an *M,* is at its highest, around 65 degrees, at 1:05 a.m. See the illustration on page 11.

Rust-red **Mars** rises in the northeast around 1:25 a.m. and is in the eastern sky before dawn.

The brightest star in the sky, **Sirius** rises in the east around 1:35 a.m. and is in the eastern sky before dawn.

The **waning crescent moon** and **Mars** appear close together in the morning sky. Look east around 2:30 a.m.

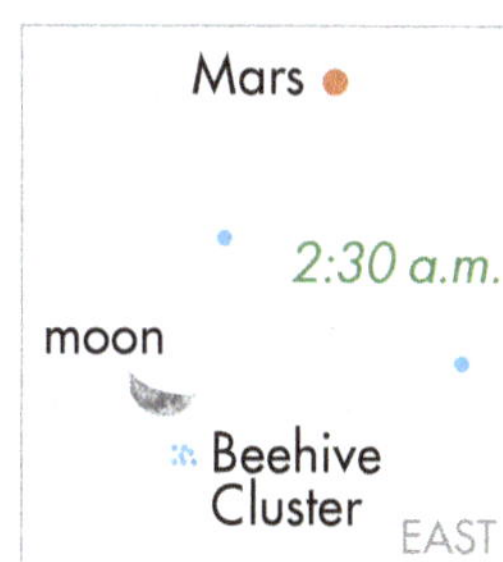

Jupiter rises in the east around 2:50 a.m. and is in the eastern sky before dawn.

The **Great Bear (Big Dipper)** rises in the northeast around 3 a.m. and is in the northeastern sky before dawn. See the illustration on page 10.

2026

Monday, October 5

TONIGHT

sunset: **6:24 p.m.**
twilight begins: **7:18 p.m.**

MOON PHASE

IN THE MORNING

twilight begins: **5:50 a.m.**
sunrise: **6:44 a.m.**

moonrise: **2:33 a.m.**
moonset: **3:44 p.m.**

IN THE SKY

Bright **Venus** sets in the southwest around dusk.

The **summer triangle**'s three bright stars are Vega, Altair, and Deneb. **Vega** sets in the northwest around 2:35 a.m. **Altair** is at its highest, around 65 degrees, around dusk and sets in the west around 2 a.m. Altair can be identified by its less bright neighboring stars, one on either side, together called the shaft of Aquila. **Deneb** is in the band of the Milky Way. See the illustration of the summer triangle on page 19.

The absence of the moon early in the evening provides excellent dark-sky viewing conditions.

Scorpius is in the western sky at dusk and sets in the southwest around 8:30 p.m. See the illustration on page 19.

The bright star **Arcturus** is in the western sky at dusk and sets in the northwest around 8:50 p.m.

Just east of Scorpius, **Sagittarius** sets in the southwest around 10:05 p.m.

Orion rises in the east around 11:50 p.m. Look for the bright star **Aldebaran** and the **Pleiades** near Orion. See the illustration on page 17.

At its closest and brightest for the year, **Saturn** is visible most of the night and is at its highest, around 60 degrees, at 12:30 a.m.

Before the moon rises, look for the **Andromeda Galaxy** with binoculars or a telescope. It is almost directly overhead at moonrise. See the guide on page 11.

Castor and Pollux, the twin stars in the constellation Gemini, rise in the northeast around 12:45 a.m. and are in the eastern sky before dawn. Read how to find Castor and Pollux on page 14.

Cassiopeia, shaped like an *M,* is at its highest, around 65 degrees, at 12:55 a.m. See the illustration on page 11.

Rust-red **Mars** rises in the northeast around 1:25 a.m. and is in the eastern sky before dawn.

The brightest star in the sky, **Sirius** rises in the east around 1:30 a.m. and is in the eastern sky before dawn.

Jupiter rises in the east around 2:45 a.m. and is in the eastern sky before dawn.

The **waning sliver-crescent moon** and **Jupiter** appear close together in the morning sky. Look east around 3:30 a.m.

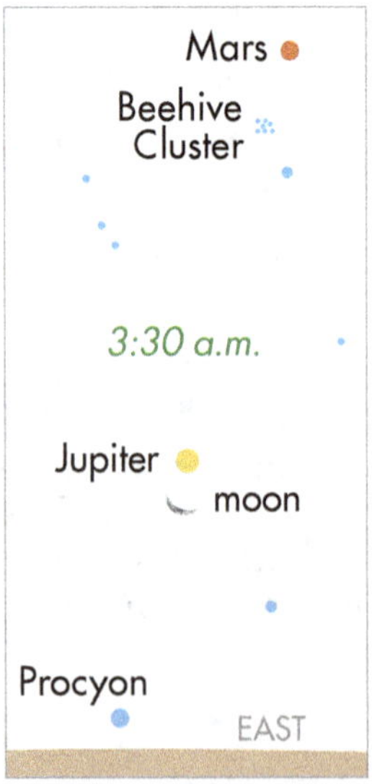

The **Great Bear (Big Dipper)** rises in the northeast around 2:55 a.m. and is in the northeastern sky before dawn. See the illustration on page 10.

2026

Tuesday, October 6

TONIGHT

sunset: **6:22 p.m.**
twilight begins: **7:17 p.m.**

IN THE MORNING

twilight begins: **5:51 a.m.**
sunrise: **6:44 a.m.**

MOON PHASE

moonrise: **3:40 a.m.**
moonset: **4:17 p.m.**

IN THE SKY

Bright **Venus** sets in the southwest around dusk.

The absence of the **moon** tonight provides excellent dark-sky viewing conditions.

Bright star **Capella** rises in the northeast around 8:05 p.m. and is in the northeastern sky at dusk and is almost directly overhead around dawn. Read about Capella on page 14.

Scorpius is in the western sky at dusk and sets in the southwest around 8:30 p.m. See the illustration on page 19.

The bright star **Arcturus** is in the western sky at dusk and sets in the northwest around 8:45 p.m.

Just east of Scorpius, **Sagittarius** sets in the southwest around 10 p.m.

Orion rises in the east around 11:45 p.m. Look for the bright star **Aldebaran** and the **Pleiades** near Orion. See the illustration on page 17.

At its closest and brightest for the year, **Saturn** is visible most of the night and is at its highest, around 60 degrees, at 12:25 a.m. Unlike bright Jupiter and Venus and ruddy Mars, Saturn is harder to distinguish from the stars. Stars twinkle and shimmer, while the planets, including Saturn, show a steady flat light.

Castor and Pollux, the twin stars in the constellation Gemini, rise in the northeast around 12:40 a.m. and are in the eastern sky before dawn. Read how to find Castor and Pollux on page 14.

Cassiopeia, shaped like an *M,* is at its highest, around 65 degrees, at 12:50 a.m. See the illustration on page 11.

Rust-red **Mars** rises in the northeast around 1:25 a.m. and is in the eastern sky before dawn.

The brightest star in the sky, **Sirius** rises in the east around 1:25 a.m. and is in the eastern sky before dawn.

The bright star **Altair** sets in the west around 1:55 a.m. Altair can be identified by its less bright neighboring stars, one on either side, together called the shaft of Aquila. See page 18.

Before the moon rises, look for the **Andromeda Galaxy** with binoculars or a telescope. It is almost directly overhead at 2:15 a.m. See the guide on page 11.

One of the brightest stars in the sky, **Vega** sets in the northwest around 2:30 a.m.

Jupiter rises in the east around 2:40 a.m. and is in the eastern sky before dawn.

The **Great Bear (Big Dipper)** rises in the northeast around 2:55 a.m. and is in the northeastern sky before dawn. See the illustration on page 10.

2026

Wednesday, October 7

TONIGHT

sunset: **6:21 p.m.**
twilight begins: **7:15 p.m.**

IN THE MORNING

twilight begins: **5:51 a.m.**
sunrise: **6:45 a.m.**

MOON PHASE

moonrise: **4:45 a.m.**
moonset: **4:46 p.m.**

IN THE SKY

The Draconids meteor shower happens tonight. A weak shower with up to 10 meteors per hour, the best time to watch is early evening. Tonight has moonless dark skies all night.

Bright Venus sets in the southwest around dusk.

The absence of the **moon** tonight provides excellent dark-sky viewing conditions.

The bright star **Arcturus** is in the western sky at dusk and sets in the northwest around 8:40 p.m.

Just east of Scorpius, **Sagittarius** sets in the southwest around 9:55 p.m.

Orion rises in the east around 11:40 p.m. Look for the bright star **Aldebaran** and the **Pleiades** near Orion. See the illustration on page 17.

At its closest and brightest for the year, Saturn is visible most of the night and is at its highest, around 60 degrees, at 12:20 a.m. Unlike bright Jupiter and Venus and ruddy Mars, Saturn is harder to distinguish from the stars. Stars twinkle and shimmer, while the planets, including Saturn, show a steady flat light.

Castor and Pollux, the twin stars in the constellation Gemini, rise in the northeast around 12:35 a.m. and are in the eastern sky before dawn. Read how to find Castor and Pollux on page 14.

Cassiopeia, shaped like an *M,* is at its highest, around 65 degrees, at 12:50 a.m. See the illustration on page 11.

The brightest star in the sky, **Sirius** rises in the east around 1:20 a.m. and is in the eastern sky before dawn.

Rust-red Mars rises in the northeast around 1:25 a.m. and is in the eastern sky before dawn.

The bright star **Altair** sets in the west around 1:50 a.m. Altair can be identified by its less bright neighboring stars, one on either side, together called the shaft of Aquila. See page 18.

Before the moon rises, around 4:40 a.m., look for the **Andromeda Galaxy** with binoculars or a telescope. It is almost directly overhead at 2:10 a.m. See the guide on page 11.

One of the brightest stars in the sky, **Vega** sets in the northwest around 2:25 a.m.

Jupiter rises in the east around 2:40 a.m. and is in the eastern sky before dawn.

Before the moon rises, around 4:40 a.m., use binoculars or a telescope to look for the **moons of Jupiter.** See the guide on page 18.

The **Great Bear (Big Dipper)** rises in the northeast around 2:50 a.m. and is in the northeastern sky before dawn. See the illustration on page 10.

2026

Thursday, October 8

TONIGHT

sunset: 6:20 p.m.
twilight begins: 7:14 p.m.

IN THE MORNING

twilight begins: 5:52 a.m.
sunrise: 6:46 a.m.

MOON PHASE

moonrise: 5:48 a.m.
moonset: 5:13 p.m.

IN THE SKY

Bright Venus sets in the southwest around dusk.

Bright star **Capella** rises in the northeast around 8 p.m. and is in the northeastern sky at dusk and is almost directly overhead around dawn. Read about Capella on page 14.

The absence of the **moon** tonight provides excellent dark-sky viewing conditions.

Scorpius is in the western sky at dusk and sets in the southwest around 8:20 p.m. See the illustration on page 19.

The bright star **Arcturus** is in the western sky at dusk and sets in the northwest around 8:40 p.m.

Just east of Scorpius, **Sagittarius** sets in the southwest around 9:50 p.m.

Orion rises in the east around 11:35 p.m. Look for the bright star **Aldebaran** and the **Pleiades** near Orion. See the illustration on page 17.

At its closest and brightest for the year, Saturn is visible most of the night and is at its highest, around 60 degrees, at 12:20 a.m. Unlike bright Jupiter and Venus and ruddy Mars, Saturn is harder to distinguish from the stars. Stars twinkle and shimmer, while the planets, including Saturn, show a steady flat light.

Castor and Pollux, the twin stars in the constellation Gemini, rise in the northeast around 12:35 a.m. and are in the eastern sky before dawn. Read how to find Castor and Pollux on page 14.

Cassiopeia, shaped like an *M,* is at its highest, around 65 degrees, at 12:45 a.m. See the illustration on page 11.

Rust-red Mars rises in the northeast around 1:20 a.m. and is in the eastern sky before dawn.

The brightest star in the sky, **Sirius** rises in the east around 1:20 a.m. and is in the eastern sky before dawn.

The bright star **Altair** sets in the west around 1:45 a.m. Altair can be identified by its less bright neighboring stars, one on either side, together called the shaft of Aquila. See page 18.

With the dark sky, look for the **Andromeda Galaxy** with binoculars or a telescope. It is almost directly overhead at 2:05 a.m. See the guide on page 11.

One of the brightest stars in the sky, **Vega** sets in the northwest around 2:20 a.m.

Jupiter rises in the east around 2:35 a.m. and is in the eastern sky before dawn.

With the dark sky, use binoculars or a telescope to look for the **moons of Jupiter.** See the guide on page 18.

The **Great Bear (Big Dipper)** rises in the northeast around 2:45 a.m. and is in the northeastern sky before dawn. See the illustration on page 10.

2026

Friday, October 9

TONIGHT

sunset: **6:18 p.m.**
twilight begins: **7:13 p.m.**

IN THE MORNING

twilight begins: **5:53 a.m.**
sunrise: **6:47 a.m.**

MOON PHASE

moonrise: **6:51 a.m.**
moonset: **5:40 p.m.**

IN THE SKY

Bright **Venus** sets in the southwest around dusk.

Bright star **Capella** rises in the northeast around 7:55 p.m. and is in the northeastern sky at dusk and is almost directly overhead at 4:50 a.m. Read about Capella on page 14.

Scorpius is in the western sky at dusk and sets in the southwest around 8:15 p.m. See the illustration on page 19.

The bright star **Arcturus** is in the western sky at dusk and sets in the northwest around 8:35 p.m.

The absence of the **moon** tonight provides excellent dark-sky viewing conditions.

Just east of Scorpius, **Sagittarius** sets in the southwest around 9:50 p.m.

At its closest and brightest for the year, **Saturn** is visible most of the night and is at its highest, around 60 degrees, at 12:15 a.m. Unlike bright Jupiter and Venus and ruddy Mars, Saturn is harder to distinguish from the stars. Stars twinkle and shimmer, while the planets, including Saturn, show a steady flat light.

Orion rises in the east around 11:35 p.m. and at its highest, around 65 degrees, around dawn. Look for the bright star **Aldebaran** and the **Pleiades** near Orion. See the illustration on page 17.

Castor and Pollux, the twin stars in the constellation Gemini, rise in the northeast around 12:30 a.m. and are in the eastern sky before dawn. Read how to find Castor and Pollux on page 14.

Cassiopeia, shaped like an *M,* is at its highest, around 65 degrees, at 12:40 a.m. See the illustration on page 11.

The brightest star in the sky, **Sirius** rises in the east around 1:15 a.m. and is in the eastern sky before dawn.

Rust-red **Mars** rises in the northeast around 1:20 a.m. and is in the eastern sky before dawn.

The bright star **Altair** sets in the west around 1:45 a.m. Altair can be identified by its less bright neighboring stars, one on either side, together called the shaft of Aquila. See page 18.

Look for the **Andromeda Galaxy** with binoculars or a telescope. It is almost directly overhead at 2:05 a.m. See the guide on page 11.

One of the brightest stars in the sky, **Vega** sets in the northwest around 2:20 a.m.

Jupiter rises in the east around 2:35 a.m. and is in the eastern sky before dawn.

Use binoculars or a telescope to look for the **moons of Jupiter.** See the guide on page 18.

The **Great Bear (Big Dipper)** rises in the northeast around 2:40 a.m. and is in the northeastern sky before dawn. See the illustration on page 10.

Saturday, October 10

MOON PHASE

TONIGHT

sunset: **6:17 p.m.**
twilight begins: **7:11 p.m.**

IN THE MORNING

twilight begins: **5:54 a.m.**
sunrise: **6:47 a.m.**

moonrise: **7:53 a.m.**
moonset: **6:08 p.m.**

IN THE SKY

Bright **Venus** sets in the southwest around dusk.

Bright star **Capella** rises in the northeast around 7:50 p.m. and is in the northeastern sky at dusk and is almost directly overhead at 4:50 a.m. Read about Capella on page 14.

Scorpius is in the western sky at dusk and sets in the southwest around 8:15 p.m. See the illustration on page 19.

The bright star **Arcturus** is in the western sky at dusk and sets in the northwest around 8:30 p.m.

The absence of the **moon** tonight provides excellent dark-sky viewing conditions.

Just east of Scorpius, **Sagittarius** sets in the southwest around 9:45 p.m.

Orion rises in the east around 11:30 p.m. and at its highest, around 65 degrees, around dawn. Look for the bright star **Aldebaran** and the **Pleiades** near Orion. See the illustration on page 17.

At its closest and brightest for the year, **Saturn** is visible most of the night and is at its highest, around 60 degrees, at 12:10 a.m. Unlike bright Jupiter and Venus and ruddy Mars, Saturn is harder to distinguish from the stars. Stars twinkle and shimmer, while the planets, including Saturn, show a steady flat light.

Castor and Pollux, the twin stars in the constellation Gemini, rise in the northeast around 12:25 a.m. and are in the eastern sky before dawn. Read how to find Castor and Pollux on page 14.

Cassiopeia, shaped like an *M*, is at its highest, around 65 degrees, at 12:35 a.m. See the illustration on page 11.

The brightest star in the sky, **Sirius** rises in the east around 1:10 a.m. and is in the eastern sky before dawn.

Rust-red **Mars** rises in the northeast around 1:20 a.m. and is in the eastern sky before dawn.

The bright star **Altair** sets in the west around 1:40 a.m. Altair can be identified by its less bright neighboring stars, one on either side, together called the shaft of Aquila. See page 18.

Look for the **Andromeda Galaxy** with binoculars or a telescope. It is almost directly overhead at 2 a.m. See the guide on page 11.

One of the brightest stars in the sky, **Vega** sets in the northwest around 2:15 a.m.

Jupiter rises in the east around 2:30 a.m. and is in the eastern sky before dawn.

Use binoculars or a telescope to look for the **moons of Jupiter.** See the guide on page 18.

The **Great Bear (Big Dipper)** rises in the northeast around 2:35 a.m. and is in the northeastern sky before dawn. See the illustration on page 10.

2026

Sunday, October 11

TONIGHT

sunset: **6:16 p.m.**
twilight begins: **7:10 p.m.**

MOON PHASE

IN THE MORNING

twilight begins: **5:54 a.m.**
sunrise: **6:48 a.m.**

moonrise: **7:53 a.m.**
moonset: **6:38 p.m.**

IN THE SKY

Look for **Mercury** in the evening sky, just after sunset in the west. It's difficult but possible to see the tiny dot in the glare of the sun. It's easier to find with the naked eye, not using binoculars or a telescope. Never look directly at the sun or point binoculars or a telescope at it.

Bright star **Capella** rises in the northeast around 7:45 p.m. and is in the northeastern sky at dusk and is almost directly overhead at 4:45 a.m. Read about Capella on page 14.

Scorpius is in the western sky at dusk and sets in the southwest around 8:10 p.m. See the illustration on page 19.

The bright star **Arcturus** is in the western sky at dusk and sets in the northwest around 8:25 p.m.

The absence of the **moon** tonight provides excellent dark-sky viewing conditions.

Just east of Scorpius, **Sagittarius** sets in the southwest around 9:40 p.m.

Orion rises in the east around 11:25 p.m. and at its highest, around 65 degrees, around dawn. Look for the bright star **Aldebaran** and the **Pleiades** near Orion. See the illustration on page 17.

At its closest and brightest for the year, **Saturn** is visible most of the night and is at its highest, around 60 degrees, at 12:05 a.m. Unlike bright Jupiter and Venus and ruddy Mars, Saturn is harder to distinguish from the stars. Stars twinkle and shimmer, while the planets, including Saturn, show a steady flat light.

Castor and Pollux, the twin stars in the constellation Gemini, rise in the northeast around 12:20 a.m. and are in the eastern sky before dawn. Read how to find Castor and Pollux on page 14.

The brightest star in the sky, **Sirius** rises in the east around 1 a.m.

Rust-red **Mars** rises in the northeast around 1:20 a.m. and is in the eastern sky before dawn.

The bright star **Altair** sets in the west around 1:35 a.m. Altair can be identified by its less bright neighboring stars, one on either side, together called the shaft of Aquila. See page 18.

Look for the **Andromeda Galaxy** with binoculars or a telescope. It is almost directly overhead at 1:55 a.m. See the guide on page 11.

One of the brightest stars in the sky, **Vega** sets in the northwest around 2:10 a.m.

Jupiter rises in the east around 2:25 a.m. and is in the eastern sky before dawn.

Use binoculars or a telescope to look for the **moons of Jupiter.** See the guide on page 18.

2026

Monday, October 12

MOON PHASE

TONIGHT

sunset: **6:15 p.m.**
twilight begins: **7:09 p.m.**

IN THE MORNING

twilight begins: **5:55 a.m.**
sunrise: **6:49 a.m.**

moonrise: **8:55 a.m.**
moonset: **7:12 p.m.**

IN THE SKY

Look for **Mercury** in the evening sky, just after sunset in the west. It's difficult but possible to see the tiny dot in the glare of the sun. It's easier to find with the naked eye, not using binoculars or a telescope. Never look directly at the sun or point binoculars or a telescope at it.

The absence of the **moon** tonight provides excellent dark-sky viewing conditions.

Bright star **Capella** rises in the northeast around 7:45 p.m. and is in the northeastern sky at dusk and is almost directly overhead at 4:40 a.m. Read about Capella on page 14.

The bright star **Arcturus** is in the western sky at dusk and sets in the northwest around 8:25 p.m.

Just east of Scorpius, **Sagittarius** sets in the southwest around 9:35 p.m.

Orion rises in the east around 11:20 p.m. and at its highest, around 65 degrees, around dawn. Look for the bright star **Aldebaran** and the **Pleiades** near Orion. See the illustration on page 17.

Saturn is visible most of the night and is at its highest, around 60 degrees, at midnight. Unlike bright Jupiter and Venus and ruddy Mars, Saturn is harder to distinguish from the stars. Stars twinkle and shimmer, while the planets, including Saturn, show a steady flat light.

Castor and Pollux, the twin stars in the constellation Gemini, rise in the northeast around 12:15 a.m. and are in the eastern sky before dawn. Read how to find Castor and Pollux on page 14.

Cassiopeia, shaped like an *M,* is at its highest, around 65 degrees, at 12:30 a.m. See the illustration on page 11.

The brightest star in the sky, **Sirius** rises in the east around 1 a.m.

The bright star **Altair** sets in the west around 1:30 a.m. Altair can be identified by its less bright neighboring stars, one on either side, together called the shaft of Aquila. See page 18.

Look for the **Andromeda Galaxy** with binoculars or a telescope. It is almost directly overhead at 1:50 a.m. See the guide on page 11.

One of the brightest stars in the sky, **Vega** sets in the northwest around 2:05 a.m.

Rust-red **Mars** rises in the northeast around 1:15 a.m. and is in the eastern sky before dawn.

Jupiter rises in the east around 2:25 a.m. and is in the eastern sky before dawn.

Use binoculars or a telescope to look for the **moons of Jupiter.** See the guide on page 18.

2026

Tuesday, October 13

TONIGHT

sunset: **6:13 p.m.**
twilight begins: **7:08 p.m.**

IN THE MORNING

twilight begins: **5:56 a.m.**
sunrise: **6:50 a.m.**

MOON PHASE

moonrise: **9:57 a.m.**
moonset: **7:50 p.m.**

IN THE SKY

Look for **Mercury** in the evening sky, just after sunset in the southwest. It's difficult but possible to see the tiny dot in the glare of the sun. It's easier to find with the naked eye, not using binoculars or a telescope. Never look directly at the sun or point binoculars or a telescope at it.

Bright star **Capella** rises in the northeast around 7:40 p.m. and is in the northeastern sky at dusk and is almost directly overhead at 4:35 a.m. Read about Capella on page 14.

Scorpius is in the western sky at dusk and sets in the southwest around 8 p.m. See the illustration on page 19.

The absence of the **moon** tonight provides excellent dark-sky viewing conditions. Explore the **dark sky** after the moon sets, from around 8:10 p.m.

The bright star **Arcturus** is in the western sky at dusk and sets in the northwest around 8:20 p.m.

Just east of Scorpius, **Sagittarius** sets in the southwest around 9:30 p.m.

Orion rises in the east around 11:20 p.m. and at its highest, around 65 degrees, around dawn. Look for the bright star **Aldebaran** and the **Pleiades** near Orion. See the illustration on page 17.

Saturn is visible most of the night and is at its highest, around 55 degrees, at 11:55 p.m. Unlike bright Jupiter and Venus and ruddy Mars, Saturn is harder to distinguish from the stars. Stars twinkle and shimmer, while the planets, including Saturn, show a steady flat light.

Castor and Pollux, the twin stars in the constellation Gemini, rise in the northeast around 12:15 a.m. and are in the eastern sky before dawn. Read how to find Castor and Pollux on page 14.

The brightest star in the sky, **Sirius** rises in the east around 12:55 a.m.

Rust-red **Mars** rises in the northeast around 1:15 a.m. and is in the eastern sky before dawn.

The bright star **Altair** sets in the west around 1:25 a.m. Altair can be identified by its less bright neighboring stars, one on either side, together called the shaft of Aquila. See page 18.

Look for the **Andromeda Galaxy** with binoculars or a telescope. It is almost directly overhead at 1:45 a.m. See the guide on page 11.

One of the brightest stars in the sky, **Vega** sets in the northwest around 2 a.m.

Jupiter rises in the east around 2:20 a.m. and is in the eastern sky before dawn.

Use binoculars or a telescope to look for the **moons of Jupiter.** See the guide on page 18.

2026

Wednesday, October 14

TONIGHT

sunset: **6:12 p.m.**
twilight begins: **7:07 p.m.**

IN THE MORNING

twilight begins: **5:57 a.m.**
sunrise: **6:51 a.m.**

MOON PHASE

moonrise: **10:57 a.m.**
moonset: **8:33 p.m.**

IN THE SKY

The **moon** is prominent in the sky early in the evening. Using a telescope or binoculars, take a look. You should be able to see rough mountains and craters along the boundary between light and dark, in astronomy called the terminator.

Bright star **Capella** is in the northeastern sky at dusk and is almost directly overhead at 4:30 a.m. Read about Capella on page 14.

Scorpius is in the western sky at dusk and sets in the southwest around 7:55 p.m. See the illustration on page 19.

The bright star **Arcturus** is in the western sky at dusk and sets in the northwest around 8:15 p.m.

Explore the **dark sky** after the moon sets, from around 8:55 p.m.

Just east of Scorpius, **Sagittarius** sets in the southwest around 9:30 p.m.

Orion rises in the east around 11:15 p.m. and at its highest, around 65 degrees, around dawn. Look for the bright star **Aldebaran** and the **Pleiades** near Orion. See the illustration on page 17.

Saturn is visible most of the night and is at its highest, around 55 degrees, at 11:55 p.m. Unlike bright Jupiter and Venus and ruddy Mars, Saturn is harder to distinguish from the stars. Stars twinkle and shimmer, while the planets, including Saturn, show a steady flat light.

Castor and Pollux, the twin stars in the constellation Gemini, rise in the northeast around 12:10 a.m. and are in the eastern sky before dawn. Read how to find Castor and Pollux on page 14.

Cassiopeia, shaped like an *M,* is at its highest, around 65 degrees, at 12:20 a.m. See the illustration on page 11.

The brightest star in the sky, **Sirius** rises in the east around 12:50 a.m.

Rust-red **Mars** rises in the northeast around 1:15 a.m. and is in the eastern sky before dawn.

The bright star **Altair** sets in the west around 1:25 a.m. Altair can be identified by its less bright neighboring stars, one on either side, together called the shaft of Aquila. See page 18.

Look for the **Andromeda Galaxy** with binoculars or a telescope. It is almost directly overhead at 1:45 a.m. See the guide on page 11.

One of the brightest stars in the sky, **Vega** sets in the northwest around 2 a.m.

Jupiter rises in the east around 2:15 a.m. and is in the eastern sky before dawn.

Use binoculars or a telescope to look for the **moons of Jupiter.** See the guide on page 18.

The **Great Bear (Big Dipper)** rises in the northeast around 2:20 a.m. and is in the northeastern sky before dawn. See the illustration on page 10.

2026

Thursday, October 15

TONIGHT

sunset: **6:11 p.m.**
twilight begins: **7:05 p.m.**

IN THE MORNING

twilight begins: **5:57 a.m.**
sunrise: **6:51 a.m.**

MOON PHASE

moonrise: **11:52 a.m.**
moonset: **9:23 p.m.**

IN THE SKY

The **moon** is prominent in the sky early in the evening. Using a telescope or binoculars, take a look. You should be able to see rough mountains and craters along the boundary between light and dark, in astronomy called the terminator.

Bright star **Capella** is in the northeastern sky at dusk and is almost directly overhead at 4:30 a.m. Read about Capella on page 14.

Scorpius is in the western sky at dusk and sets in the southwest around 7:55 p.m. See the illustration on page 19.

The bright star **Arcturus** is in the western sky at dusk and sets in the northwest around 8:10 p.m.

Just east of Scorpius, **Sagittarius** sets in the southwest around 9:25 p.m.

Explore the **dark sky** after the moon sets, from around 9:45 p.m.

Cassiopeia, shaped like an *M,* is at its highest, around 65 degrees, at 12:15 a.m. See the illustration on page 11.

Orion rises in the east around 11:10 p.m. and at its highest, around 65 degrees, around dawn. Look for the bright star **Aldebaran** and the **Pleiades** near Orion. See the illustration on page 17.

Saturn is visible most of the night and is at its highest, around 55 degrees, at 11:50 p.m. Unlike bright Jupiter and Venus and ruddy Mars, Saturn is harder to distinguish from the stars. Stars twinkle and shimmer, while the planets, including Saturn, show a steady flat light.

Castor and Pollux, the twin stars in the constellation Gemini, rise in the northeast around 12:05 a.m. and are in the eastern sky before dawn. Read how to find Castor and Pollux on page 14.

The brightest star in the sky, **Sirius** rises in the east around 12:45 a.m.

Rust-red **Mars** rises in the northeast around 1:15 a.m. and is in the eastern sky before dawn.

The bright star **Altair** sets in the west around 1:20 a.m. Altair can be identified by its less bright neighboring stars, one on either side, together called the shaft of Aquila. See page 18.

Look for the **Andromeda Galaxy** with binoculars or a telescope. It is almost directly overhead at 1:40 a.m. See the guide on page 11.

One of the brightest stars in the sky, **Vega** sets in the northwest around 1:55 a.m.

Jupiter rises in the east around 2:15 a.m. and is in the eastern sky before dawn.

Use binoculars or a telescope to look for the **moons of Jupiter.** See the guide on page 18.

2026

Friday, October 16

TONIGHT

sunset: **6:10 p.m.**
twilight begins: **7:04 p.m.**

MOON PHASE

IN THE MORNING

twilight begins: **5:58 a.m.**
sunrise: **6:52 a.m.**

moonrise: **12:43 p.m.**
moonset: **10:17 p.m.**

IN THE SKY

The **moon** is prominent in the sky early in the evening. Using a telescope or binoculars, take a look. You should be able to see rough mountains and craters along the boundary between light and dark, in astronomy called the terminator.

Bright star **Capella** is in the northeastern sky at dusk and is almost directly overhead at 4:25 a.m. Read about Capella on page 14.

Scorpius is in the western sky at dusk and sets in the southwest around 7:50 p.m. See the illustration on page 19.

The bright star **Arcturus** is in the western sky at dusk and sets in the northwest around 8:05 p.m.

Just east of Scorpius, **Sagittarius** sets in the southwest around 9:20 p.m.

Explore the **dark sky** after the moon sets, from around 10:35 p.m.

After the moon sets, from around 10:35 p.m., look for the **Andromeda Galaxy** with binoculars or a telescope. It is almost directly overhead at 1:35 a.m. See the guide on page 11.

Orion rises in the east around 11:05 p.m. and at its highest, around 65 degrees, around dawn. Look for the bright star **Aldebaran** and the **Pleiades** near Orion. See the illustration on page 17.

Saturn is visible most of the night and is at its highest, around 55 degrees, at 11:45 p.m. Unlike bright Jupiter and Venus and ruddy Mars, Saturn is harder to distinguish from the stars. Stars twinkle and shimmer, while the planets, including Saturn, show a steady flat light.

Castor and Pollux, the twin stars in the constellation Gemini, rise in the northeast around midnight and are in the eastern sky before dawn. Read how to find Castor and Pollux on page 14.

Cassiopeia, shaped like an *M,* is at its highest, around 65 degrees, at 12:10 a.m. See the illustration on page 11.

The brightest star in the sky, **Sirius** rises in the east around 12:45 a.m.

Rust-red **Mars** rises in the northeast around 1:10 a.m. and is in the eastern sky before dawn.

The bright star **Altair** sets in the west around 1:15 a.m. Altair can be identified by its less bright neighboring stars, one on either side, together called the shaft of Aquila. See page 18.

One of the brightest stars in the sky, **Vega** sets in the northwest around 1:50 a.m.

Jupiter rises in the east around 2:10 a.m. and is in the eastern sky before dawn.

Use binoculars or a telescope to look for the **moons of Jupiter.** See the guide on page 18.

2026

Saturday, October 17

TONIGHT

sunset: **6:08 p.m.**
twilight begins: **7:03 p.m.**

IN THE MORNING

twilight begins: **5:59 a.m.**
sunrise: **6:53 a.m.**

MOON PHASE

moonrise: **1:27 p.m.**
moonset: **11:14 p.m.**

IN THE SKY

The **moon** is prominent in the sky early in the evening. Using a telescope or binoculars, take a look. You should be able to see rough mountains and craters along the boundary between light and dark, in astronomy called the terminator.

Scorpius is in the western sky at dusk and sets in the southwest around 7:45 p.m. See the illustration on page 19.

Bright star **Capella** is in the northeastern sky at dusk and is almost directly overhead at 4:20 a.m. Read about Capella on page 14.

The bright star **Arcturus** is in the western sky at dusk and sets in the northwest around 8:05 p.m.

Just east of Scorpius, **Sagittarius** sets in the southwest around 9:15 p.m. Tonight the **moon** is in Sagittarius, making it easy to locate the constellation.

Orion rises in the east around 11 p.m. and at its highest, around 65 degrees, at 5 a.m. Look for the bright star **Aldebaran** and the **Pleiades** near Orion. See the illustration on page 17.

Explore the **dark sky** after the moon sets, from around 11:35 p.m.

After the moon sets, from around 11:35 p.m., look for the **Andromeda Galaxy** with binoculars or a telescope. It is almost directly overhead at 1:30 a.m. See the guide on page 11.

Saturn is visible most of the night and is at its highest, around 55 degrees, at 11:40 p.m. Unlike bright Jupiter and Venus and ruddy Mars, Saturn is harder to distinguish from the stars. Stars twinkle and shimmer, while the planets, including Saturn, show a steady flat light.

Castor and Pollux, the twin stars in the constellation Gemini, rise in the northeast around 11:55 p.m. Read how to find Castor and Pollux on page 14.

Cassiopeia, shaped like an *M,* is at its highest, around 65 degrees, at 12:10 a.m. See the illustration on page 11.

The brightest star in the sky, **Sirius** rises in the east around 12:40 a.m.

Rust-red **Mars** rises in the northeast around 1:10 a.m. and is in the eastern sky before dawn.

One of the brightest stars in the sky, **Vega** sets in the northwest around 1:45 a.m.

Jupiter rises in the east around 2:10 a.m. and is in the eastern sky before dawn.

Use binoculars or a telescope to look for the **moons of Jupiter.** See the guide on page 18.

2026

Sunday, October 18

TONIGHT

sunset: **6:07 p.m.**
twilight begins: **7:02 p.m.**

IN THE MORNING

twilight begins: **6 a.m.**
sunrise: **6:54 a.m.**

MOON PHASE

moonrise: **2:06 p.m.**
moonset: **12:14 a.m.**

IN THE SKY

The **moon** is prominent in the sky early in the evening. Using a telescope or binoculars, take a look. You should be able to see rough mountains and craters along the boundary between light and dark, in astronomy called the terminator.

Bright star **Capella** is in the northeastern sky at dusk and is almost directly overhead at 4:15 a.m. Read about Capella on page 14.

Scorpius sets in the southwest around 7:40 p.m. See the illustration on page 19.

The bright star **Arcturus** is in the western sky at dusk and sets in the northwest around 8 p.m.

Just east of Scorpius, **Sagittarius** sets in the southwest around 9:10 p.m.

Orion rises in the east around 11 p.m. and at its highest, around 65 degrees, at 4:55 a.m. Look for the bright star **Aldebaran** and the **Pleiades** near Orion. See the illustration on page 17.

Saturn is visible most of the night and is at its highest, around 55 degrees, at 11:35 p.m. Unlike bright Jupiter and Venus and ruddy Mars, Saturn is harder to distinguish from the stars. Stars twinkle and shimmer, while the planets, including Saturn, show a steady flat light.

Castor and Pollux, the twin stars in the constellation Gemini, rise in the northeast around 11:55 p.m. Read how to find Castor and Pollux on page 14.

Cassiopeia, shaped like an *M,* is at its highest, around 65 degrees, at 12:05 a.m. See the illustration on page 11.

Explore the **dark sky** after the moon sets, from around 12:35 a.m.

After the moon sets, from around 12:35 a.m., look for the **Andromeda Galaxy** with binoculars or a telescope. It is almost directly overhead at 1:25 a.m. See the guide on page 11.

The brightest star in the sky, **Sirius** rises in the east around 12:35 a.m.

Rust-red **Mars** rises in the northeast around 1:10 a.m.

The bright star **Altair** sets in the west around 1:10 a.m. Altair can be identified by its less bright neighboring stars, one on either side, together called the shaft of Aquila. See page 18.

One of the brightest stars in the sky, **Vega** sets in the northwest around 1:45 a.m.

Jupiter rises in the east around 2:05 a.m. and is in the eastern sky before dawn.

Use binoculars or a telescope to look for the **moons of Jupiter.** See the guide on page 18.

The **Great Bear (Big Dipper)** rises in the northeast around 2:05 a.m. and is in the northeastern sky before dawn. See the illustration on page 10.

2026

Monday, October 19

TONIGHT

sunset: **6:06 p.m.**
twilight begins: **7:01 p.m.**

MOON PHASE

IN THE MORNING

twilight begins: **6 a.m.**
sunrise: **6:55 a.m.**

moonrise: **2:39 p.m.**
moonset: **1:14 a.m.**

IN THE SKY

The **moon** is prominent in the sky early in the evening. Using a telescope or binoculars, take a look. You should be able to see rough mountains and craters along the boundary between light and dark, in astronomy called the terminator.

Scorpius sets in the southwest around 7:35 p.m. See the illustration on page 19.

The bright star **Arcturus** is in the western sky at dusk and sets in the northwest around 7:55 p.m.

Bright star **Capella** is in the northeastern sky at dusk and is almost directly overhead at 4:10 a.m. Read about Capella on page 14.

Just east of Scorpius, **Sagittarius** sets in the southwest around 9:10 p.m.

Orion rises in the east around 10:55 p.m. and at its highest, around 65 degrees, at 4:50 a.m. Look for the bright star **Aldebaran** and the **Pleiades** near Orion. See the illustration on page 17.

Saturn is visible most of the night and is at its highest, around 55 degrees, at 11:30 p.m. Unlike bright Jupiter and Venus and ruddy Mars, Saturn is harder to distinguish from the stars. Stars twinkle and shimmer, while the planets, including Saturn, show a steady flat light.

Castor and Pollux, the twin stars in the constellation Gemini, rise in the northeast around 11:50 p.m. Read how to find Castor and Pollux on page 14.

Cassiopeia, shaped like an *M,* is at its highest, around 65 degrees, at midnight. See the illustration on page 11.

The brightest star in the sky, **Sirius** rises in the east around 12:30 a.m.

Rust-red **Mars** rises in the northeast around 1:05 a.m.

The bright star **Altair** sets in the west around 1:05 a.m. Altair can be identified by its less bright neighboring stars, one on either side, together called the shaft of Aquila. See page 18.

One of the brightest stars in the sky, **Vega** sets in the northwest around 1:40 a.m.

Jupiter rises in the east around 2 a.m. and is in the eastern sky before dawn.

Use binoculars or a telescope to look for the **moons of Jupiter.** See the guide on page 18.

The **Great Bear (Big Dipper)** rises in the northeast around 2 a.m. and is in the northeastern sky before dawn. See the illustration on page 10.

2026

Tuesday, October 20

TONIGHT

sunset: **6:05 p.m.**
twilight begins: **7 p.m.**

IN THE MORNING

twilight begins: **6:01 a.m.**
sunrise: **6:56 a.m.**

MOON PHASE

moonrise: **3:09 p.m.**
moonset: **2:14 a.m.**

IN THE SKY

The **Orionids meteor shower** happens tonight. A good shower with up to 20 meteors per hour, the best time to watch is after midnight. Tonight the moonlight interferes until the moon sets, around 2:25 a.m.

Scorpius sets in the southwest around 7:35 p.m. See the illustration on page 19.

The bright star **Arcturus** is in the western sky at dusk and sets in the northwest around 7:50 p.m.

Bright star **Capella** is in the northeastern sky at dusk and is almost directly overhead at 4:10 a.m. Read about Capella on page 14.

Just east of Scorpius, **Sagittarius** sets in the southwest around 9:05 p.m.

Orion rises in the east around 10:50 p.m. and at its highest, around 65 degrees, at 4:45 a.m. Look for the bright star **Aldebaran** and the **Pleiades** near Orion. See the illustration on page 17.

Saturn is visible most of the night and is at its highest, around 55 degrees, at 11:30 p.m. and sets in the west around 5:30 a.m. Unlike bright Jupiter and Venus and ruddy Mars, Saturn is harder to distinguish from the stars. Stars twinkle and shimmer, while the planets, including Saturn, show a steady flat light.

Castor and Pollux, the twin stars in the constellation Gemini, rise in the northeast around 11:45 p.m. Read how to find Castor and Pollux on page 14.

Cassiopeia, shaped like an *M,* is at its highest, around 65 degrees, at 11:55 p.m. See the illustration on page 11.

The brightest star in the sky, **Sirius** rises in the east around 12:25 a.m. and at its highest in the south, around 40 degrees, around dawn.

The bright star **Altair** sets in the west around 1 a.m. Altair can be identified by its less bright neighboring stars, one on either side, together called the shaft of Aquila. See page 18.

Rust-red **Mars** rises in the northeast around 1:05 a.m.

One of the brightest stars in the sky, **Vega** sets in the northwest around 1:35 a.m.

Jupiter rises in the east around 2 a.m. and is in the eastern sky before dawn.

Use binoculars or a telescope to look for the **moons of Jupiter.** See the guide on page 18.

The **Great Bear (Big Dipper)** rises in the northeast around 1:55 a.m. and is in the northeastern sky before dawn. See the illustration on page 10.

2026

Wednesday, October 21

TONIGHT

sunset: **6:04 p.m.**
twilight begins: **6:58 p.m.**

IN THE MORNING

twilight begins: **6:02 a.m.**
sunrise: **6:56 a.m.**

MOON PHASE

moonrise: **3:37 p.m.**
moonset: **3:15 a.m.**

IN THE SKY

The **Orionids meteor shower** happens tonight. A good shower with up to 20 meteors per hour, the best time to watch is after midnight. Tonight unfortunately moonlight washes out the shower.

Scorpius sets in the southwest around 7:30 p.m. See the illustration on page 19.

The bright star **Arcturus** is in the western sky at dusk and sets in the northwest around 7:45 p.m.

Bright star **Capella** is in the northeastern sky at dusk and is almost directly overhead at 4:05 a.m. Read about Capella on page 14.

Just east of Scorpius, **Sagittarius** sets in the southwest around 9 p.m.

Orion rises in the east around 10:45 p.m. and at its highest, around 65 degrees, at 4:40 a.m. Look for the bright star **Aldebaran** and the **Pleiades** near Orion. See the illustration on page 17.

Saturn is visible most of the night and is at its highest, around 55 degrees, at 11:25 p.m. and sets in the west around 5:25 a.m. Unlike bright Jupiter and Venus and ruddy Mars, Saturn is harder to distinguish from the stars. Stars twinkle and shimmer, while the planets, including Saturn, show a steady flat light.

Castor and Pollux, the twin stars in the constellation Gemini, rise in the northeast around 11:40 p.m. Read how to find Castor and Pollux on page 14.

Cassiopeia, shaped like an *M,* is at its highest, around 65 degrees, at 11:55 p.m. See the illustration on page 11.

The brightest star in the sky, **Sirius** rises in the east around 12:25 a.m. and at its highest in the south, around 40 degrees, around dawn.

The bright star **Altair** sets in the west around 12:55 a.m. Altair can be identified by its less bright neighboring stars, one on either side, together called the shaft of Aquila. See page 18.

Rust-red **Mars** rises in the east around 1:05 a.m.

One of the brightest stars in the sky, **Vega** sets in the northwest around 1:30 a.m.

The **Great Bear (Big Dipper)** rises in the northeast around 1:50 a.m. and is in the northeastern sky before dawn. See the illustration on page 10.

Jupiter rises in the east around 1:55 a.m. and is in the eastern sky before dawn.

2026
Thursday, October 22

TONIGHT

sunset: **6:02 p.m.**
twilight begins: **6:57 p.m.**

IN THE MORNING

twilight begins: **6:03 a.m.**
sunrise: **6:57 a.m.**

MOON PHASE

moonrise: **4:03 p.m.**
moonset: **4:17 a.m.**

IN THE SKY

The **Orionids meteor shower** happens tonight. A good shower with up to 20 meteors per hour, the best time to watch is after midnight. Tonight unfortunately moonlight washes out the shower.

Scorpius sets in the southwest around 7:25 p.m. See the illustration on page 19.

The bright star **Arcturus** is in the western sky at dusk and sets in the northwest around 7:45 p.m.

Just east of Scorpius, **Sagittarius** is in the western sky at dusk and sets in the southwest around 8:55 p.m.

Bright star **Capella** is in the northeastern sky at dusk and is almost directly overhead at 4 a.m. Read about Capella on page 14.

The nearly **full moon** tonight acts like a spotlight. Look around at the desert landscape at how much illumination the moon provides.

Orion rises in the east around 10:40 p.m. and at its highest, around 65 degrees, at 4:40 a.m. Look for the bright star **Aldebaran** and the **Pleiades** near Orion. See the illustration on page 17.

Saturn is visible most of the night and is at its highest, around 55 degrees, at 11:20 p.m. and sets in the west around 5:20 a.m. Unlike bright Jupiter and Venus and ruddy Mars, Saturn is harder to distinguish from the stars. Stars twinkle and shimmer, while the planets, including Saturn, show a steady flat light.

Castor and Pollux, the twin stars in the constellation Gemini, rise in the northeast around 11:40 p.m. Read how to find Castor and Pollux on page 14.

Cassiopeia, shaped like an *M,* is at its highest, around 65 degrees, at 11:50 p.m. See the illustration on page 11.

The brightest star in the sky, **Sirius** rises in the east around 12:20 a.m. and at its highest in the south, around 40 degrees, around dawn.

The bright star **Altair** sets in the west around 12:50 a.m. Altair can be identified by its less bright neighboring stars, one on either side, together called the shaft of Aquila. See page 18.

Rust-red **Mars** rises in the east around 1:05 a.m.

One of the brightest stars in the sky, **Vega** sets in the northwest around 1:25 a.m.

The **Great Bear (Big Dipper)** rises in the northeast around 1:45 a.m. and is in the northeastern sky before dawn. See the illustration on page 10.

Jupiter rises in the east around 1:50 a.m. and is in the eastern sky before dawn.

2026

Friday, October 23

TONIGHT

sunset: **6:01 p.m.**
twilight begins: **6:56 p.m.**

IN THE MORNING

twilight begins: **6:03 a.m.**
sunrise: **6:58 a.m.**

MOON PHASE

moonrise: **4:30 p.m.**
moonset: **5:21 a.m.**

IN THE SKY

Scorpius sets in the southwest around 7:20 p.m. See the illustration on page 19.

The bright star **Arcturus** is in the western sky at dusk and sets in the northwest around 7:40 p.m.

Just east of Scorpius, **Sagittarius** is in the western sky at dusk and sets in the southwest around 8:55 p.m.

Bright star **Capella** is in the northeastern sky at dusk and is almost directly overhead at 3:55 a.m. Read about Capella on page 14.

The **full moon** tonight acts like a spotlight. Look around at the desert landscape at how much illumination the moon provides.

Orion rises in the east around 10:40 p.m. and at its highest, around 65 degrees, at 4:35 a.m. Look for the bright star **Aldebaran** and the **Pleiades** near Orion. See the illustration on page 17.

Saturn is visible most of the night and is at its highest, around 55 degrees, at 11:15 p.m. and sets in the west around 5:15 a.m. Unlike bright Jupiter and Venus and ruddy Mars, Saturn is harder to distinguish from the stars. Stars twinkle and shimmer, while the planets, including Saturn, show a steady flat light.

Castor and Pollux, the twin stars in the constellation Gemini, rise in the northeast around 11:35 p.m. Read how to find Castor and Pollux on page 14.

Cassiopeia, shaped like an *M,* is at its highest, around 65 degrees, at 11:45 p.m. See the illustration on page 11.

The brightest star in the sky, **Sirius** rises in the east around 12:15 a.m. and at its highest in the south, around 40 degrees, around dawn.

The bright star **Altair** sets in the west around 12:45 a.m. Altair can be identified by its less bright neighboring stars, one on either side, together called the shaft of Aquila. See page 18.

Rust-red Mars rises in the east around 1 a.m.

One of the brightest stars in the sky, **Vega** sets in the northwest around 1:25 a.m.

The **Great Bear (Big Dipper)** rises in the northeast around 1:40 a.m. and is in the northeastern sky before dawn. See the illustration on page 10.

Jupiter rises in the east around 1:50 a.m. and is in the eastern sky before dawn.

2026

Saturday, October 24

TONIGHT

sunset: **6 p.m.**
twilight begins: **6:55 p.m.**

IN THE MORNING

twilight begins: **6:04 a.m.**
sunrise: **6:59 a.m.**

MOON PHASE

moonrise: **4:59 p.m.**
moonset: **6:28 a.m.**

IN THE SKY

Scorpius sets in the southwest around 7:20 p.m. See the illustration on page 19.

The bright star **Arcturus** is in the western sky at dusk and sets in the northwest around 7:35 p.m.

Just east of Scorpius, **Sagittarius** is in the western sky at dusk and sets in the southwest around 8:50 p.m.

Bright star **Capella** is in the northeastern sky at dusk and is almost directly overhead at 3:55 a.m. Read about Capella on page 14.

The **full moon** tonight acts like a spotlight. Look around at the desert landscape at how much illumination the moon provides.

Orion rises in the east around 10:35 p.m. and at its highest, around 65 degrees, at 4:30 a.m. Look for the bright star **Aldebaran** and the **Pleiades** near Orion. See the illustration on page 17.

Saturn is visible most of the night and is at its highest, around 55 degrees, at 11:10 p.m. and sets in the west around 5:10 a.m. Unlike bright Jupiter and Venus and ruddy Mars, Saturn is harder to distinguish from the stars. Stars twinkle and shimmer, while the planets, including Saturn, show a steady flat light.

Castor and Pollux, the twin stars in the constellation Gemini, rise in the northeast around 11:30 p.m. Read how to find Castor and Pollux on page 14.

Cassiopeia, shaped like an *M,* is at its highest, around 65 degrees, at 11:40 p.m. See the illustration on page 11.

The brightest star in the sky, **Sirius** rises in the east around 12:10 a.m. and at its highest in the south, around 40 degrees, around dawn.

The bright star **Altair** sets in the west around 12:40 a.m. Altair can be identified by its less bright neighboring stars, one on either side, together called the shaft of Aquila. See page 18.

Rust-red **Mars** rises in the east around 1 a.m.

One of the brightest stars in the sky, **Vega** sets in the northwest around 1:20 a.m.

The **Great Bear (Big Dipper)** rises in the northeast around 1:40 a.m. and is in the northeastern sky before dawn. See the illustration on page 10.

Jupiter rises in the east around 1:45 a.m. and is in the eastern sky before dawn.

2026

Sunday, October 25

TONIGHT

sunset: **5:59 p.m.**
twilight begins: **6:54 p.m.**

IN THE MORNING

twilight begins: **6:05 a.m.**
sunrise: **7 a.m.**

MOON PHASE

moonrise: **5:31 p.m.**
moonset: **7:39 a.m.**

IN THE SKY

Scorpius sets in the southwest around 7:15 p.m. See the illustration on page 19.

The bright star **Arcturus** is in the western sky at dusk and sets in the northwest around 7:30 p.m.

Just east of Scorpius, **Sagittarius** is in the western sky at dusk and sets in the southwest around 8:45 p.m.

Bright star **Capella** is in the northeastern sky at dusk and is almost directly overhead at 3:50 a.m. Read about Capella on page 14.

The **full moon** tonight acts like a spotlight. Look around at the desert landscape at how much illumination the moon provides.

Orion rises in the east around 10:30 p.m. and at its highest, around 65 degrees, at 4:25 a.m. Look for the bright star **Aldebaran** and the **Pleiades** near Orion. See the illustration on page 17.

Saturn is visible most of the night and is at its highest, around 55 degrees, at 11:05 p.m. and sets in the west around 5:05 a.m. Unlike bright Jupiter and Venus and ruddy Mars, Saturn is harder to distinguish from the stars. Stars twinkle and shimmer, while the planets, including Saturn, show a steady flat light.

Castor and Pollux, the twin stars in the constellation Gemini, rise in the northeast around 11:25 p.m. Read how to find Castor and Pollux on page 14.

Cassiopeia, shaped like an *M,* is at its highest, around 65 degrees, at 11:35 p.m. See the illustration on page 11.

The brightest star in the sky, **Sirius** rises in the east around 12:05 a.m. and at its highest in the south, around 40 degrees, around dawn.

The bright star **Altair** sets in the west around 12:35 a.m. Altair can be identified by its less bright neighboring stars, one on either side, together called the shaft of Aquila. See page 18.

Rust-red **Mars** rises in the east around 1 a.m.

One of the brightest stars in the sky, **Vega** sets in the northwest around 1:15 a.m.

The **Great Bear (Big Dipper)** rises in the northeast around 1:35 a.m. and is in the northeastern sky before dawn. See the illustration on page 10.

Jupiter rises in the east around 1:40 a.m. and is in the eastern sky before dawn.

2026

Monday, October 26

TONIGHT

sunset: **5:58 p.m.**
twilight begins: **6:53 p.m.**

MOON PHASE

IN THE MORNING

twilight begins: **6:06 a.m.**
sunrise: **7:01 a.m.**

moonrise: **6:09 p.m.**
moonset: **7:39 a.m.**

IN THE SKY

The bright star **Arcturus** is in the western sky at dusk and sets in the northwest around 7:30 p.m.

Sagittarius is in the western sky at dusk and sets in the southwest around 8:40 p.m.

Bright star **Capella** is in the northeastern sky at dusk and is almost directly overhead at 3:45 a.m. Read about Capella on page 14.

The **full moon** tonight acts like a spotlight. Look around at the desert landscape at how much illumination the moon provides.

Orion rises in the east around 10:25 p.m. and at its highest, around 65 degrees, at 4:25 a.m. Look for the bright star **Aldebaran** and the **Pleiades** near Orion. See the illustration on page 17.

Saturn is visible most of the night and is at its highest, around 55 degrees, at 11 p.m. and sets in the west around 5 a.m. Unlike bright Jupiter and Venus and ruddy Mars, Saturn is harder to distinguish from the stars. Stars twinkle and shimmer, while the planets, including Saturn, show a steady flat light.

Castor and Pollux, the twin stars in the constellation Gemini, rise in the northeast around 11:20 p.m. Read how to find Castor and Pollux on page 14.

Cassiopeia, shaped like an *M,* is at its highest, around 65 degrees, at 11:35 p.m. See the illustration on page 11.

The brightest star in the sky, **Sirius** rises in the east around 12:05 a.m. and at its highest in the south, around 40 degrees, around dawn.

The bright star **Altair** sets in the west around 12:30 a.m. Altair can be identified by its less bright neighboring stars, one on either side, together called the shaft of Aquila. See page 18.

Rust-red **Mars** rises in the east around 12:55 a.m.

One of the brightest stars in the sky, **Vega** sets in the northwest around 1:10 a.m.

The **Great Bear (Big Dipper)** rises in the northeast around 1:30 a.m. and is in the northeastern sky before dawn. See the illustration on page 10.

Jupiter rises in the east around 1:40 a.m. and is in the eastern sky before dawn.

2026

Tuesday, October 27

TONIGHT

sunset: **5:57 p.m.**
twilight begins: **6:52 p.m.**

IN THE MORNING

twilight begins: **6:07 a.m.**
sunrise: **7:02 a.m.**

MOON PHASE

moonrise: **6:55 p.m.**
moonset: **8:53 a.m.**

IN THE SKY

The bright star **Arcturus** is in the western sky at dusk and sets in the northwest around 7:25 p.m.

Sagittarius is in the western sky at dusk and sets in the southwest around 8:35 p.m.

Bright star **Capella** is in the northeastern sky at dusk and is almost directly overhead at 3:40 a.m. Read about Capella on page 14.

The **full moon** tonight acts like a spotlight. Look around at the desert landscape at how much illumination the moon provides.

Orion rises in the east around 10:25 p.m. and at its highest, around 65 degrees, at 4:20 a.m. Look for the bright star **Aldebaran** and the **Pleiades** near Orion. See the illustration on page 17.

Saturn is visible most of the night and is at its highest, around 55 degrees, at 11 p.m. and sets in the west around 5 a.m. Unlike bright Jupiter and Venus and ruddy Mars, Saturn is harder to distinguish from the stars. Stars twinkle and shimmer, while the planets, including Saturn, show a steady flat light.

Castor and Pollux, the twin stars in the constellation Gemini, rise in the northeast around 11:20 p.m. Read how to find Castor and Pollux on page 14.

Cassiopeia, shaped like an *M,* is at its highest, around 65 degrees, at 11:30 p.m. See the illustration on page 11.

The brightest star in the sky, **Sirius** rises in the east around midnight and at its highest in the south, around 40 degrees, around dawn.

The bright star **Altair** sets in the west around 12:30 a.m. Altair can be identified by its less bright neighboring stars, one on either side, together called the shaft of Aquila. See page 18.

Rust-red **Mars** rises in the east around 12:55 a.m.

One of the brightest stars in the sky, **Vega** sets in the northwest around 1:05 a.m.

The **Great Bear (Big Dipper)** rises in the northeast around 1:25 a.m. and is in the northeastern sky before dawn. See the illustration on page 10.

Jupiter rises in the east around 1:35 a.m. and is in the eastern sky before dawn.

2026

Wednesday, October 28

TONIGHT

sunset: **5:56 p.m.**
twilight begins: **6:51 p.m.**

IN THE MORNING

twilight begins: **6:07 a.m.**
sunrise: **7:02 a.m.**

MOON PHASE

moonrise: **7:50 p.m.**
moonset: **10:06 a.m.**

IN THE SKY

The bright star **Arcturus** sets in the northwest around 7:20 p.m. See page 13 for an easy way to locate the star.

Sagittarius is in the western sky at dusk and sets in the southwest around 8:35 p.m.

Bright star **Capella** is in the northeastern sky at dusk and is almost directly overhead at 3:35 a.m. Read about Capella on page 14.

Orion rises in the east around 10:20 p.m. and at its highest, around 65 degrees, at 4:15 a.m. Look for the bright star **Aldebaran** and the **Pleiades** near Orion. See the illustration on page 17.

Saturn is visible most of the night and is at its highest, around 55 degrees, at 10:55 p.m. and sets in the west around 4:55 a.m. Unlike bright Jupiter and Venus and ruddy Mars, Saturn is harder to distinguish from the stars. Stars twinkle and shimmer, while the planets, including Saturn, show a steady flat light.

Castor and Pollux, the twin stars in the constellation Gemini, rise in the northeast around 11:15 p.m. Read how to find Castor and Pollux on page 14.

Cassiopeia, shaped like an *M,* is at its highest, around 65 degrees, at 11:25 p.m. See the illustration on page 11.

The brightest star in the sky, **Sirius** rises in the east around 11:55 p.m. and at its highest in the south, around 40 degrees, at 5:05 a.m.

The bright star **Altair** sets in the west around 12:25 a.m. Altair can be identified by its less bright neighboring stars, one on either side, together called the shaft of Aquila. See page 18.

Rust-red **Mars** rises in the east around 12:55 a.m.

One of the brightest stars in the sky, **Vega** sets in the northwest around 1:05 a.m.

The **Great Bear (Big Dipper)** rises in the northeast around 1:25 a.m. and is in the northeastern sky before dawn. See the illustration on page 10.

Jupiter rises in the east around 1:30 a.m. and is in the eastern sky before dawn.

Bright **Venus** rises in the east around dawn.

2026

Thursday, October 29

TONIGHT

sunset: **5:55 p.m.**
twilight begins: **6:50 p.m.**

IN THE MORNING

twilight begins: **6:08 a.m.**
sunrise: **7:03 a.m.**

MOON PHASE

moonrise: **8:54 p.m.**
moonset: **11:15 a.m.**

IN THE SKY

Sagittarius is in the western sky at dusk and sets in the southwest around 8:30 p.m.

Bright star **Capella** is in the northeastern sky at dusk and is almost directly overhead at 3:35 a.m. Read about Capella on page 14.

Explore the **dark sky** after twilight and before the moon rises, around 8:35 p.m.

Before the moon rises, until around 8:35 p.m., look for the **Andromeda Galaxy** with binoculars or a telescope. It is almost directly overhead at 12:40 a.m. See the guide on page 11.

Orion rises in the east around 10:15 p.m. and at its highest, around 65 degrees, at 4:10 a.m. Look for the bright star **Aldebaran** and the **Pleiades** near Orion. See the illustration on page 17.

Saturn is visible most of the night and is at its highest, around 55 degrees, at 10:50 p.m. and sets in the west around 4:50 a.m. Unlike bright Jupiter and Venus and ruddy Mars, Saturn is harder to distinguish from the stars. Stars twinkle and shimmer, while the planets, including Saturn, show a steady flat light.

Castor and Pollux, the twin stars in the constellation Gemini, rise in the northeast around 11:10 p.m. Read how to find Castor and Pollux on page 14.

Cassiopeia, shaped like an *M,* is at its highest, around 65 degrees, at 11:20 p.m. See the illustration on page 11.

The brightest star in the sky, **Sirius** rises in the east around 11:50 p.m. and at its highest in the south, around 40 degrees, at 5 a.m.

The bright star **Altair** sets in the west around 12:20 a.m. Altair can be identified by its less bright neighboring stars, one on either side, together called the shaft of Aquila. See page 18.

Rust-red **Mars** rises in the east around 12:55 a.m.

One of the brightest stars in the sky, **Vega** sets in the northwest around 1 a.m.

The **Great Bear (Big Dipper)** rises in the northeast around 1:20 a.m. and is in the northeastern sky before dawn. See the illustration on page 10.

Jupiter rises in the east around 1:30 a.m. and is in the eastern sky before dawn.

The bright star **Arcturus** rises in the northeast around 5:45 a.m. and is in the eastern sky before dawn. See page 13 for an easy way to locate the star.

Bright **Venus** rises in the east around dawn.

2026

Friday, October 30

TONIGHT

sunset: **5:54 p.m.**
twilight begins: **6:50 p.m.**

IN THE MORNING

twilight begins: **6:09 a.m.**
sunrise: **7:04 a.m.**

MOON PHASE

moonrise: **10:04 p.m.**
moonset: **12:14 p.m.**

IN THE SKY

Sagittarius is in the western sky at dusk and sets in the southwest around 8:25 p.m.

Bright star **Capella** is in the northeastern sky at dusk and is almost directly overhead at 3:30 a.m. Read about Capella on page 14.

Explore the **dark sky** after twilight and before the moon rises, around 9:45 p.m.

Orion rises in the east around 10:10 p.m. and at its highest, around 65 degrees, at 4:05 a.m. Look for the bright star **Aldebaran** and the **Pleiades** near Orion. See the illustration on page 17.

Saturn is visible most of the night and is at its highest, around 55 degrees, at 10:45 p.m. and sets in the west around 4:45 a.m. Unlike bright Jupiter and Venus and ruddy Mars, Saturn is harder to distinguish from the stars. Stars twinkle and shimmer, while the planets, including Saturn, show a steady flat light.

Castor and Pollux, the twin stars in the constellation Gemini, rise in the northeast around 11:05 p.m. Read how to find Castor and Pollux on page 14.

Cassiopeia, shaped like an *M,* is at its highest, around 65 degrees, at 11:15 p.m. See the illustration on page 11.

The brightest star in the sky, **Sirius** rises in the east around 11:50 p.m. and at its highest in the south, around 40 degrees, at 4:55 a.m.

The bright star **Altair** sets in the west around 12:15 a.m. Altair can be identified by its less bright neighboring stars, one on either side, together called the shaft of Aquila. See page 18.

Rust-red **Mars** rises in the east around 12:50 a.m.

One of the brightest stars in the sky, **Vega** sets in the northwest around 12:55 a.m.

The **Great Bear (Big Dipper)** rises in the northeast around 1:15 a.m. and is in the northeastern sky before dawn. See the illustration on page 10.

Jupiter rises in the east around 1:25 a.m. and is in the eastern sky before dawn.

The bright star **Arcturus** rises in the northeast around 5:40 a.m. and is in the eastern sky before dawn. See page 13 for an easy way to locate the star.

Bright **Venus** rises in the east around 6:30 a.m.

2026

Saturday, October 31

TONIGHT

sunset: **5:53 p.m.**
twilight begins: **6:49 p.m.**

IN THE MORNING

twilight begins: **5:10 a.m.**
sunrise: **6:05 a.m.**

MOON PHASE

moonrise: **11:16 p.m.**
moonset: **1:04 p.m.**

IN THE SKY

The bright star **Arcturus** sets in the northwest around 7:10 p.m. See page 13 for an easy way to locate the star.

Sagittarius is in the western sky at dusk and sets in the southwest around 8:20 p.m.

Bright star **Capella** is in the northeastern sky at dusk and is almost directly overhead at 3:25 a.m. Read about Capella on page 14.

Orion rises in the east around 10:05 p.m. and at its highest, around 65 degrees, at 4:05 a.m. Look for the bright star **Aldebaran** and the **Pleiades** near Orion. See the illustration on page 17.

Saturn is visible most of the night and is at its highest, around 55 degrees, at 10:40 p.m. Unlike bright Jupiter and Venus and ruddy Mars, Saturn is harder to distinguish from the stars. Stars twinkle and shimmer, while the planets, including Saturn, show a steady flat light.

Explore the **dark sky** after twilight and before the moon rises, around 10:55 p.m.

Before the moon rises, until around 10:55 p.m., look for the **Andromeda Galaxy** with binoculars or a telescope. See the guide on page 11.

Castor and Pollux, the twin stars in the constellation Gemini, rise in the northeast around 11 p.m. Read how to find Castor and Pollux on page 14.

Cassiopeia, shaped like an *M,* is at its highest, around 65 degrees, at 11:15 p.m. See the illustration on page 11.

The brightest star in the sky, **Sirius** rises in the east around 11:45 p.m.

The bright star **Altair** sets in the west around 12:10 a.m. Altair can be identified by its less bright neighboring stars, one on either side, together called the shaft of Aquila. See page 18.

Rust-red **Mars** rises in the east around 12:50 a.m.

One of the brightest stars in the sky, **Vega** sets in the northwest around 12:50 a.m.

The **Great Bear (Big Dipper)** rises in the northeast around 1:10 a.m. and is in the northeastern sky before dawn. See the illustration on page 10.

Jupiter rises in the east around 1:20 a.m. and is in the eastern sky before dawn.

Bright **Venus** rises in the east around 5:20 a.m.

2026

Sunday, November 1

TONIGHT

sunset: **4:52 p.m.**
twilight begins: **5:48 p.m.**

IN THE MORNING

twilight begins: **5:11 a.m.**
sunrise: **6:06 a.m.**

MOON PHASE

moonrise: **11:26 p.m.**
moonset: **12:45 p.m.**

IN THE SKY

Sagittarius is in the western sky at dusk and sets in the southwest around 7:15 p.m.

Saturn is visible most of the night and is at its highest, around 55 degrees, at 9:35 p.m. and sets in the west around 3:35 a.m. Unlike bright Jupiter and Venus and ruddy Mars, Saturn is harder to distinguish from the stars. Stars twinkle and shimmer, while the planets, including Saturn, show a steady flat light.

Orion rises in the east around 9:05 p.m. and at its highest, around 65 degrees, at 3 a.m. Look for the bright star **Aldebaran** and the **Pleiades** near Orion. See the illustration on page 17.

Castor and Pollux, the twin stars in the constellation Gemini, rise in the northeast around 10 p.m. and are almost directly overhead around dawn. Read how to find Castor and Pollux on page 14.

Cassiopeia, shaped like an *M,* is at its highest, around 65 degrees, at 10:10 p.m. See the illustration on page 11.

The brightest star in the sky, **Sirius** rises in the east around 10:40 p.m. and at its highest in the south, around 40 degrees, at 3:50 a.m.

The bright star **Altair** sets in the west around 11:10 p.m. Altair can be identified by its less bright neighboring stars, one on either side, together called the shaft of Aquila. See page 18.

Before the moon rises, around 11:20 p.m., look for the **Andromeda Galaxy** with binoculars or a telescope. It is almost directly overhead at moonrise. See the guide on page 11.

One of the brightest stars in the sky, **Vega** sets in the northwest around 11:45 p.m.

Rust-red **Mars** rises in the east around 11:50 p.m.

The **Great Bear (Big Dipper)** rises in the northeast around 12:05 a.m. and is in the northeastern sky before dawn. See the illustration on page 10.

Jupiter rises in the east around 12:20 a.m. and is in the eastern sky before dawn.

The **waning crescent moon, Mars,** and **Jupiter** appear close together in the midnight sky. Look east around 12:30 a.m.

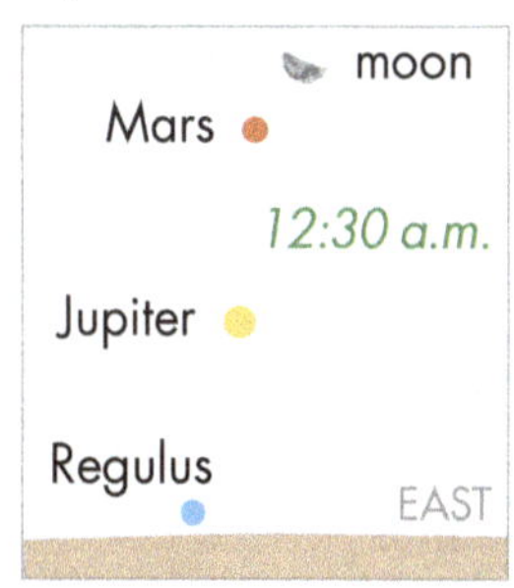

Bright star **Capella** is in the northeastern sky at dusk and is almost directly overhead at 2:20 a.m. Read about Capella on page 14.

The bright star **Arcturus** rises in the northeast around 4:30 a.m. and is in the eastern sky before dawn. See page 13 for an easy way to locate the star.

Bright **Venus** rises in the east around 5:20 a.m.

2026

Monday, November 2

TONIGHT

sunset: **4:51 p.m.**
twilight begins: **5:47 p.m.**

IN THE MORNING

twilight begins: **5:12 a.m.**
sunrise: **6:07 a.m.**

MOON PHASE

moonrise: **12:33 a.m.**
moonset: **1:19 p.m.**

IN THE SKY

Sagittarius is in the western sky at dusk and sets in the southwest around 7:15 p.m.

Bright star **Capella** is in the northeastern sky at dusk and is almost directly overhead at 2:15 a.m. Read about Capella on page 14.

Orion rises in the east around 9 p.m. and at its highest, around 65 degrees, at 2:55 a.m. Look for the bright star **Aldebaran** and the **Pleiades** near Orion. See the illustration on page 17.

Saturn is visible most of the night and is at its highest, around 55 degrees, at 9:35 p.m. It sets in the west around 3:30 a.m. Unlike bright Jupiter and Venus and ruddy Mars, Saturn is harder to distinguish from the stars. Stars twinkle and shimmer, while the planets, including Saturn, show a steady flat light.

Castor and Pollux, the twin stars in the constellation Gemini, rise in the northeast around 9:55 p.m. and are almost directly overhead around dawn. Read how to find Castor and Pollux on page 14.

Cassiopeia, shaped like an *M,* is at its highest, around 65 degrees, at 10:05 p.m. See the illustration on page 11.

The brightest star in the sky, **Sirius** rises in the east around 10:35 p.m. and at its highest in the south, around 40 degrees, at 3:45 a.m.

The bright star **Altair** sets in the west around 11:05 p.m. Altair can be identified by its less bright neighboring stars, one on either side, together called the shaft of Aquila. See page 18.

One of the brightest stars in the sky, **Vega** sets in the northwest around 11:40 p.m.

Before the moon rises, around 12:15 a.m., look for the **Andromeda Galaxy** with binoculars or a telescope. It is almost directly overhead at 11:25 p.m. See the guide on page 11.

Rust-red **Mars** rises in the east around 11:45 p.m.

The **Great Bear (Big Dipper)** rises in the northeast around 12:05 a.m. and is in the northeastern sky before dawn. See the illustration on page 10.

Jupiter rises in the east around 12:15 a.m. and is in the eastern sky before dawn.

The bright star **Arcturus** rises in the northeast around 4:30 a.m. and is in the eastern sky before dawn. See page 13 for an easy way to locate the star.

Bright **Venus** rises in the east around 5:10 a.m.

2026

Tuesday, November 3

TONIGHT

sunset: **4:50 p.m.**
twilight begins: **5:46 p.m.**

IN THE MORNING

twilight begins: **5:12 a.m.**
sunrise: **6:08 a.m.**

MOON PHASE

moonrise: **1:37 a.m.**
moonset: **1:49 p.m.**

IN THE SKY

Sagittarius is in the western sky at dusk and sets in the southwest around 7:10 p.m.

Bright star **Capella** is in the northeastern sky at dusk and is almost directly overhead at 2:15 a.m. Read about Capella on page 14.

Orion rises in the east around 8:55 p.m. and at its highest, around 65 degrees, at 2:50 a.m. Look for the bright star **Aldebaran** and the **Pleiades** near Orion. See the illustration on page 17.

Saturn is visible most of the night and is at its highest, around 55 degrees, at 9:30 p.m. and sets in the west around 3:30 a.m. Unlike bright Jupiter and Venus and ruddy Mars, Saturn is harder to distinguish from the stars. Stars twinkle and shimmer, while the planets, including Saturn, show a steady flat light.

Castor and Pollux, the twin stars in the constellation Gemini, rise in the northeast around 9:50 p.m. and are almost directly overhead around dawn. Read how to find Castor and Pollux on page 14.

Cassiopeia, shaped like an *M,* is at its highest, around 65 degrees, at 10 p.m. See the illustration on page 11.

The brightest star in the sky, **Sirius** rises in the east around 10:30 p.m. and at its highest in the south, around 40 degrees, at 3:40 a.m.

The bright star **Altair** sets in the west around 11 p.m. Altair can be identified by its less bright neighboring stars, one on either side, together called the shaft of Aquila. See page 18.

One of the brightest stars in the sky, **Vega** sets in the northwest around 11:35 p.m.

Before the moon rises, around 12:30 a.m., look for the **Andromeda Galaxy** with binoculars or a telescope. It is almost directly overhead at 11:20 p.m. See the guide on page 11.

Rust-red **Mars** rises in the east around 11:45 p.m.

The **Great Bear (Big Dipper)** rises in the northeast around midnight. See the illustration on page 10.

Jupiter rises in the east around 12:10 a.m. and is in the eastern sky before dawn.

The bright star **Arcturus** rises in the northeast around 4:25 a.m. and is in the eastern sky before dawn. See page 13 for an easy way to locate the star.

Bright **Venus** rises in the east around 5:05 a.m.

2026

Wednesday, November 4

TONIGHT

sunset: **4:49 p.m.**
twilight begins: **5:45 p.m.**

MOON PHASE

IN THE MORNING

twilight begins: **5:13 a.m.**
sunrise: **6:09 a.m.**

moonrise: **2:39 a.m.**
moonset: **2:17 p.m.**

IN THE SKY

The Taurids meteor shower happens tonight. A weak shower with up to 10 meteors per hour, the best time to watch is after midnight. Tonight the moon will wash out the shower from when it rises, around 2:30 a.m.

Sagittarius is in the western sky at dusk and sets in the southwest around 7:05 p.m.

The absence of the **moon** tonight provides excellent dark-sky viewing conditions.

Orion rises in the east around 8:50 p.m. and at its highest, around 65 degrees, at 2:45 a.m. Look for the bright star **Aldebaran** and the **Pleiades** near Orion. See the illustration on page 17.

Saturn is at its highest, around 55 degrees, at 9:25 p.m. and sets in the west around 3:25 a.m. Unlike bright Jupiter and Venus and ruddy Mars, Saturn is harder to distinguish from the stars. Stars twinkle and shimmer, while the planets, including Saturn, show a steady flat light.

Castor and Pollux, the twin stars in the constellation Gemini, rise in the northeast around 9:45 p.m. and are almost directly overhead around dawn. Read how to find Castor and Pollux on page 14.

Cassiopeia, shaped like an *M,* is at its highest, around 65 degrees, at 10 p.m. See the illustration on page 11.

The brightest star in the sky, **Sirius** rises in the east around 10:30 p.m. and at its highest in the south, around 40 degrees, at 3:35 a.m.

The bright star **Altair** sets in the west around 10:55 p.m. Altair can be identified by its less bright neighboring stars, one on either side, together called the shaft of Aquila. See page 18.

Before the moon rises, around 2:30 a.m., look for the **Andromeda Galaxy** with binoculars or a telescope. It is almost directly overhead at 11:15 p.m. See the guide on page 11.

One of the brightest stars in the sky, **Vega** sets in the northwest around 11:30 p.m.

Rust-red Mars rises in the east around 11:45 p.m.

The **Great Bear (Big Dipper)** rises in the northeast around 11:55 p.m. See the illustration on page 10.

Jupiter rises in the east around 12:05 a.m.

Before the moon rises, around 2:30 a.m., use binoculars or a telescope to look for the **moons of Jupiter.** See the guide on page 18.

Jupiter and Mars appear close together in the midnight sky. Look east near the horizon around 12:50 a.m.

Bright star **Capella** is almost directly overhead at 2:10 a.m. Read about Capella on page 14.

Bright Venus rises in the east around 5 a.m.

2026
Thursday, November 5

TONIGHT

sunset: 4:48 p.m.
twilight begins: 5:45 p.m.

IN THE MORNING

twilight begins: 5:14 a.m.
sunrise: 6:10 a.m.

MOON PHASE

moonrise: 3:41 a.m.
moonset: 2:43 p.m.

IN THE SKY

Sagittarius is in the western sky at dusk and sets in the southwest around 7 p.m.

The absence of the **moon** tonight provides excellent dark-sky viewing conditions.

Orion rises in the east around 8:45 p.m. and at its highest, around 65 degrees, at 2:45 a.m. Look for the bright star **Aldebaran** and the **Pleiades** near Orion. See the illustration on page 17.

Saturn is at its highest, around 55 degrees, at 9:20 p.m. and sets in the west around 3:20 a.m. Unlike bright Jupiter and Venus and ruddy Mars, Saturn is harder to distinguish from the stars. Stars twinkle and shimmer, while the planets, including Saturn, show a steady flat light.

Castor and Pollux, the twin stars in the constellation Gemini, rise in the northeast around 9:45 p.m. and are almost directly overhead around dawn. Read how to find Castor and Pollux on page 14.

Cassiopeia, shaped like an *M,* is at its highest, around 65 degrees, at 9:55 p.m. See the illustration on page 11.

The brightest star in the sky, **Sirius** rises in the east around 10:25 p.m. and at its highest in the south, around 40 degrees, at 3:35 a.m.

The bright star **Altair** sets in the west around 10:55 p.m. Altair can be identified by its less bright neighboring stars, one on either side, together called the shaft of Aquila. See page 18.

With the dark sky, look for the **Andromeda Galaxy** with binoculars or a telescope. It is almost directly overhead at 11:15 p.m. See the guide on page 11.

One of the brightest stars in the sky, **Vega** sets in the northwest around 11:30 p.m.

Rust-red **Mars** rises in the east around 11:40 p.m.

The **Great Bear (Big Dipper)** rises in the northeast around 11:50 p.m. See the illustration on page 10.

Jupiter rises in the east around midnight.

With the dark sky, use binoculars or a telescope to look for the **moons of Jupiter.** See the guide on page 18.

Jupiter and **Mars** appear close together in the midnight sky. Look east near the horizon around 12:50 a.m.

Bright star **Capella** is almost directly overhead at 2:05 a.m. Read about Capella on page 14.

The bright star **Arcturus** rises in the northeast around 4:15 a.m. and is in the eastern sky before dawn. See page 13 for an easy way to locate the star.

Bright **Venus** rises in the east around 4:55 a.m.

2026

Friday, November 6

TONIGHT

sunset: **4:47 p.m.**
twilight begins: **5:44 p.m.**

IN THE MORNING

twilight begins: **5:15 a.m.**
sunrise: **6:11 a.m.**

MOON PHASE

moonrise: **4:42 a.m.**
moonset: **3:10 p.m.**

IN THE SKY

Sagittarius is in the western sky at dusk and sets in the southwest around 7 p.m.

The absence of the **moon** tonight provides excellent dark-sky viewing conditions.

Orion rises in the east around 8:45 p.m. and at its highest, around 65 degrees, at 2:40 a.m. Look for the bright star **Aldebaran** and the **Pleiades** near Orion. See the illustration on page 17.

Saturn is at its highest, around 55 degrees, at 9:15 p.m. and sets in the west around 3:15 a.m. Unlike bright Jupiter and Venus and ruddy Mars, Saturn is harder to distinguish from the stars. Stars twinkle and shimmer, while the planets, including Saturn, show a steady flat light.

Castor and Pollux, the twin stars in the constellation Gemini, rise in the northeast around 9:40 p.m. and are almost directly overhead around dawn. Read how to find Castor and Pollux on page 14.

Cassiopeia, shaped like an *M,* is at its highest, around 65 degrees, at 9:50 p.m. See the illustration on page 11.

The brightest star in the sky, **Sirius** rises in the east around 10:20 p.m. and at its highest in the south, around 40 degrees, at 3:30 a.m.

The bright star **Altair** sets in the west around 10:50 p.m. Altair can be identified by its less bright neighboring stars, one on either side, together called the shaft of Aquila. See page 18.

With the dark sky, look for the **Andromeda Galaxy** with binoculars or a telescope. It is almost directly overhead at 11:10 p.m. See the guide on page 11.

One of the brightest stars in the sky, **Vega** sets in the northwest around 11:25 p.m.

Rust-red **Mars** rises in the east around 11:40 p.m.

The **Great Bear (Big Dipper)** rises in the northeast around 11:45 p.m. See the illustration on page 10.

Jupiter rises in the east around midnight.

With the dark sky, use binoculars or a telescope to look for the **moons of Jupiter.** See the guide on page 18.

Bright star **Capella** is almost directly overhead at 2 a.m. Read about Capella on page 14.

The bright star **Arcturus** rises in the northeast around 4:10 a.m. and is in the eastern sky before dawn. See page 13 for an easy way to locate the star.

Bright **Venus** rises in the east around 4:50 a.m.

The **new moon** and **Venus** appear close together in the morning sky. Look east around 5 a.m.

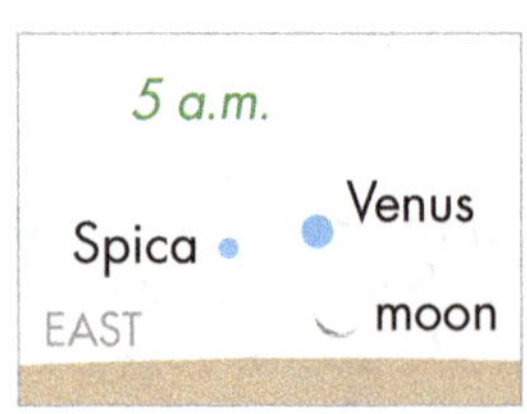

2026

Saturday, November 7

TONIGHT

sunset: **4:47 p.m.**
twilight begins: **5:43 p.m.**

IN THE MORNING

twilight begins: **5:16 a.m.**
sunrise: **6:12 a.m.**

MOON PHASE

moonrise: **5:44 a.m.**
moonset: **3:39 p.m.**

IN THE SKY

Sagittarius is in the western sky at dusk and sets in the southwest around 6:55 p.m.

The absence of the **moon** tonight provides excellent dark-sky viewing conditions.

Orion rises in the east around 8:40 p.m. and at its highest, around 65 degrees, at 2:35 a.m. Look for the bright star **Aldebaran** and the **Pleiades** near Orion. See the illustration on page 17.

Saturn is at its highest, around 55 degrees, at 9:15 p.m. and sets in the west around 3:10 a.m. Unlike bright Jupiter and Venus and ruddy Mars, Saturn is harder to distinguish from the stars. Stars twinkle and shimmer, while the planets, including Saturn, show a steady flat light.

Castor and Pollux, the twin stars in the constellation Gemini, rise in the northeast around 9:35 p.m. and are almost directly overhead around dawn. Read how to find Castor and Pollux on page 14.

Cassiopeia, shaped like an *M,* is at its highest, around 65 degrees, at 9:45 p.m. See the illustration on page 11.

The brightest star in the sky, **Sirius** rises in the east around 10:15 p.m. and at its highest in the south, around 40 degrees, at 3:25 a.m.

With the dark sky, look for the **Andromeda Galaxy** with binoculars or a telescope. It is almost directly overhead at 11:05 p.m. See the guide on page 11.

One of the brightest stars in the sky, **Vega** sets in the northwest around 11:20 p.m.

Rust-red **Mars** rises in the east around 11:40 p.m.

The **Great Bear (Big Dipper)** rises in the northeast around 11:45 p.m. See the illustration on page 10.

Jupiter rises in the east around 11:55 p.m.

With the dark sky, use binoculars or a telescope to look for the **moons of Jupiter.** See the guide on page 18.

Jupiter and **Mars** appear close together in the midnight sky. Look east near the horizon around 12:50 a.m.

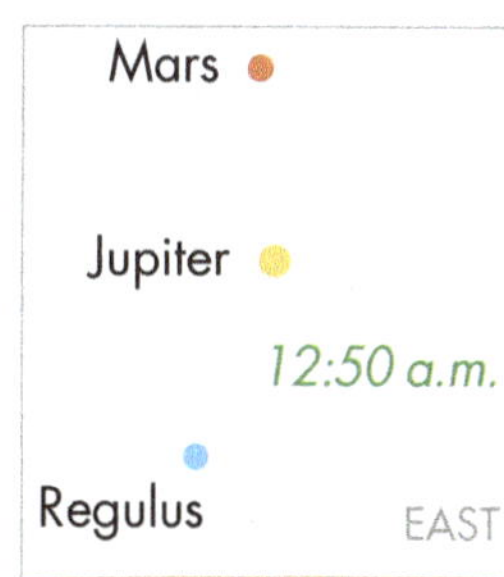

Bright star **Capella** is almost directly overhead at 2 a.m. Read about Capella on page 14.

The bright star **Arcturus** rises in the northeast around 4:10 a.m. and is in the eastern sky before dawn. See page 13 for an easy way to locate the star.

Bright **Venus** rises in the east around 4:45 a.m. and is in the eastern sky before dawn.

2026

Sunday, November 8

TONIGHT

sunset: **4:46 p.m.**
twilight begins: **5:42 p.m.**

IN THE MORNING

twilight begins: **5:17 a.m.**
sunrise: **6:13 a.m.**

MOON PHASE

moonrise: **6:45 a.m.**
moonset: **4:11 p.m.**

IN THE SKY

Sagittarius is in the western sky at dusk and sets in the southwest around 6:50 p.m.

The absence of the **moon** tonight provides excellent dark-sky viewing conditions.

Orion rises in the east around 8:35 p.m. and at its highest, around 65 degrees, at 2:30 a.m. Look for the bright star **Aldebaran** and the **Pleiades** near Orion. See the illustration on page 17.

Saturn is at its highest, around 55 degrees, at 9:10 p.m. and sets in the west around 3:05 a.m. Unlike bright Jupiter and Venus and ruddy Mars, Saturn is harder to distinguish from the stars. Stars twinkle and shimmer, while the planets, including Saturn, show a steady flat light.

Castor and Pollux, the twin stars in the constellation Gemini, rise in the northeast around 9:30 p.m. and are almost directly overhead around dawn. Read how to find Castor and Pollux on page 14.

Cassiopeia, shaped like an *M,* is at its highest, around 65 degrees, at 9:40 p.m. See the illustration on page 11.

The brightest star in the sky, **Sirius** rises in the east around 10:10 p.m. and at its highest in the south, around 40 degrees, at 3:20 a.m.

The bright star **Altair** sets in the west around 10:40 p.m. Altair can be identified by its less bright neighboring stars, one on either side, together called the shaft of Aquila. See page 18.

Look for the **Andromeda Galaxy** with binoculars or a telescope. It is almost directly overhead at 11 p.m. See the guide on page 11.

One of the brightest stars in the sky, **Vega** sets in the northwest around 11:15 p.m.

Rust-red **Mars** rises in the east around 11:35 p.m.

The **Great Bear (Big Dipper)** rises in the northeast around 11:40 p.m. See the illustration on page 10.

Jupiter rises in the east around 11:50 p.m.

Use binoculars or a telescope to look for the **moons of Jupiter.** See the guide on page 18.

Jupiter and **Mars** appear close together in the midnight sky. Look east near the horizon around 12:40 a.m.

Bright star **Capella** is almost directly overhead at 1:55 a.m. Read about Capella on page 14.

The bright star **Arcturus** rises in the northeast around 4:05 a.m. and is in the eastern sky before dawn. See page 13 for an easy way to locate the star.

Bright **Venus** rises in the east around 4:35 a.m. and is in the eastern sky before dawn.

2026

Monday, November 9

TONIGHT

sunset: **4:45 p.m.**
twilight begins: **5:42 p.m.**

IN THE MORNING

twilight begins: **5:17 a.m.**
sunrise: **6:14 a.m.**

MOON PHASE

moonrise: **6:45 a.m.**
moonset: **4:47 p.m.**

IN THE SKY

Sagittarius is in the western sky at dusk and sets in the southwest around 6:45 p.m.

The absence of the **moon** tonight provides excellent dark-sky viewing conditions.

Orion rises in the east around 8:30 p.m. and at its highest, around 65 degrees, at 2:30 a.m. Look for the bright star **Aldebaran** and the **Pleiades** near Orion. See the illustration on page 17.

Saturn is at its highest, around 55 degrees, at 9:05 p.m. and sets in the west around 3:05 a.m. Unlike bright Jupiter and Venus and ruddy Mars, Saturn is harder to distinguish from the stars. Stars twinkle and shimmer, while the planets, including Saturn, show a steady flat light.

Castor and Pollux, the twin stars in the constellation Gemini, rise in the northeast around 9:25 p.m. and are almost directly overhead around dawn. Read how to find Castor and Pollux on page 14.

Cassiopeia, shaped like an *M*, is at its highest, around 65 degrees, at 9:40 p.m. See the illustration on page 11.

The brightest star in the sky, **Sirius** rises in the east around 10:10 p.m. and at its highest in the south, around 40 degrees, at 3:15 a.m.

The bright star **Altair** sets in the west around 10:35 p.m. Altair can be identified by its less bright neighboring stars, one on either side, together called the shaft of Aquila. See page 18.

Look for the **Andromeda Galaxy** with binoculars or a telescope. It is almost directly overhead at 10:55 p.m. See the guide on page 11.

One of the brightest stars in the sky, **Vega** sets in the northwest around 11:10 p.m.

Rust-red Mars rises in the east around 11:35 p.m.

The **Great Bear (Big Dipper)** rises in the northeast around 11:35 p.m. See the illustration on page 10.

Jupiter rises in the east around 11:50 p.m.

Use binoculars or a telescope to look for the **moons of Jupiter.** See the guide on page 18.

Jupiter and Mars appear close together in the midnight sky. Look east near the horizon around 12:30 a.m.

Bright star **Capella** is almost directly overhead at 1:50 a.m. Read about Capella on page 14.

The bright star **Arcturus** rises in the northeast around 4 a.m. and is in the eastern sky before dawn. See page 13 for an easy way to locate the star.

Bright Venus rises in the east around 4:30 a.m. and is in the eastern sky before dawn.

2026

Tuesday, November 10

TONIGHT

sunset: **4:44 p.m.**
twilight begins: **5:41 p.m.**

IN THE MORNING

twilight begins: **5:18 a.m.**
sunrise: **6:14 a.m.**

MOON PHASE

moonrise: **7:46 a.m.**
moonset: **5:29 p.m.**

IN THE SKY

Sagittarius is in the western sky at dusk and sets in the southwest around 6:40 p.m.

The absence of the **moon** tonight provides excellent dark-sky viewing conditions.

Orion rises in the east around 8:30 p.m. and at its highest, around 65 degrees, at 2:25 a.m. Look for the bright star **Aldebaran** and the **Pleiades** near Orion. See the illustration on page 17.

Saturn is at its highest, around 55 degrees, at 9 p.m. and sets in the west around 3 a.m. Unlike bright Jupiter and Venus and ruddy Mars, Saturn is harder to distinguish from the stars. Stars twinkle and shimmer, while the planets, including Saturn, show a steady flat light.

Castor and Pollux, the twin stars in the constellation Gemini, rise in the northeast around 9:25 p.m. and are almost directly overhead at 4:15 a.m. Read how to find Castor and Pollux on page 14.

Cassiopeia, shaped like an *M,* is at its highest, around 65 degrees, at 9:35 p.m. See the illustration on page 11.

The brightest star in the sky, **Sirius** rises in the east around 10:05 p.m. and at its highest in the south, around 40 degrees, at 3:15 a.m.

The bright star **Altair** sets in the west around 10:35 p.m. Altair can be identified by its less bright neighboring stars, one on either side, together called the shaft of Aquila. See page 18.

Look for the **Andromeda Galaxy** with binoculars or a telescope. It is almost directly overhead at 10:55 p.m. See the guide on page 11.

One of the brightest stars in the sky, **Vega** sets in the northwest around 11:10 p.m.

The **Great Bear (Big Dipper)** rises in the northeast around 11:30 p.m. See the illustration on page 10.

Rust-red **Mars** rises in the east around 11:35 p.m.

Jupiter rises in the east around 11:45 p.m.

Use binoculars or a telescope to look for the **moons of Jupiter.** See the guide on page 18.

Jupiter and **Mars** appear close together in the midnight sky. Look east near the horizon around 12:30 a.m.

Bright star **Capella** is almost directly overhead at 1:45 a.m. Read about Capella on page 14.

The bright star **Arcturus** rises in the northeast around 3:55 a.m. and is in the eastern sky before dawn. See page 13 for an easy way to locate the star.

Bright **Venus** rises in the east around 4:25 a.m. and is in the eastern sky before dawn.

2026

Wednesday, November 11

TONIGHT

sunset: **4:44 p.m.**
twilight begins: **5:40 p.m.**

MOON PHASE

IN THE MORNING

twilight begins: **5:19 a.m.**
sunrise: **6:15 a.m.**

moonrise: **8:43 a.m.**
moonset: **6:16 p.m.**

IN THE SKY

The absence of the **moon** tonight provides excellent dark-sky viewing conditions. Explore the **dark sky** after the moon sets, from around 6:35 p.m.

Orion rises in the east around 8:25 p.m. and at its highest, around 65 degrees, at 2:20 a.m. Look for the bright star **Aldebaran** and the **Pleiades** near Orion. See the illustration on page 17.

Saturn is at its highest, around 55 degrees, at 8:55 p.m. and sets in the west around 2:55 a.m. Unlike bright Jupiter and Venus and ruddy Mars, Saturn is harder to distinguish from the stars. Stars twinkle and shimmer, while the planets, including Saturn, show a steady flat light.

Castor and Pollux, the twin stars in the constellation Gemini, rise in the northeast around 9:20 p.m. and are almost directly overhead at 4:10 a.m. Read how to find Castor and Pollux on page 14.

Cassiopeia, shaped like an *M,* is at its highest, around 65 degrees, at 9:30 p.m. See the illustration on page 11.

The brightest star in the sky, **Sirius** rises in the east around 10 p.m. and at its highest in the south, around 40 degrees, at 3:10 a.m.

The bright star **Altair** sets in the west around 10:30 p.m. Altair can be identified by its less bright neighboring stars, one on either side, together called the shaft of Aquila. See page 18.

Look for the **Andromeda Galaxy** with binoculars or a telescope. It is almost directly overhead at 10:50 p.m. See the guide on page 11.

One of the brightest stars in the sky, **Vega** sets in the northwest around 11:05 p.m.

Rust-red **Mars** rises in the east around 11:30 p.m.

The **Great Bear (Big Dipper)** rises in the northeast around 11:30 p.m. See the illustration on page 10.

Jupiter rises in the east around 11:40 p.m.

Use binoculars or a telescope to look for the **moons of Jupiter.** See the guide on page 18.

Jupiter and **Mars** appear close together in the midnight sky. Look east near the horizon around 12:30 a.m.

Bright star **Capella** is almost directly overhead at 1:40 a.m. Read about Capella on page 14.

The bright star **Arcturus** rises in the northeast around 3:55 a.m. and is in the eastern sky before dawn. See page 13 for an easy way to locate the star.

Bright **Venus** rises in the east around 4:20 a.m. and is in the eastern sky before dawn.

2026

Thursday, November 12

TONIGHT

sunset: **4:43 p.m.**
twilight begins: **5:40 p.m.**

IN THE MORNING

twilight begins: **5:20 a.m.**
sunrise: **6:16 a.m.**

MOON PHASE

moonrise: **9:36 a.m.**
moonset: **7:08 p.m.**

IN THE SKY

The absence of the **moon** tonight provides excellent dark-sky viewing conditions. Explore the **dark sky** after the moon sets, from around 7:30 p.m.

After the moon sets, from around 7:30 p.m., look for the **Andromeda Galaxy** with binoculars or a telescope. It is almost directly overhead at 10:45 p.m. See the guide on page 11.

Orion rises in the east around 8:20 p.m. and at its highest, around 65 degrees, at 2:15 a.m. Look for the bright star **Aldebaran** and the **Pleiades** near Orion. See the illustration on page 17.

Saturn is at its highest, around 55 degrees, at 8:50 p.m. and sets in the west around 2:50 a.m. Unlike bright Jupiter and Venus and ruddy Mars, Saturn is harder to distinguish from the stars. Stars twinkle and shimmer, while the planets, including Saturn, show a steady flat light.

Castor and Pollux, the twin stars in the constellation Gemini, rise in the northeast around 9:15 p.m. and are almost directly overhead at 4:05 a.m. Read how to find Castor and Pollux on page 14.

Cassiopeia, shaped like an *M,* is at its highest, around 65 degrees, at 9:25 p.m. See the illustration on page 11.

The brightest star in the sky, **Sirius** rises in the east around 9:55 p.m. and at its highest in the south, around 40 degrees, at 3:05 a.m.

The bright star **Altair** sets in the west around 10:25 p.m. Altair can be identified by its less bright neighboring stars, one on either side, together called the shaft of Aquila. See page 18.

One of the brightest stars in the sky, **Vega** sets in the northwest around 11 p.m.

The **Great Bear (Big Dipper)** rises in the northeast around 11:25 p.m. See the illustration on page 10.

Rust-red Mars rises in the east around 11:30 p.m.

Jupiter rises in the east around 11:35 p.m.

Use binoculars or a telescope to look for the **moons of Jupiter.** See the guide on page 18.

Jupiter and Mars appear close together in the midnight sky. Look east near the horizon around 12:30 a.m.

Bright star **Capella** is almost directly overhead at 1:40 a.m. Read about Capella on page 14.

The bright star **Arcturus** rises in the northeast around 3:50 a.m. and is in the eastern sky before dawn. See page 13 for an easy way to locate the star.

Bright Venus rises in the east around 4:20 a.m. and is in the eastern sky before dawn.

2026
Friday, November 13

TONIGHT

sunset: **4:42 p.m.**
twilight begins: **5:39 p.m.**

IN THE MORNING

twilight begins: **5:21 a.m.**
sunrise: **6:17 a.m.**

MOON PHASE

moonrise: **10:22 a.m.**
moonset: **8:05 p.m.**

IN THE SKY

The **moon** is prominent in the sky early in the evening. Using a telescope or binoculars, take a look. You should be able to see rough mountains and craters along the boundary between light and dark, in astronomy called the terminator.

Explore the **dark sky** after the moon sets, from around 8:25 p.m.

Orion rises in the east around 8:15 p.m. and at its highest, around 65 degrees, at 2:10 a.m. Look for the bright star **Aldebaran** and the **Pleiades** near Orion. See the illustration on page 17.

After the moon sets, from around 8:25 p.m., look for the **Andromeda Galaxy** with binoculars or a telescope. It is almost directly overhead at 10:40 p.m. See the guide on page 11.

Saturn is at its highest, around 55 degrees, at 8:50 p.m. and sets in the west around 2:45 a.m. Unlike bright Jupiter and Venus and ruddy Mars, Saturn is harder to distinguish from the stars. Stars twinkle and shimmer, while the planets, including Saturn, show a steady flat light.

Castor and Pollux, the twin stars in the constellation Gemini, rise in the northeast around 9:10 p.m. and are almost directly overhead at 4 a.m. Read how to find Castor and Pollux on page 14.

Cassiopeia, shaped like an *M,* is at its highest, around 65 degrees, at 9:20 p.m. See the illustration on page 11.

The brightest star in the sky, **Sirius** rises in the east around 9:55 p.m. and at its highest in the south, around 40 degrees, at 3 a.m.

The bright star **Altair** sets in the west around 10:20 p.m. Altair can be identified by its less bright neighboring stars, one on either side, together called the shaft of Aquila. See page 18.

One of the brightest stars in the sky, **Vega** sets in the northwest around 10:55 p.m.

The **Great Bear (Big Dipper)** rises in the northeast around 11:20 p.m. See the illustration on page 10.

Rust-red **Mars** rises in the east around 11:30 p.m.

Jupiter rises in the east around 11:35 p.m.

Use binoculars or a telescope to look for the **moons of Jupiter.** See the guide on page 18.

Jupiter and **Mars** appear close together in the midnight sky. Look east near the horizon around 12:30 a.m.

Bright star **Capella** is almost directly overhead at 1:35 a.m. Read about Capella on page 14.

The bright star **Arcturus** rises in the northeast around 3:45 a.m. and is in the eastern sky before dawn. See page 13 for an easy way to locate the star.

Bright **Venus** rises in the east around 4:15 a.m. and is in the eastern sky before dawn.

2026

Saturday, November 14

TONIGHT

sunset: **4:42 p.m.**
twilight begins: **5:39 p.m.**

MOON PHASE

IN THE MORNING

twilight begins: **5:22 a.m.**
sunrise: **6:18 a.m.**

moonrise: **11:02 a.m.**
moonset: **9:03 p.m.**

IN THE SKY

The **moon** is prominent in the sky early in the evening. Using a telescope or binoculars, take a look. You should be able to see rough mountains and craters along the boundary between light and dark, in astronomy called the terminator.

Orion rises in the east around 8:10 p.m. and at its highest, around 65 degrees, at 2:10 a.m. Look for the bright star **Aldebaran** and the **Pleiades** near Orion. See the illustration on page 17.

Saturn is at its highest, around 55 degrees, at 8:45 p.m. and sets in the west around 2:40 a.m. Unlike bright Jupiter and Venus and ruddy Mars, Saturn is harder to distinguish from the stars. Stars twinkle and shimmer, while the planets, including Saturn, show a steady flat light.

Castor and Pollux, the twin stars in the constellation Gemini, rise in the northeast around 9:05 p.m. and are almost directly overhead at 4 a.m. Read how to find Castor and Pollux on page 14.

Cassiopeia, shaped like an *M,* is at its highest, around 65 degrees, at 9:20 p.m. See the illustration on page 11.

Explore the **dark sky** after the moon sets, from around 9:25 p.m.

The brightest star in the sky, **Sirius** rises in the east around 9:50 p.m. and at its highest in the south, around 40 degrees, at 3 a.m.

The bright star **Altair** sets in the west around 10:15 p.m. Altair can be identified by its less bright neighboring stars, one on either side, together called the shaft of Aquila. See page 18.

After the moon sets, from around 9:25 p.m., look for the **Andromeda Galaxy** with binoculars or a telescope. It is almost directly overhead at 10:35 p.m. See the guide on page 11.

One of the brightest stars in the sky, **Vega** sets in the northwest around 10:50 p.m.

Rust-red **Mars** rises in the east around 11:25 p.m.

Jupiter rises in the east around 11:30 p.m.

Use binoculars or a telescope to look for the **moons of Jupiter.** See the guide on page 18.

Jupiter and **Mars** appear close together in the midnight sky. Look east near the horizon around 12:30 a.m.

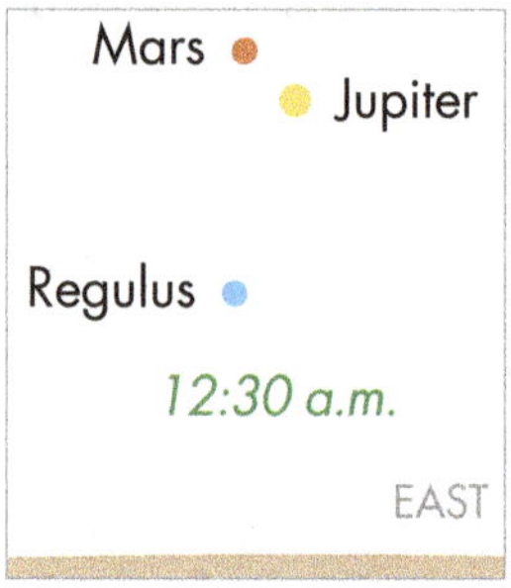

Bright star **Capella** is almost directly overhead at 1:30 a.m. Read about Capella on page 14.

Bright **Venus** rises in the east around 4:10 a.m. and is in the eastern sky before dawn.

2026
Sunday, November 15

TONIGHT

sunset: **4:41 p.m.**
twilight begins: **5:38 p.m.**

IN THE MORNING

twilight begins: **5:22 a.m.**
sunrise: **6:19 a.m.**

MOON PHASE

moonrise: **11:37 a.m.**
moonset: **10:02 p.m.**

IN THE SKY

The **moon** is prominent in the sky early in the evening. Using a telescope or binoculars, take a look. You should be able to see rough mountains and craters along the boundary between light and dark, in astronomy called the terminator.

Orion rises in the east around 8:10 p.m. and at its highest, around 65 degrees, at 2:05 a.m. Look for the bright star **Aldebaran** and the **Pleiades** near Orion. See the illustration on page 17.

Saturn is at its highest, around 55 degrees, at 8:40 p.m. and sets in the west around 2:40 a.m. Unlike bright Jupiter and Venus and ruddy Mars, Saturn is harder to distinguish from the stars. Stars twinkle and shimmer, while the planets, including Saturn, show a steady flat light.

Castor and Pollux, the twin stars in the constellation Gemini, rise in the northeast around 9:05 p.m. and are almost directly overhead at 3:55 a.m. Read how to find Castor and Pollux on page 14.

Cassiopeia, shaped like an *M,* is at its highest, around 65 degrees, at 9:15 p.m. See the illustration on page 11.

The brightest star in the sky, **Sirius** rises in the east around 9:45 p.m. and at its highest in the south, around 40 degrees, at 2:55 a.m.

The bright star **Altair** sets in the west around 10:15 p.m. Altair can be identified by its less bright neighboring stars, one on either side, together called the shaft of Aquila. See page 18.

Explore the **dark sky** after the moon sets, from around 10:20 p.m.

After the moon sets, from around 10:20 p.m., look for the **Andromeda Galaxy** with binoculars or a telescope. It is almost directly overhead at 10:35 p.m. See the guide on page 11.

One of the brightest stars in the sky, **Vega** sets in the northwest around 10:50 p.m.

The **Great Bear (Big Dipper)** rises in the northeast around 11:10 p.m. See the illustration on page 10.

Rust-red **Mars** rises in the east around 11:25 p.m.

Jupiter rises in the east around 11:25 p.m.

Use binoculars or a telescope to look for the **moons of Jupiter.** See the guide on page 18.

Jupiter and **Mars** appear close together in the midnight sky. Look east near the horizon around 12:30 a.m.

Bright star **Capella** is almost directly overhead at 1:25 a.m. Read about Capella on page 14.

Bright **Venus** rises in the east around 4:05 a.m. and is in the eastern sky before dawn.

2026

Monday, November 16

TONIGHT

sunset: **4:40 p.m.**
twilight begins: **5:38 p.m.**

MOON PHASE

IN THE MORNING

twilight begins: **5:23 a.m.**
sunrise: **6:20 a.m.**

moonrise: **12:08 p.m.**
moonset: **11:01 p.m.**

IN THE SKY

The **Leonids meteor shower** happens tonight. A weak shower with up to 15 meteors per hour, the best time to watch is after midnight. Tonight moonlight interferes before the moon sets, around 11:10 p.m., but afterward the viewing conditions are good.

The **moon** is prominent in the sky early in the evening. Using a telescope or binoculars, take a look. You should be able to see rough mountains and craters along the boundary between light and dark, in astronomy called the terminator.

Orion rises in the east around 8:05 p.m. and at its highest, around 65 degrees, at 2 a.m. Look for the bright star **Aldebaran** and the **Pleiades** near Orion. See the illustration on page 17.

Saturn is at its highest, around 55 degrees, at 8:35 p.m. and sets in the west around 2:35 a.m. Unlike bright Jupiter and Venus and ruddy Mars, Saturn is harder to distinguish from the stars. Stars twinkle and shimmer, while the planets, including Saturn, show a steady flat light.

Cassiopeia, shaped like an *M,* is at its highest, around 65 degrees, at 9:10 p.m. See the illustration on page 11.

The brightest star in the sky, **Sirius** rises in the east around 9:40 p.m. and at its highest in the south, around 40 degrees, at 2:50 a.m.

The bright star **Altair** sets in the west around 10:10 p.m. Altair can be identified by its less bright neighboring stars, one on either side, together called the shaft of Aquila. See page 18.

One of the brightest stars in the sky, **Vega** sets in the northwest around 10:45 p.m.

The **Great Bear (Big Dipper)** rises in the northeast around 11:10 p.m. See the illustration on page 10.

Explore the **dark sky** after the moon sets, from around 11:20 p.m.

After the moon sets, from around 11:20 p.m., look for the **Andromeda Galaxy** with binoculars or a telescope. It is in the northwest at midnight. See the guide on page 11.

Rust-red **Mars** rises in the east around 11:25 p.m.

Jupiter rises in the east around 11:25 p.m.

Use binoculars or a telescope to look for the **moons of Jupiter.** See the guide on page 18.

Jupiter and **Mars** appear close together in the midnight sky. Look east near the horizon around 12:30 a.m.

Bright star **Capella** is almost directly overhead at 1:20 a.m. Read about Capella on page 14.

Bright **Venus** rises in the east around 4 a.m. and is in the eastern sky before dawn.

2026

Tuesday, November 17

TONIGHT

sunset: **4:40 p.m.**
twilight begins: **5:37 p.m.**

IN THE MORNING

twilight begins: **5:24 a.m.**
sunrise: **6:21 a.m.**

MOON PHASE

moonrise: **12:36 p.m.**
moonset: **midnight**

IN THE SKY

The **Leonids meteor shower** happens tonight. A weak shower with up to 15 meteors per hour, the best time to watch is after midnight. Tonight the moonlight interferes until the moon sets, around 12:10 a.m.

The **moon** is prominent in the sky early in the evening. Using a telescope or binoculars, take a look. You should be able to see rough mountains and craters along the boundary between light and dark, in astronomy called the terminator.

Orion is visible most of the night and rises in the east around 8 p.m. It is at its highest, around 65 degrees, at 1:55 a.m. Look for the bright star **Aldebaran** and the **Pleiades** near Orion. See the illustration on page 17.

Saturn is at its highest, around 55 degrees, at 8:30 p.m. and sets in the west around 2:30 a.m. Unlike bright Jupiter and Venus and ruddy Mars, Saturn is harder to distinguish from the stars. Stars twinkle and shimmer, while the planets, including Saturn, show a steady flat light.

Castor and Pollux, the twin stars in the constellation Gemini, rise in the northeast around 8:55 p.m. and are almost directly overhead at 3:45 a.m. Read how to find Castor and Pollux on page 14.

The brightest star in the sky, **Sirius** rises in the east around 9:35 p.m. and at its highest in the south, around 40 degrees, at 2:45 a.m.

The bright star **Altair** sets in the west around 10:05 p.m. Altair can be identified by its less bright neighboring stars, one on either side, together called the shaft of Aquila. See page 18.

One of the brightest stars in the sky, **Vega** sets in the northwest around 10:40 p.m.

The **Great Bear (Big Dipper)** rises in the northeast around 11:05 p.m. See the illustration on page 10.

Rust-red **Mars** rises in the east around 11:20 p.m.

Jupiter rises in the east around 11:20 p.m.

Explore the **dark sky** after the moon sets, from around 12:20 a.m.

After the moon sets, from around 12:20 a.m., look for the **Andromeda Galaxy** with binoculars or a telescope. It is in the northwest after midnight. See the guide on page 11.

After the moon sets, from around 12:20 a.m., use binoculars or a telescope to look for the **moons of Jupiter.** See the guide on page 18.

Jupiter and **Mars** appear close together in the midnight sky. Look east near the horizon around 12:30 a.m.

Bright star **Capella** is almost directly overhead at 1:20 a.m. Read about Capella on page 14.

Bright **Venus** rises in the east around 4 a.m. and is in the eastern sky before dawn.

2026

Wednesday, November 18

TONIGHT

sunset: **4:39 p.m.**
twilight begins: **5:37 p.m.**

MOON PHASE

IN THE MORNING

twilight begins: **5:25 a.m.**
sunrise: **6:22 a.m.**

moonrise: **1:02 p.m.**
moonset: **12:59 a.m.**

IN THE SKY

The **Leonids meteor shower** happens tonight. A weak shower with up to 15 meteors per hour, the best time to watch is after midnight. Tonight the moonlight interferes until the moon sets, around 1:10 a.m.

The **moon** is prominent in the sky early in the evening. Using a telescope or binoculars, take a look. You should be able to see rough mountains and craters along the boundary between light and dark, in astronomy called the terminator.

Orion is visible most of the night and rises in the east around 7:55 p.m. It is at its highest, around 65 degrees, at 1:50 a.m. Look for the bright star **Aldebaran** and the **Pleiades** near Orion. See the illustration on page 17.

Saturn is at its highest, around 55 degrees, at 8:25 p.m. and sets in the west around 2:25 a.m. Unlike bright Jupiter and Venus and ruddy Mars, Saturn is harder to distinguish from the stars. Stars twinkle and shimmer, while the planets, including Saturn, show a steady flat light.

Castor and Pollux, the twin stars in the constellation Gemini, rise in the northeast around 8:50 p.m. and are almost directly overhead at 3:40 a.m. Read how to find Castor and Pollux on page 14.

Cassiopeia, shaped like an *M,* is at its highest, around 65 degrees, at 9:05 p.m. See the illustration on page 11.

The brightest star in the sky, **Sirius** rises in the east around 9:35 p.m. and at its highest in the south, around 40 degrees, at 2:40 a.m.

The bright star **Altair** sets in the west around 10 p.m. Altair can be identified by its less bright neighboring stars, one on either side, together called the shaft of Aquila. See page 18.

One of the brightest stars in the sky, **Vega** sets in the northwest around 10:35 p.m.

The **Great Bear (Big Dipper)** rises in the northeast around 11 p.m. See the illustration on page 10.

Jupiter rises in the east around 11:15 p.m.

Rust-red **Mars** rises in the east around 11:20 p.m.

Jupiter and **Mars** appear close together in the midnight sky. Look east near the horizon around 12:30 a.m.

Bright star **Capella** is almost directly overhead at 1:15 a.m. Read about Capella on page 14.

The bright star **Arcturus** rises in the northeast around 3:25 a.m. and is in the eastern sky before dawn. See page 13 for an easy way to locate the star.

Bright **Venus** rises in the east around 3:55 a.m. and is in the eastern sky before dawn.

2026

Thursday, November 19

TONIGHT

sunset: **4:39 p.m.**
twilight begins: **5:37 p.m.**

IN THE MORNING

twilight begins: **5:26 a.m.**
sunrise: **6:23 a.m.**

MOON PHASE

moonrise: **1:28 p.m.**
moonset: **2:01 a.m.**

IN THE SKY

Orion is visible most of the night and rises in the east around 7:50 p.m. It is at its highest, around 65 degrees, at 1:50 a.m. Look for the bright star **Aldebaran** and the **Pleiades** near Orion. See the illustration on page 17.

Saturn is at its highest, around 55 degrees, at 8:25 p.m. and sets in the west around 2:20 a.m. Unlike bright Jupiter and Venus and ruddy Mars, Saturn is harder to distinguish from the stars. Stars twinkle and shimmer, while the planets, including Saturn, show a steady flat light.

Castor and Pollux, the twin stars in the constellation Gemini, rise in the northeast around 8:50 p.m. and are almost directly overhead at 3:40 a.m. Read how to find Castor and Pollux on page 14.

Cassiopeia, shaped like an *M,* is at its highest, around 65 degrees, at 9 p.m. See the illustration on page 11.

The brightest star in the sky, **Sirius** rises in the east around 9:30 p.m. and is at its highest in the south, around 40 degrees, at 2:40 a.m.

The bright star **Altair** sets in the west around 10 p.m. Altair can be identified by its less bright neighboring stars, one on either side, together called the shaft of Aquila. See page 18.

One of the brightest stars in the sky, **Vega** sets in the northwest around 10:35 p.m.

The **Great Bear (Big Dipper)** rises in the northeast around 10:55 p.m. See the illustration on page 10.

Jupiter rises in the east around 11:15 p.m.

Rust-red **Mars** rises in the east around 11:20 p.m.

Jupiter and **Mars** appear close together in the midnight sky. Look east near the horizon around 12:30 a.m.

Bright star **Capella** is almost directly overhead at 1:10 a.m. Read about Capella on page 14.

The bright star **Arcturus** rises in the northeast around 3:20 a.m. and is in the eastern sky before dawn. See page 13 for an easy way to locate the star.

Bright **Venus** rises in the east around 3:50 a.m. and is in the eastern sky before dawn.

Look for **Mercury** in the morning sky, just before sunrise in the east. It's difficult but possible to see the tiny dot in the glare of the sun. It's easier to find with the naked eye, not using binoculars or a telescope. Never look directly at the sun or point binoculars or a telescope at it.

2026

Friday, November 20

TONIGHT

sunset: **4:38** p.m.
twilight begins: **5:36** p.m.

IN THE MORNING

twilight begins: **5:27** a.m.
sunrise: **6:24** a.m.

MOON PHASE

moonrise: **1:55** p.m.
moonset: **3:05** a.m.

IN THE SKY

The **waxing gibbous moon** and **Saturn** appear close together in the evening sky. Look southeast at about 45 degrees elevation around 5:40 p.m.

Orion is visible most of the night and rises in the east around 7:50 p.m. It is at its highest, around 65 degrees, at 1:45 a.m. Look for the bright star **Aldebaran** and the **Pleiades** near Orion. See the illustration on page 17.

Saturn is at its highest, around 55 degrees, at 8:20 p.m. and sets in the west around 2:15 a.m. Unlike bright Jupiter and Venus and ruddy Mars, Saturn is harder to distinguish from the stars. Stars twinkle and shimmer, while the planets, including Saturn, show a steady flat light.

Castor and Pollux, the twin stars in the constellation Gemini, rise in the northeast around 8:45 p.m. and are almost directly overhead at 3:35 a.m. Read how to find Castor and Pollux on page 14.

Cassiopeia, shaped like an *M,* is at its highest, around 65 degrees, at 8:55 p.m. See the illustration on page 11.

The brightest star in the sky, **Sirius** rises in the east around 9:25 p.m. and at its highest in the south, around 40 degrees, at 2:35 a.m.

The bright star **Altair** sets in the west around 9:55 p.m. Altair can be identified by its less bright neighboring stars, one on either side, together called the shaft of Aquila. See page 18.

One of the brightest stars in the sky, **Vega** sets in the northwest around 10:30 p.m.

The **Great Bear (Big Dipper)** rises in the northeast around 10:50 p.m. See the illustration on page 10.

Jupiter rises in the east around 11:10 p.m.

Rust-red **Mars** rises in the east around 11:15 p.m.

Bright star **Capella** is almost directly overhead at 1:05 a.m. Read about Capella on page 14.

The bright star **Arcturus** rises in the northeast around 3:15 a.m. and is in the eastern sky before dawn. See page 13 for an easy way to locate the star.

Bright **Venus** rises in the east around 3:50 a.m. and is in the eastern sky before dawn.

Look for **Mercury** in the morning sky, just before sunrise in the east. It's difficult but possible to see the tiny dot in the glare of the sun. It's easier to find with the naked eye, not using binoculars or a telescope. Never look directly at the sun or point binoculars or a telescope at it.

2026

Saturday, November 21

TONIGHT

sunset: **4:38 p.m.**
twilight begins: **5:36 p.m.**

IN THE MORNING

twilight begins: **5:28 a.m.**
sunrise: **6:25 a.m.**

moonrise: **2:25 p.m.**
moonset: **4:14 a.m.**

IN THE SKY

The **full moon** tonight acts like a spotlight. Look around at the desert landscape at how much illumination the moon provides.

Orion is visible most of the night and rises in the east around 7:45 p.m. It is at its highest, around 65 degrees, at 1:40 a.m. Look for the bright star **Aldebaran** and the **Pleiades** near Orion. See the illustration on page 17.

Saturn is at its highest, around 55 degrees, at 8:15 p.m. and sets in the west around 2:15 a.m. Unlike bright Jupiter and Venus and ruddy Mars, Saturn is harder to distinguish from the stars. Stars twinkle and shimmer, while the planets, including Saturn, show a steady flat light.

Castor and Pollux, the twin stars in the constellation Gemini, rise in the northeast around 8:40 p.m. and are almost directly overhead at 3:30 a.m. Read how to find Castor and Pollux on page 14.

Cassiopeia, shaped like an *M,* is at its highest, around 65 degrees, at 8:50 p.m. See the illustration on page 11.

The brightest star in the sky, **Sirius** rises in the east around 9:20 p.m. and at its highest in the south, around 40 degrees, at 2:30 a.m.

The bright star **Altair** sets in the west around 9:50 p.m. Altair can be identified by its less bright neighboring stars, one on either side, together called the shaft of Aquila. See page 18.

One of the brightest stars in the sky, **Vega** sets in the northwest around 10:25 p.m.

The **Great Bear (Big Dipper)** rises in the northeast around 10:50 p.m. See the illustration on page 10.

Jupiter rises in the east around 11:05 p.m.

Rust-red **Mars** rises in the east around 11:15 p.m.

Jupiter and **Mars** appear close together in the midnight sky. Look east near the horizon around midnight.

Bright **Venus** rises in the east around 3:45 a.m. and is in the eastern sky before dawn.

Look for **Mercury** in the morning sky, just before sunrise in the east. It's difficult but possible to see the tiny dot in the glare of the sun. It's easier to find with the naked eye, not using binoculars or a telescope. Never look directly at the sun or point binoculars or a telescope at it.

2026

Sunday, November 22

TONIGHT

sunset: **4:37 p.m.**
twilight begins: **5:36 p.m.**

MOON PHASE

IN THE MORNING

twilight begins: **5:28 a.m.**
sunrise: **6:26 a.m.**

moonrise: **3:01 p.m.**
moonset: **5:27 a.m.**

IN THE SKY

The **full moon** tonight acts like a spotlight. Look around at the desert landscape at how much illumination the moon provides.

Orion is visible most of the night and rises in the east around 7:40 p.m. It is at its highest, around 65 degrees, at 1:35 a.m. Look for the bright star **Aldebaran** and the **Pleiades** near Orion. See the illustration on page 17.

Saturn is at its highest, around 55 degrees, at 8:10 p.m. and sets in the west around 2:10 a.m. Unlike bright Jupiter and Venus and ruddy Mars, Saturn is harder to distinguish from the stars. Stars twinkle and shimmer, while the planets, including Saturn, show a steady flat light.

Castor and Pollux, the twin stars in the constellation Gemini, rise in the northeast around 8:35 p.m. and are almost directly overhead at 3:25 a.m. Read how to find Castor and Pollux on page 14.

Cassiopeia, shaped like an *M,* is at its highest, around 65 degrees, at 8:45 p.m. See the illustration on page 11.

The brightest star in the sky, **Sirius** rises in the east around 9:15 p.m. and at its highest in the south, around 40 degrees, at 2:25 a.m.

The bright star **Altair** sets in the west around 9:45 p.m. Altair can be identified by its less bright neighboring stars, one on either side, together called the shaft of Aquila. See page 18.

One of the brightest stars in the sky, **Vega** sets in the northwest around 10:20 p.m.

The **Great Bear (Big Dipper)** rises in the northeast around 10:45 p.m. See the illustration on page 10.

Jupiter rises in the east around 11 p.m.

Rust-red **Mars** rises in the east around 11:10 p.m.

Jupiter and **Mars** appear close together in the midnight sky. Look east near the horizon around midnight.

Bright star **Capella** is almost directly overhead at 1 a.m. Read about Capella on page 14.

The bright star **Arcturus** rises in the northeast around 3:10 a.m. and is in the eastern sky before dawn. See page 13 for an easy way to locate the star.

Bright **Venus** rises in the east around 3:40 a.m. and is in the eastern sky before dawn.

2026

Monday, November 23

TONIGHT

sunset: **4:37 p.m.**
twilight begins: **5:35 p.m.**

IN THE MORNING

twilight begins: **5:29 a.m.**
sunrise: **6:27 a.m.**

MOON PHASE

moonrise: **3:43 p.m.**
moonset: **6:42 a.m.**

IN THE SKY

The **full moon** tonight acts like a spotlight. Look around at the desert landscape at how much illumination the moon provides.

Orion is visible most of the night and rises in the east around 7:35 p.m. It is at its highest, around 65 degrees, at 1:35 a.m. Look for the bright star **Aldebaran** and the **Pleiades** near Orion. See the illustration on page 17.

Castor and Pollux, the twin stars in the constellation Gemini, rise in the northeast around 8:30 p.m. and are almost directly overhead at 3:25 a.m. Read how to find Castor and Pollux on page 14.

Cassiopeia, shaped like an *M,* is at its highest, around 65 degrees, at 8:45 p.m. See the illustration on page 11.

Saturn is at its highest, around 55 degrees, at 8:05 p.m. and sets in the west around 2:05 a.m. Unlike bright Jupiter and Venus and ruddy Mars, Saturn is harder to distinguish from the stars. Stars twinkle and shimmer, while the planets, including Saturn, show a steady flat light.

The brightest star in the sky, **Sirius** rises in the east around 9:15 p.m. and at its highest in the south, around 40 degrees, at 2:20 a.m.

The bright star **Altair** sets in the west around 9:40 p.m. Altair can be identified by its less bright neighboring stars, one on either side, together called the shaft of Aquila. See page 18.

One of the brightest stars in the sky, **Vega** sets in the northwest around 10:15 p.m.

The **Great Bear (Big Dipper)** rises in the northeast around 10:40 p.m. See the illustration on page 10.

Jupiter rises in the east around 11 p.m.

Rust-red **Mars** rises in the east around 11:10 p.m.

Jupiter and **Mars** appear close together in the midnight sky. Look east near the horizon around midnight.

Bright star **Capella** is almost directly overhead at 12:55 a.m. Read about Capella on page 14.

The bright star **Arcturus** rises in the northeast around 3:05 a.m. and is in the eastern sky before dawn. See page 13 for an easy way to locate the star.

Bright **Venus** rises in the east around 3:40 a.m. and is in the eastern sky before dawn.

2026

Tuesday, November 24

TONIGHT

sunset: **4:37 p.m.**
twilight begins: **5:35 p.m.**

IN THE MORNING

twilight begins: **5:30 a.m.**
sunrise: **6:28 a.m.**

MOON PHASE

moonrise: **4:35 p.m.**
moonset: **6:42 a.m.**

IN THE SKY

The **full moon** tonight acts like a spotlight. Look around at the desert landscape at how much illumination the moon provides.

Orion is visible most of the night and rises in the east around 7:35 p.m. It is at its highest, around 65 degrees, at 1:30 a.m. Look for the bright star **Aldebaran** and the **Pleiades** near Orion. See the illustration on page 17.

Saturn is at its highest, around 55 degrees, at 8:05 p.m. and sets in the west around 2 a.m. Unlike bright Jupiter and Venus and ruddy Mars, Saturn is harder to distinguish from the stars. Stars twinkle and shimmer, while the planets, including Saturn, show a steady flat light.

Castor and Pollux, the twin stars in the constellation Gemini, rise in the northeast around 8:30 p.m. and are almost directly overhead at 3:20 a.m. Read how to find Castor and Pollux on page 14.

Cassiopeia, shaped like an *M,* is at its highest, around 65 degrees, at 8:40 p.m. See the illustration on page 11.

The brightest star in the sky, **Sirius** rises in the east around 9:10 p.m. and at its highest in the south, around 40 degrees, at 2:20 a.m.

The bright star **Altair** sets in the west around 9:40 p.m. Altair can be identified by its less bright neighboring stars, one on either side, together called the shaft of Aquila. See page 18.

One of the brightest stars in the sky, **Vega** sets in the northwest around 10:15 p.m.

The **Great Bear (Big Dipper)** rises in the northeast around 10:35 p.m. See the illustration on page 10.

Jupiter rises in the east around 10:55 p.m.

Rust-red **Mars** rises in the east around 11:10 p.m.

Jupiter and **Mars** appear close together in the midnight sky. Look east near the horizon around midnight.

Bright star **Capella** is almost directly overhead at 12:50 a.m. Read about Capella on page 14.

The bright star **Arcturus** rises in the northeast around 3 a.m. and is in the eastern sky before dawn. See page 13 for an easy way to locate the star.

At its brightest, **Venus** rises in the east around 3:35 a.m.

2026

Wednesday, November 25

TONIGHT

sunset: 4:36 p.m.
twilight begins: 5:35 p.m.

IN THE MORNING

twilight begins: 5:31 a.m.
sunrise: 6:29 a.m.

MOON PHASE

moonrise: 5:37 p.m.
moonset: 7:55 a.m.

IN THE SKY

The **full moon** tonight acts like a spotlight. Look around at the desert landscape at how much illumination the moon provides.

Orion is visible most of the night and rises in the east around 7:30 p.m. It is at its highest, around 65 degrees, at 1:25 a.m. Look for the bright star **Aldebaran** and the **Pleiades** near Orion. See the illustration on page 17.

Saturn is at its highest, around 55 degrees, at 8 p.m. and sets in the west around 1:55 a.m. Unlike bright Jupiter and Venus and ruddy Mars, Saturn is harder to distinguish from the stars. Stars twinkle and shimmer, while the planets, including Saturn, show a steady flat light.

Castor and Pollux, the twin stars in the constellation Gemini, rise in the northeast around 8:25 p.m. and are almost directly overhead at 3:15 a.m. Read how to find Castor and Pollux on page 14.

Cassiopeia, shaped like an *M,* is at its highest, around 65 degrees, at 8:35 p.m. and is in the northwestern sky before dawn. Shaped like a *Σ,* it sets in the northwest around 5:05 a.m. See the illustration on page 11.

The brightest star in the sky, **Sirius** rises in the east around 9:05 p.m. and at its highest in the south, around 40 degrees, at 2:15 a.m.

The bright star **Altair** sets in the west around 9:35 p.m. Altair can be identified by its less bright neighboring stars, one on either side, together called the shaft of Aquila. See page 18.

One of the brightest stars in the sky, **Vega** sets in the northwest around 10:10 p.m.

The **Great Bear (Big Dipper)** rises in the northeast around 10:35 p.m. See the illustration on page 10.

Jupiter rises in the east around 10:50 p.m.

Rust-red and bright, **Mars** rises in the east around 11:05 p.m.

Jupiter and **Mars** appear close together in the midnight sky. Look east near the horizon around 11:40 p.m.

Bright star **Capella** is almost directly overhead at 12:45 a.m. Read about Capella on page 14.

The bright star **Arcturus** rises in the northeast around 3 a.m. and is in the eastern sky before dawn. See page 13 for an easy way to locate the star.

At its brightest, **Venus** rises in the east around 3:35 a.m.

2026

Thursday, November 26

TONIGHT

sunset: **4:36 p.m.**
twilight begins: **5:35 p.m.**

MOON PHASE

IN THE MORNING

twilight begins: **5:32 a.m.**
sunrise: **6:29 a.m.**

moonrise: **6:47 p.m.**
moonset: **9:01 a.m.**

IN THE SKY

The nearly **full moon** tonight acts like a spotlight. Look around at the desert landscape at how much illumination the moon provides.

Orion is visible most of the night. It rises in the east around 7:25 p.m. and is at its highest, around 65 degrees, at 1:20 a.m. Look for the bright star **Aldebaran** and the **Pleiades** near Orion. See the illustration on page 17.

Saturn is at its highest, around 55 degrees, at 7:55 p.m. and sets in the west around 1:55 a.m. Unlike bright Jupiter and Venus and ruddy Mars, Saturn is harder to distinguish from the stars. Stars twinkle and shimmer, while the planets, including Saturn, show a steady flat light.

Castor and Pollux, the twin stars in the constellation Gemini, rise in the northeast around 8:20 p.m. and are almost directly overhead at 3:10 a.m. Read how to find Castor and Pollux on page 14.

Cassiopeia, shaped like an *M,* is at its highest, around 65 degrees, at 8:30 p.m. and is in the northwestern sky before dawn. Shaped like a *Σ,* it sets in the northwest around 5 a.m. See the illustration on page 11.

The brightest star in the sky, **Sirius** rises in the east around 9 p.m. and at its highest in the south, around 40 degrees, at 2:10 a.m.

The bright star **Altair** sets in the west around 9:30 p.m. Altair can be identified by its less bright neighboring stars, one on either side, together called the shaft of Aquila. See page 18.

One of the brightest stars in the sky, **Vega** sets in the northwest around 10:05 p.m.

The **Great Bear (Big Dipper)** rises in the northeast around 10:30 p.m. See the illustration on page 10.

Jupiter rises in the east around 10:45 p.m.

Rust-red and bright, **Mars** rises in the east around 11:05 p.m.

Jupiter and **Mars** appear close together in the midnight sky. Look east near the horizon around 11:40 p.m.

Bright star **Capella** is almost directly overhead at 12:45 a.m. Read about Capella on page 14.

The bright star **Arcturus** rises in the northeast around 2:55 a.m. and is in the eastern sky before dawn. See page 13 for an easy way to locate the star.

At its brightest, **Venus** rises in the east around 3:35 a.m.

2026

Friday, November 27

TONIGHT

sunset: **4:36 p.m.**
twilight begins: **5:34 p.m.**

IN THE MORNING

twilight begins: **5:33 a.m.**
sunrise: **6:30 a.m.**

MOON PHASE

moonrise: **8:02 p.m.**
moonset: **9:57 a.m.**

IN THE SKY

Orion is visible most of the night and rises in the east around 7:20 p.m. It is at its highest, around 65 degrees, at 1:15 a.m. Look for the bright star **Aldebaran** and the **Pleiades** near Orion. See the illustration on page 17.

Saturn is at its highest, around 55 degrees, at 7:50 p.m. and sets in the west around 1:50 a.m. Unlike bright Jupiter and Venus and ruddy Mars, Saturn is harder to distinguish from the stars. Stars twinkle and shimmer, while the planets, including Saturn, show a steady flat light.

Castor and Pollux, the twin stars in the constellation Gemini, rise in the northeast around 8:15 p.m. and are almost directly overhead at 3:05 a.m. Read how to find Castor and Pollux on page 14.

Cassiopeia, shaped like an *M,* is at its highest, around 65 degrees, at 8:25 p.m. and is in the northwestern sky before dawn. Shaped like a *Σ,* it sets in the northwest around 5 a.m. See the illustration on page 11.

The brightest star in the sky, **Sirius** rises in the east around 9 p.m. and at its highest in the south, around 40 degrees, at 2:05 a.m.

The bright star **Altair** sets in the west around 9:25 p.m. Altair can be identified by its less bright neighboring stars, one on either side, together called the shaft of Aquila. See page 18.

One of the brightest stars in the sky, **Vega** sets in the northwest around 10 p.m.

The **Great Bear (Big Dipper)** rises in the northeast around 10:25 p.m. See the illustration on page 10.

Jupiter rises in the east around 10:45 p.m.

Rust-red and bright, **Mars** rises in the east around 11 p.m.

Jupiter and **Mars** appear close together in the midnight sky. Look east near the horizon around 11:40 p.m.

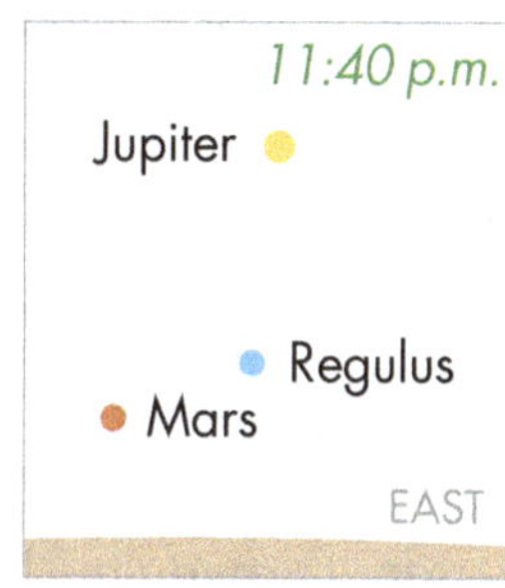

Bright star **Capella** is almost directly overhead at 12:40 a.m. Read about Capella on page 14.

The bright star **Arcturus** rises in the northeast around 2:50 a.m. and is in the eastern sky before dawn. See page 13 for an easy way to locate the star.

At its brightest, **Venus** rises in the east around 3:30 a.m.

2026

Saturday, November 28

TONIGHT

sunset: **4:36 p.m.**
twilight begins: **5:34 p.m.**

MOON PHASE

IN THE MORNING

twilight begins: **5:33 a.m.**
sunrise: **6:31 a.m.**

moonrise: **9:15 p.m.**
moonset: **10:43 a.m.**

IN THE SKY

Explore the **dark sky** after twilight and before the moon rises, around 8:55 p.m.

Before the moon rises, until around 8:55 p.m., look for the **Andromeda Galaxy** with binoculars or a telescope. It is almost directly overhead at 8:30 p.m. See the guide on page 11.

Orion is visible most of the night and rises in the east around 7:15 p.m. It is at its highest, around 65 degrees, at 1:15 a.m. Look for the bright star **Aldebaran** and the **Pleiades** near Orion. See the illustration on page 17.

Saturn is at its highest, around 55 degrees, at 7:45 p.m. and sets in the west around 1:45 a.m. Unlike bright Jupiter and Venus and ruddy Mars, Saturn is harder to distinguish from the stars. Stars twinkle and shimmer, while the planets, including Saturn, show a steady flat light.

Castor and Pollux, the twin stars in the constellation Gemini, rise in the northeast around 8:10 p.m. and are almost directly overhead at 3:05 a.m. Read how to find Castor and Pollux on page 14.

Cassiopeia, shaped like an *M,* is at its highest, around 65 degrees, at 8:25 p.m. and is in the northwestern sky before dawn. Shaped like a *Σ,* it sets in the northwest around 4:55 a.m. See the illustration on page 11.

The brightest star in the sky, **Sirius** rises in the east around 8:55 p.m. and at its highest in the south, around 40 degrees, at 2:05 a.m.

The bright star **Altair** sets in the west around 9:20 p.m. Altair can be identified by its less bright neighboring stars, one on either side, together called the shaft of Aquila. See page 18.

One of the brightest stars in the sky, **Vega** sets in the northwest around 9:55 p.m.

The **Great Bear (Big Dipper)** rises in the northeast around 10:20 p.m. See the illustration on page 10.

Jupiter rises in the east around 10:40 p.m. and at its highest, around 70 degrees, around dawn.

Rust-red and bright, **Mars** rises in the east around 11 p.m.

Jupiter and **Mars** appear close together in the midnight sky. Look east near the horizon around 11:40 p.m.

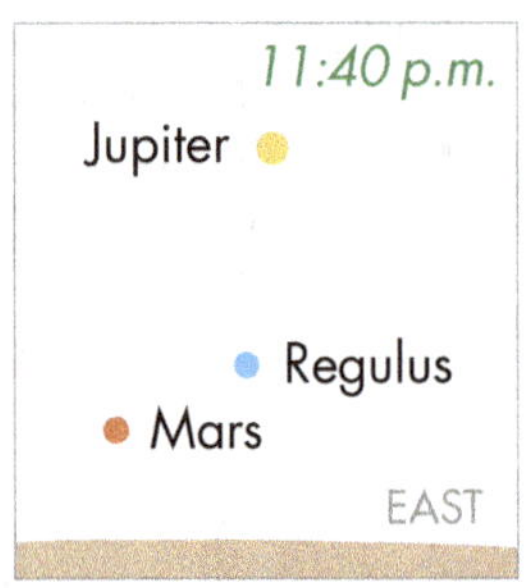

Bright star **Capella** is almost directly overhead at 12:35 a.m. Read about Capella on page 14.

At its brightest, **Venus** rises in the east around 3:30 a.m.

2026

Sunday, November 29

TONIGHT

sunset: 4:36 p.m.
twilight begins: 5:34 p.m.

IN THE MORNING

twilight begins: 5:34 a.m.
sunrise: 6:32 a.m.

MOON PHASE

moonrise: 10:25 p.m.
moonset: 11:20 a.m.

IN THE SKY

Explore the **dark sky** after twilight and before the moon rises, around 10:05 p.m.

Orion is visible most of the night and rises in the east around 7:15 p.m. It is at its highest, around 65 degrees, at 1:10 a.m. Look for the bright star **Aldebaran** and the **Pleiades** near Orion. See the illustration on page 17.

Saturn is at its highest, around 55 degrees, at 7:45 p.m. and sets in the west around 1:40 a.m. Unlike bright Jupiter and Venus and ruddy Mars, Saturn is harder to distinguish from the stars. Stars twinkle and shimmer, while the planets, including Saturn, show a steady flat light.

Castor and Pollux, the twin stars in the constellation Gemini, rise in the northeast around 8:10 p.m. Read how to find Castor and Pollux on page 14.

Cassiopeia, shaped like an *M,* is at its highest, around 65 degrees, at 8:20 p.m. and is in the northwestern sky before dawn. Shaped like a *Σ,* it sets in the northwest around 4:50 a.m. See the illustration on page 11.

The brightest star in the sky, **Sirius** rises in the east around 8:50 p.m. and at its highest in the south, around 40 degrees, at 2 a.m.

The bright star **Altair** sets in the west around 9:20 p.m. Altair can be identified by its less bright neighboring stars, one on either side, together called the shaft of Aquila. See page 18.

One of the brightest stars in the sky, **Vega** sets in the northwest around 9:55 p.m.

Before the moon rises, until around 10:05 p.m., look for the **Andromeda Galaxy** with binoculars or a telescope. It is almost directly overhead at 9:40 p.m. See the guide on page 11.

The **Great Bear (Big Dipper)** rises in the northeast around 10:15 p.m. See the illustration on page 10.

Jupiter rises in the east around 10:35 p.m. and at its highest, around 70 degrees, around dawn.

Rust-red and bright, Mars rises in the east around 11 p.m.

The **waning gibbous moon,** Jupiter, and Mars appear close together in the midnight sky. Look east around 11:40 p.m.

Bright star **Capella** is almost directly overhead at 12:30 a.m. Read about Capella on page 14.

The bright star **Arcturus** rises in the northeast around 2:40 a.m. and is in the eastern sky before dawn. See page 13 for an easy way to locate the star.

Bright Venus rises in the east around 3:25 a.m.

2026

Monday, November 30

TONIGHT

sunset: **4:35 p.m.**
twilight begins: **5:34 p.m.**

IN THE MORNING

twilight begins: **5:35 a.m.**
sunrise: **6:33 a.m.**

MOON PHASE

moonrise: **11:31 p.m.**
moonset: **11:52 a.m.**

IN THE SKY

Explore the **dark sky** after twilight and before the moon rises, around 11:10 p.m.

Orion is visible most of the night and rises in the east around 7:10 p.m. It is at its highest, around 65 degrees, at 1:05 a.m. Look for the bright star **Aldebaran** and the **Pleiades** near Orion. See the illustration on page 17.

Saturn is at its highest, around 55 degrees, at 7:40 p.m. and sets in the west around 1:35 a.m. Unlike bright Jupiter and Venus and ruddy Mars, Saturn is harder to distinguish from the stars. Stars twinkle and shimmer, while the planets, including Saturn, show a steady flat light.

Castor and Pollux, the twin stars in the constellation Gemini, rise in the northeast around 8:05 p.m. and are almost directly overhead at 2:55 a.m. Read how to find Castor and Pollux on page 14.

Cassiopeia, shaped like an *M,* is at its highest, around 65 degrees, at 8:15 p.m. and is in the northwestern sky before dawn. Shaped like a *Σ,* it sets in the northwest around 4:45 a.m. See the illustration on page 11.

The brightest star in the sky, **Sirius** rises in the east around 8:45 p.m. and at its highest in the south, around 40 degrees, at 1:55 a.m.

The bright star **Altair** sets in the west around 9:15 p.m. Altair can be identified by its less bright neighboring stars, one on either side, together called the shaft of Aquila. See page 18.

Before the moon rises, until around 11:10 p.m., look for the **Andromeda Galaxy** with binoculars or a telescope. It is almost directly overhead at 9:35 p.m. See the guide on page 11.

One of the brightest stars in the sky, **Vega** sets in the northwest around 9:50 p.m.

The **Great Bear (Big Dipper)** rises in the northeast around 10:15 p.m. See the illustration on page 10.

Jupiter rises in the east around 10:30 p.m. and at its highest, around 70 degrees, around dawn.

Rust-red and bright, **Mars** rises in the east around 10:55 p.m.

Bright star **Capella** is almost directly overhead at 12:25 a.m. Read about Capella on page 14.

The bright star **Arcturus** rises in the northeast around 2:40 a.m. and is in the eastern sky before dawn. See page 13 for an easy way to locate the star.

Bright **Venus** rises in the east around 3:25 a.m.

2026

Tuesday, December 1

TONIGHT

sunset: **4:35 p.m.**
twilight begins: **5:34 p.m.**

IN THE MORNING

twilight begins: **5:36 a.m.**
sunrise: **6:34 a.m.**

MOON PHASE

moonrise: **12:34 a.m.**
moonset: **12:20 p.m.**

IN THE SKY

Saturn is at its highest, around 55 degrees, at 7:35 p.m. and sets in the west around 1:35 a.m. Unlike bright Jupiter and Venus and ruddy Mars, Saturn is harder to distinguish from the stars. Stars twinkle and shimmer, while the planets, including Saturn, show a steady flat light.

Castor and Pollux, the twin stars in the constellation Gemini, rise in the northeast around 8 p.m. and are almost directly overhead at 2:50 a.m. Read how to find Castor and Pollux on page 14.

Cassiopeia, shaped like an *M,* is at its highest, around 65 degrees, at 8:10 p.m. and is in the northwestern sky before dawn. Shaped like a *Σ,* it sets in the northwest around 4:45 a.m. See the illustration on page 11.

The brightest star in the sky, **Sirius** rises in the east around 8:40 p.m. and at its highest in the south, around 40 degrees, at 1:50 a.m.

The bright star **Altair** sets in the west around 9:10 p.m. Altair can be identified by its less bright neighboring stars, one on either side, together called the shaft of Aquila. See page 18.

Before the moon rises, around 12:30 a.m., look for the **Andromeda Galaxy** with binoculars or a telescope. It is almost directly overhead at 9:30 p.m. and is in the northwestern sky before dawn. See the guide on page 11.

One of the brightest stars in the sky, **Vega** sets in the northwest around 9:45 p.m.

The **Great Bear (Big Dipper)** rises in the northeast around 10:10 p.m. See the illustration on page 10.

Jupiter rises in the east around 10:30 p.m. and at its highest, around 70 degrees, around dawn.

Rust-red and bright, Mars rises in the east around 10:55 p.m.

Before the moon rises, around 12:30 a.m., use binoculars or a telescope to look for the **moons of Jupiter.** See the guide on page 18.

Bright star **Capella** is almost directly overhead at 12:25 a.m. Read about Capella on page 14.

Orion is visible most of the night and is at its highest, around 65 degrees, at 1 a.m. Look for the bright star **Aldebaran** and the **Pleiades** near Orion. See the illustration on page 17.

The bright star **Arcturus** rises in the northeast around 2:35 a.m. and is in the eastern sky before dawn. See page 13 for an easy way to locate the star.

Bright Venus rises in the east around 3:25 a.m.

2026

Wednesday, December 2

TONIGHT

sunset: **4:35 p.m.**
twilight begins: **5:34 p.m.**

IN THE MORNING

twilight begins: **5:36 a.m.**
sunrise: **6:35 a.m.**

MOON PHASE

moonrise: **1:35 a.m.**
moonset: **12:47 p.m.**

IN THE SKY

Saturn is at its highest, around 55 degrees, at 7:30 p.m. and sets in the west around 1:30 a.m. Unlike bright Jupiter and Venus and ruddy Mars, Saturn is harder to distinguish from the stars. Stars twinkle and shimmer, while the planets, including Saturn, show a steady flat light.

Castor and Pollux, the twin stars in the constellation Gemini, rise in the northeast around 7:55 p.m. and are almost directly overhead at 2:45 a.m. Read how to find Castor and Pollux on page 14.

Cassiopeia, shaped like an *M,* is at its highest, around 65 degrees, at 8:05 p.m. Shaped like a *Σ,* it sets in the northwest around 4:40 a.m. See the illustration on page 11.

The brightest star in the sky, **Sirius** rises in the east around 8:40 p.m. and at its highest in the south, around 40 degrees, at 1:45 a.m.

The bright star **Altair** sets in the west around 9:05 p.m. Altair can be identified by its less bright neighboring stars, one on either side, together called the shaft of Aquila. See page 18.

Before the moon rises, around 1:30 a.m., look for the **Andromeda Galaxy** with binoculars or a telescope. It is almost directly overhead at 9:25 p.m. and is in the northwestern sky before dawn. See the guide on page 11.

One of the brightest stars in the sky, **Vega** sets in the northwest around 9:40 p.m.

The **Great Bear (Big Dipper)** rises in the northeast around 10:05 p.m. See the illustration on page 10.

Jupiter rises in the east around 10:25 p.m. and at its highest, around 70 degrees, around dawn.

Before the moon rises, around 1:30 a.m., use binoculars or a telescope to look for the **moons of Jupiter.** See the guide on page 18.

Rust-red and bright, **Mars** rises in the east around 10:50 p.m.

Bright star **Capella** is almost directly overhead at 12:20 a.m. Read about Capella on page 14.

Orion is visible most of the night and is at its highest, around 65 degrees, at 12:55 a.m. Look for the bright star **Aldebaran** and the **Pleiades** near Orion. See the illustration on page 17.

The bright star **Arcturus** rises in the northeast around 2:30 a.m. and is in the eastern sky before dawn. See page 13 for an easy way to locate the star.

Bright **Venus** rises in the east around 3:25 a.m.

2026

Thursday, December 3

TONIGHT

sunset: **4:35 p.m.**
twilight begins: **5:34 p.m.**

IN THE MORNING

twilight begins: **5:37 a.m.**
sunrise: **6:36 a.m.**

MOON PHASE

moonrise: **2:36 a.m.**
moonset: **1:13 p.m.**

IN THE SKY

The absence of the **moon** tonight provides excellent dark-sky viewing conditions.

Saturn is at its highest, around 55 degrees, at 7:25 p.m. and sets in the west around 1:25 a.m. Unlike bright Jupiter and Venus and ruddy Mars, Saturn is harder to distinguish from the stars. Stars twinkle and shimmer, while the planets, including Saturn, show a steady flat light.

Castor and Pollux, the twin stars in the constellation Gemini, rise in the northeast around 7:55 p.m. and are almost directly overhead at 2:45 a.m. Read how to find Castor and Pollux on page 14.

Cassiopeia, shaped like an *M,* is at its highest, around 65 degrees, at 8:05 p.m. and is in the northwestern sky before dawn. Shaped like a *Σ,* it sets in the northwest around 4:35 a.m. See the illustration on page 11.

The brightest star in the sky, **Sirius** rises in the east around 8:35 p.m. and at its highest in the south, around 40 degrees, at 1:45 a.m.

The bright star **Altair** sets in the west around 9:05 p.m. Altair can be identified by its less bright neighboring stars, one on either side, together called the shaft of Aquila. See page 18.

Before the moon rises, around 2:30 a.m., look for the **Andromeda Galaxy** with binoculars or a telescope. It is almost directly overhead at 9:20 p.m. and is in the northwestern sky before dawn. See the guide on page 11.

One of the brightest stars in the sky, **Vega** sets in the northwest around 9:40 p.m.

The **Great Bear (Big Dipper)** rises in the northeast around 10 p.m. See the illustration on page 10.

Jupiter rises in the east around 10:20 p.m. and at its highest, around 70 degrees, around dawn.

Before the moon rises, around 2:30 a.m., use binoculars or a telescope to look for the **moons of Jupiter.** See the guide on page 18.

Rust-red and bright, **Mars** rises in the east around 10:50 p.m.

Bright star **Capella** is almost directly overhead at 12:15 a.m. Read about Capella on page 14.

Orion is visible most of the night and is at its highest, around 65 degrees, at 12:55 a.m. Look for the bright star **Aldebaran** and the **Pleiades** near Orion. See the illustration on page 17.

The bright star **Arcturus** rises in the northeast around 2:25 a.m. and is in the eastern sky before dawn. See page 13 for an easy way to locate the star.

Bright **Venus** rises in the east around 3:20 a.m.

2026

Friday, December 4

TONIGHT

sunset: **4:35 p.m.**
twilight begins: **5:34 p.m.**

MOON PHASE

IN THE MORNING

twilight begins: **5:38 a.m.**
sunrise: **6:36 a.m.**

moonrise: **3:36 a.m.**
moonset: **1:41 p.m.**

IN THE SKY

The absence of the **moon** tonight provides excellent dark-sky viewing conditions.

Saturn is at its highest, around 55 degrees, at 7:25 p.m. and sets in the west around 1:20 a.m. Unlike bright Jupiter and Venus and ruddy Mars, Saturn is harder to distinguish from the stars. Stars twinkle and shimmer, while the planets, including Saturn, show a steady flat light.

Castor and Pollux, the twin stars in the constellation Gemini, rise in the northeast around 7:50 p.m. and are almost directly overhead at 2:40 a.m. Read how to find Castor and Pollux on page 14.

Cassiopeia, shaped like an *M,* is at its highest, around 65 degrees, at 8 p.m. Shaped like a *Σ,* it sets in the northwest around 4:30 a.m. See the illustration on page 11.

The brightest star in the sky, **Sirius** rises in the east around 8:30 p.m. and at its highest in the south, around 40 degrees, at 1:40 a.m.

The bright star **Altair** sets in the west around 9 p.m. Altair can be identified by its less bright neighboring stars, one on either side, together called the shaft of Aquila. See page 18.

Before the moon rises, around 3:30 a.m., look for the **Andromeda Galaxy** with binoculars or a telescope. It is almost directly overhead at 9:20 p.m. and is in the northwestern sky before dawn. See the guide on page 11.

One of the brightest stars in the sky, **Vega** sets in the northwest around 9:35 p.m.

The **Great Bear (Big Dipper)** rises in the northeast around 9:55 p.m. See the illustration on page 10.

Rust-red and bright, **Mars** rises in the east around 10:45 p.m. and at its highest, around 70 degrees, around dawn.

Jupiter rises in the east around 10:15 p.m. and at its highest, around 70 degrees, around dawn.

Before the moon rises, around 3:30 a.m., use binoculars or a telescope to look for the **moons of Jupiter.** See the guide on page 18.

Bright star **Capella** is almost directly overhead at 12:10 a.m. Read about Capella on page 14.

Orion is visible most of the night and is at its highest, around 65 degrees, at 12:50 a.m. Look for the bright star **Aldebaran** and the **Pleiades** near Orion. See the illustration on page 17.

The bright star **Arcturus** rises in the northeast around 2:20 a.m. and is in the eastern sky before dawn. See page 13 for an easy way to locate the star.

Bright **Venus** rises in the east around 3:20 a.m.

2026

Saturday, December 5

TONIGHT

sunset: **4:35 p.m.**
twilight begins: **5:34 p.m.**

IN THE MORNING

twilight begins: **5:39 a.m.**
sunrise: **6:37 a.m.**

MOON PHASE

moonrise: **4:37 a.m.**
moonset: **2:12 p.m.**

IN THE SKY

The absence of the **moon** tonight provides excellent dark-sky viewing conditions.

Saturn is at its highest, around 55 degrees, at 7:20 p.m. and sets in the west around 1:15 a.m. Unlike bright Jupiter and Venus and ruddy Mars, Saturn is harder to distinguish from the stars. Stars twinkle and shimmer, while the planets, including Saturn, show a steady flat light.

Castor and Pollux, the twin stars in the constellation Gemini, rise in the northeast around 7:45 p.m. and are almost directly overhead at 2:35 a.m. Read how to find Castor and Pollux on page 14.

Cassiopeia, shaped like an *M,* is at its highest, around 65 degrees, at 7:55 p.m. and is in the northwestern sky before dawn. Shaped like a *Σ,* it sets in the northwest around 4:25 a.m. See the illustration on page 11.

The brightest star in the sky, **Sirius** rises in the east around 8:25 p.m. and at its highest in the south, around 40 degrees, at 1:35 a.m.

The bright star **Altair** sets in the west around 8:55 p.m. Altair can be identified by its less bright neighboring stars, one on either side, together called the shaft of Aquila. See page 18.

One of the brightest stars in the sky, **Vega** sets in the northwest around 9:30 p.m.

The **Great Bear (Big Dipper)** rises in the northeast around 9:55 p.m. See the illustration on page 10.

Jupiter rises in the east around 10:15 p.m. and at its highest, around 70 degrees, around dawn.

With the dark sky, use binoculars or a telescope to look for the **moons of Jupiter.** See the guide on page 18.

With the dark sky, look for the **Andromeda Galaxy** with binoculars or a telescope. It is in the northwestern sky before dawn. It is a good target until around 4:15 a.m. See the guide on page 11.

Bright star **Capella** is almost directly overhead at 12:10 a.m. Read about Capella on page 14.

Rust-red and bright, Mars rises in the east around 10:45 p.m. and at its highest, around 70 degrees, around dawn.

Orion is visible most of the night and is at its highest, around 65 degrees, at 12:45 a.m. Look for the bright star **Aldebaran** and the **Pleiades** near Orion. See the illustration on page 17.

The bright star **Arcturus** rises in the northeast around 2:20 a.m. and is in the eastern sky before dawn. See page 13 for an easy way to locate the star.

Bright Venus rises in the east around 3:20 a.m.

2026

Sunday, December 6

TONIGHT

sunset: **4:35 p.m.**
twilight begins: **5:34 p.m.**

MOON PHASE

IN THE MORNING

twilight begins: **5:40 a.m.**
sunrise: **6:38 a.m.**

moonrise: **5:37 a.m.**
moonset: **2:46 p.m.**

IN THE SKY

The absence of the **moon** tonight provides excellent dark-sky viewing conditions.

Saturn is at its highest, around 55 degrees, at 7:15 p.m. and sets in the west around 1:15 a.m. Unlike bright Jupiter and Venus and ruddy Mars, Saturn is harder to distinguish from the stars. Stars twinkle and shimmer, while the planets, including Saturn, show a steady flat light.

Castor and Pollux, the twin stars in the constellation Gemini, rise in the northeast around 7:40 p.m. and are almost directly overhead at 2:30 a.m. Read how to find Castor and Pollux on page 14.

Cassiopeia, shaped like an *M,* is at its highest, around 65 degrees, at 7:50 p.m. and is in the northwestern sky before dawn. Shaped like a *Σ,* it sets in the northwest around 4:25 a.m. See the illustration on page 11.

The brightest star in the sky, **Sirius** rises in the east around 8:20 p.m. and at its highest in the south, around 40 degrees, at 1:30 a.m.

The bright star **Altair** sets in the west around 8:50 p.m. Altair can be identified by its less bright neighboring stars, one on either side, together called the shaft of Aquila. See page 18.

One of the brightest stars in the sky, **Vega** sets in the northwest around 9:25 p.m.

The **Great Bear (Big Dipper)** rises in the northeast around 9:50 p.m. See the illustration on page 10.

Jupiter rises in the east around 10:10 p.m. and at its highest, around 70 degrees, around dawn.

With the dark sky, use binoculars or a telescope to look for the **moons of Jupiter.** See the guide on page 18.

Rust-red and bright, **Mars** rises in the east around 10:40 p.m. and at its highest, around 70 degrees, around dawn.

With the dark sky, look for the **Andromeda Galaxy** with binoculars or a telescope. It is in the northwestern sky before dawn. It is a good target until around 4:15 a.m. See the guide on page 11.

Bright star **Capella** is almost directly overhead at 12:05 a.m. Read about Capella on page 14.

Orion is visible most of the night and is at its highest, around 65 degrees, at 12:40 a.m. Look for the bright star **Aldebaran** and the **Pleiades** near Orion. See the illustration on page 17.

The bright star **Arcturus** rises in the northeast around 2:15 a.m. and is in the eastern sky before dawn. See page 13 for an easy way to locate the star.

Bright **Venus** rises in the east around 3:20 a.m.

2026

Monday, December 7

TONIGHT

sunset: **4:35 p.m.**
twilight begins: **5:34 p.m.**

IN THE MORNING

twilight begins: **5:40 a.m.**
sunrise: **6:39 a.m.**

MOON PHASE

moonrise: **6:36 a.m.**
moonset: **3:26 p.m.**

IN THE SKY

The absence of the **moon** tonight provides excellent dark-sky viewing conditions.

Saturn is at its highest, around 55 degrees, at 7:10 p.m. and sets in the west around 1:10 a.m. Unlike bright Jupiter and Venus and ruddy Mars, Saturn is harder to distinguish from the stars. Stars twinkle and shimmer, while the planets, including Saturn, show a steady flat light.

Castor and Pollux, the twin stars in the constellation Gemini, rise in the northeast around 7:35 p.m. and are almost directly overhead at 2:30 a.m. Read how to find Castor and Pollux on page 14.

Cassiopeia, shaped like an *M,* is at its highest, around 65 degrees, at 7:50 p.m. and is in the northwestern sky before dawn. Shaped like a *Σ,* it sets in the northwest around 4:20 a.m. See the illustration on page 11.

The brightest star in the sky, **Sirius** rises in the east around 8:20 p.m. and at its highest in the south, around 40 degrees, at 1:25 a.m.

The bright star **Altair** sets in the west around 8:45 p.m. Altair can be identified by its less bright neighboring stars, one on either side, together called the shaft of Aquila. See page 18.

One of the brightest stars in the sky, **Vega** sets in the northwest around 9:20 p.m.

The **Great Bear (Big Dipper)** rises in the northeast around 9:45 p.m. See the illustration on page 10.

Jupiter rises in the east around 10:05 p.m. and at its highest, around 70 degrees, at 4:40 a.m.

With the dark sky, use binoculars or a telescope to look for the **moons of Jupiter.** See the guide on page 18.

Rust-red and bright, Mars rises in the east around 10:40 p.m. and at its highest, around 70 degrees, around dawn.

Bright star **Capella** is almost directly overhead at 11:55 p.m. Read about Capella on page 14.

With the dark sky, look for the **Andromeda Galaxy** with binoculars or a telescope. It is in the northwestern sky before dawn. It is a good target until around 4:10 a.m. See the guide on page 11.

Orion is visible most of the night and is at its highest, around 65 degrees, at 12:40 a.m. Look for the bright star **Aldebaran** and the **Pleiades** near Orion. See the illustration on page 17.

The bright star **Arcturus** rises in the northeast around 2:10 a.m. and is in the eastern sky before dawn. See page 13 for an easy way to locate the star.

Bright Venus rises in the east around 3:20 a.m.

2026

Tuesday, December 8

TONIGHT

sunset: **4:35 p.m.**
twilight begins: **5:35 p.m.**

MOON PHASE

IN THE MORNING

twilight begins: **5:41 a.m.**
sunrise: **6:40 a.m.**

moonrise: **6:36 a.m.**
moonset: **4:11 p.m.**

IN THE SKY

The absence of the **moon** tonight provides excellent dark-sky viewing conditions.

Saturn is at its highest, around 55 degrees, at 7:05 p.m. and sets in the west around 1:05 a.m. Unlike bright Jupiter and Venus and ruddy Mars, Saturn is harder to distinguish from the stars. Stars twinkle and shimmer, while the planets, including Saturn, show a steady flat light.

Castor and Pollux, the twin stars in the constellation Gemini, rise in the northeast around 7:35 p.m. and are almost directly overhead at 2:25 a.m. Read how to find Castor and Pollux on page 14.

Cassiopeia, shaped like an *M,* is at its highest, around 65 degrees, at 7:45 p.m. and is in the northwestern sky before dawn. Shaped like a *Σ,* it sets in the northwest around 4:15 a.m. See the illustration on page 11.

The brightest star in the sky, **Sirius** rises in the east around 8:15 p.m. and at its highest in the south, around 40 degrees, at 1:25 a.m.

The bright star **Altair** sets in the west around 8:45 p.m. Altair can be identified by its less bright neighboring stars, one on either side, together called the shaft of Aquila. See page 18.

One of the brightest stars in the sky, **Vega** sets in the northwest around 9:20 p.m.

The **Great Bear (Big Dipper)** rises in the northeast around 9:40 p.m. See the illustration on page 10.

Jupiter rises in the east around 10 p.m. and at its highest, around 70 degrees, at 4:35 a.m.

Use binoculars or a telescope to look for the **moons of Jupiter.** See the guide on page 18.

Rust-red and bright, **Mars** rises in the east around 10:35 p.m. and at its highest, around 70 degrees, around dawn.

Look for the **Andromeda Galaxy** with binoculars or a telescope. It is in the northwestern sky before dawn. It is a good target until around 4:05 a.m. See the guide on page 11.

Bright star **Capella** is almost directly overhead at 11:50 p.m. Read about Capella on page 14.

Orion is visible most of the night and is at its highest, around 65 degrees, at 12:35 a.m. Look for the bright star **Aldebaran** and the **Pleiades** near Orion. See the illustration on page 17.

The bright star **Arcturus** rises in the northeast around 2:05 a.m. and is in the eastern sky before dawn. See page 13 for an easy way to locate the star.

Bright **Venus** rises in the east around 3:20 a.m.

2026

Wednesday, December 9

TONIGHT

sunset: **4:35 p.m.**
twilight begins: **5:35 p.m.**

IN THE MORNING

twilight begins: **5:42 a.m.**
sunrise: **6:40 a.m.**

MOON PHASE

moonrise: **7:30 a.m.**
moonset: **5:02 p.m.**

IN THE SKY

The absence of the **moon** tonight provides excellent dark-sky viewing conditions.

Saturn is at its highest, around 55 degrees, at 7:05 p.m. and sets in the west around 1 a.m. Unlike bright Jupiter and Venus and ruddy Mars, Saturn is harder to distinguish from the stars. Stars twinkle and shimmer, while the planets, including Saturn, show a steady flat light.

Castor and Pollux, the twin stars in the constellation Gemini, rise in the northeast around 7:30 p.m. and are almost directly overhead at 2:20 a.m. Read how to find Castor and Pollux on page 14.

Cassiopeia, shaped like an *M,* is at its highest, around 65 degrees, at 7:40 p.m. and is in the northwestern sky before dawn. Shaped like a *Σ,* it sets in the northwest around 4:10 a.m. See the illustration on page 11.

The brightest star in the sky, **Sirius** rises in the east around 8:10 p.m. and at its highest in the south, around 40 degrees, at 1:20 a.m.

The bright star **Altair** sets in the west around 8:40 p.m. Altair can be identified by its less bright neighboring stars, one on either side, together called the shaft of Aquila. See page 18.

One of the brightest stars in the sky, **Vega** sets in the northwest around 9:15 p.m.

The **Great Bear (Big Dipper)** rises in the northeast around 9:35 p.m. See the illustration on page 10.

Jupiter rises in the east around 10 p.m. and at its highest, around 70 degrees, at 4:30 a.m.

Use binoculars or a telescope to look for the **moons of Jupiter.** See the guide on page 18.

Rust-red and bright, **Mars** rises in the east around 10:35 p.m. and at its highest, around 70 degrees, around dawn.

Look for the **Andromeda Galaxy** with binoculars or a telescope. It is in the northwestern sky before dawn. It is a good target until around 4 a.m. See the guide on page 11.

Bright star **Capella** is almost directly overhead at 11:50 p.m. Read about Capella on page 14.

Orion is visible most of the night and is at its highest, around 65 degrees, at 12:30 a.m. Look for the bright star **Aldebaran** and the **Pleiades** near Orion. See the illustration on page 17.

The bright star **Arcturus** rises in the northeast around 2 a.m. and is in the eastern sky before dawn. See page 13 for an easy way to locate the star.

Bright **Venus** rises in the east around 3:15 a.m.

2026

Thursday, December 10

TONIGHT

sunset: **4:36 p.m.**
twilight begins: **5:35 p.m.**

MOON PHASE

IN THE MORNING

twilight begins: **5:42 a.m.**
sunrise: **6:41 a.m.**

moonrise: **8:18 a.m.**
moonset: **5:58 p.m.**

IN THE SKY

The absence of the **moon** tonight provides excellent dark-sky viewing conditions.

Saturn is at its highest, around 55 degrees, at 7 p.m. and sets in the west around 12:55 a.m. Unlike bright Jupiter and Venus and ruddy Mars, Saturn is harder to distinguish from the stars. Stars twinkle and shimmer, while the planets, including Saturn, show a steady flat light.

Castor and Pollux, the twin stars in the constellation Gemini, rise in the northeast around 7:25 p.m. and are almost directly overhead at 2:15 a.m. Read how to find Castor and Pollux on page 14.

Cassiopeia, shaped like an *M,* is at its highest, around 65 degrees, at 7:35 p.m. and is in the northwestern sky before dawn. Shaped like a *Σ,* it sets in the northwest around 4:05 a.m. See the illustration on page 11.

The brightest star in the sky, **Sirius** is visible most of the night and rises in the east around 8:05 p.m. and at its highest in the south, around 40 degrees, at 1:15 a.m.

The bright star **Altair** sets in the west around 8:35 p.m. Altair can be identified by its less bright neighboring stars, one on either side, together called the shaft of Aquila. See page 18.

One of the brightest stars in the sky, **Vega** sets in the northwest around 9:10 p.m.

The **Great Bear (Big Dipper)** rises in the northeast around 9:35 p.m. See the illustration on page 10.

Jupiter rises in the east around 9:55 p.m. and at its highest, around 70 degrees, at 4:25 a.m.

Use binoculars or a telescope to look for the **moons of Jupiter.** See the guide on page 18.

Rust-red and bright, **Mars** rises in the east around 10:30 p.m. and at its highest, around 70 degrees, around dawn.

After the moon sets, from around 6:20 p.m., look for the **Andromeda Galaxy** with binoculars or a telescope. It is in the northwestern sky before dawn. It is a good target until around 4 a.m. See the guide on page 11.

Bright star **Capella** is almost directly overhead at 11:45 p.m. Read about Capella on page 14.

Orion is visible most of the night and is at its highest, around 65 degrees, at 12:25 a.m. Look for the bright star **Aldebaran** and the **Pleiades** near Orion. See the illustration on page 17.

The bright star **Arcturus** rises in the northeast around 2 a.m. and is in the eastern sky before dawn. See page 13 for an easy way to locate the star.

Bright **Venus** rises in the east around 3:15 a.m.

2026

Friday, December 11

MOON PHASE

TONIGHT

sunset: 4:36 p.m.
twilight begins: 5:35 p.m.

IN THE MORNING

twilight begins: 5:43 a.m.
sunrise: 6:42 a.m.

moonrise: 9 a.m.
moonset: 6:55 p.m.

IN THE SKY

Saturn is at its highest, around 55 degrees, at 6:55 p.m. and sets in the west around 12:55 a.m. Unlike bright Jupiter and Venus and ruddy Mars, Saturn is harder to distinguish from the stars. Stars twinkle and shimmer, while the planets, including Saturn, show a steady flat light.

The absence of the **moon** tonight provides excellent dark-sky viewing conditions. Explore the **dark sky** after the moon sets, from around 7:15 p.m.

After the moon sets, from around 7:15 p.m., look for the **Andromeda Galaxy** with binoculars or a telescope. It is in the northwestern sky before dawn. It is a good target until around 3:55 a.m. See the guide on page 11.

Castor and Pollux, the twin stars in the constellation Gemini, rise in the northeast around 7:20 p.m. and are almost directly overhead at 2:10 a.m. Read how to find Castor and Pollux on page 14.

Cassiopeia, shaped like an *M,* is at its highest, around 65 degrees, at 7:30 p.m. and is in the northwestern sky before dawn. Shaped like a *Σ,* it sets in the northwest around 4:05 a.m. See the illustration on page 11.

The brightest star in the sky, **Sirius** is visible most of the night and rises in the east around 8:05 p.m. and at its highest in the south, around 40 degrees, at 1:10 a.m.

The bright star **Altair** sets in the west around 8:30 p.m. Altair can be identified by its less bright neighboring stars, one on either side, together called the shaft of Aquila. See page 18.

One of the brightest stars in the sky, **Vega** sets in the northwest around 9:05 p.m.

The **Great Bear (Big Dipper)** rises in the northeast around 9:30 p.m. See the illustration on page 10.

Jupiter rises in the east around 9:50 p.m. and at its highest, around 70 degrees, at 4:25 a.m.

Use binoculars or a telescope to look for the **moons of Jupiter.** See the guide on page 18.

Rust-red and bright, Mars rises in the east around 10:30 p.m. and at its highest, around 70 degrees, around dawn.

Bright star **Capella** is almost directly overhead at 11:40 p.m. Read about Capella on page 14.

Orion is visible most of the night and is at its highest, around 65 degrees, at 12:20 a.m. Look for the bright star **Aldebaran** and the **Pleiades** near Orion. See the illustration on page 17.

The bright star **Arcturus** rises in the northeast around 1:55 a.m. and is in the eastern sky before dawn. See page 13 for an easy way to locate the star.

Bright Venus rises in the east around 3:15 a.m.

2026

Saturday, December 12

TONIGHT

sunset: **4:36 p.m.**
twilight begins: **5:36 p.m.**

IN THE MORNING

twilight begins: **5:44 a.m.**
sunrise: **6:42 a.m.**

MOON PHASE

moonrise: **9:37 a.m.**
moonset: **7:54 p.m.**

IN THE SKY

The Geminids meteor shower happens tonight. An exceptional shower with up to 100 meteors per hour, the best time to watch is after midnight. Tonight moonlight interferes before the moon sets, around 8:05 p.m., but afterward the viewing conditions are good.

Saturn is at its highest, around 55 degrees, at 6:50 p.m. and sets in the west around 12:50 a.m. Unlike bright Jupiter and Venus and ruddy Mars, Saturn is harder to distinguish from the stars. Stars twinkle and shimmer, while the planets, including Saturn, show a steady flat light.

Castor and Pollux, the twin stars in the constellation Gemini, rise in the northeast around 7:15 p.m. and are almost directly overhead at 2:10 a.m. Read how to find Castor and Pollux on page 14.

Cassiopeia, shaped like an *M,* is at its highest, around 65 degrees, at 7:30 p.m. and is in the northwestern sky before dawn. Shaped like a *Σ,* it sets in the northwest around 4 a.m. See the illustration on page 11.

The brightest star in the sky, **Sirius** is visible most of the night and rises in the east around 8 p.m. and at its highest in the south, around 40 degrees, at 1:05 a.m.

The absence of the **moon** tonight provides excellent dark-sky viewing conditions. Explore the **dark sky** after the moon sets, from around 8:15 p.m.

After the moon sets, from around 8:15 p.m., look for the **Andromeda Galaxy** with binoculars or a telescope. It is in the northwestern sky before dawn. It is a good target until around 3:50 a.m. See the guide on page 11.

The bright star **Altair** sets in the west around 8:25 p.m. Altair can be identified by its less bright neighboring stars, one on either side, together called the shaft of Aquila. See page 18.

One of the brightest stars in the sky, **Vega** sets in the northwest around 9 p.m.

Jupiter rises in the east around 9:45 p.m. and at its highest, around 70 degrees, at 4:20 a.m.

Use binoculars or a telescope to look for the **moons of Jupiter.** See the guide on page 18.

Rust-red and bright, Mars rises in the east around 10:25 p.m. and at its highest, around 70 degrees, around dawn.

Bright star **Capella** is almost directly overhead at 11:35 p.m. Read about Capella on page 14.

Orion is visible most of the night and is at its highest, around 65 degrees, at 12:20 a.m. It sets in the west around 5:20 a.m. Look for the bright star **Aldebaran** and the **Pleiades** near Orion. See the illustration on page 17.

Bright Venus rises in the east around 3:15 a.m.

2026

Sunday, December 13

TONIGHT

sunset: 4:36 p.m.
twilight begins: 5:36 p.m.

IN THE MORNING

twilight begins: 5:44 a.m.
sunrise: 6:43 a.m.

MOON PHASE

moonrise: 10:09 a.m.
moonset: 8:52 p.m.

IN THE SKY

The **Geminids meteor shower** happens tonight. An exceptional shower with up to 100 meteors per hour, the best time to watch is after midnight. Tonight moonlight interferes before the moon sets, around 9 p.m., but afterward the viewing conditions are good.

The **moon** is prominent in the sky early in the evening. Using a telescope or binoculars, take a look. You should be able to see rough mountains and craters along the boundary between light and dark, in astronomy called the terminator.

Saturn is at its highest, around 55 degrees, at 6:45 p.m. and sets in the west around 12:45 a.m. Unlike bright Jupiter and Venus and ruddy Mars, Saturn is harder to distinguish from the stars. Stars twinkle and shimmer, while the planets, including Saturn, show a steady flat light.

Castor and Pollux, the twin stars in the constellation Gemini, rise in the northeast around 7:15 p.m. and are almost directly overhead at 2:05 a.m. Read how to find Castor and Pollux on page 14.

The brightest star in the sky, **Sirius** is visible most of the night and rises in the east around 7:55 p.m. and at its highest in the south, around 40 degrees, at 1:05 a.m.

The bright star **Altair** sets in the west around 8:25 p.m. Altair can be identified by its less bright neighboring stars, one on either side, together called the shaft of Aquila. See page 18.

One of the brightest stars in the sky, **Vega** sets in the northwest around 9 p.m.

Explore the **dark sky** after the moon sets, from around 9:10 p.m.

After the moon sets, from around 9:10 p.m., look for the **Andromeda Galaxy** with binoculars or a telescope. It is in the northwestern sky before dawn. It is a good target until around 3:45 a.m. See the guide on page 11.

The **Great Bear (Big Dipper)** rises in the northeast around 9:20 p.m. See the illustration on page 10.

Jupiter rises in the east around 9:40 p.m. and at its highest, around 70 degrees, at 4:15 a.m.

Use binoculars or a telescope to look for the **moons of Jupiter.** See the guide on page 18.

Rust-red and bright, **Mars** rises in the east around 10:25 p.m. and at its highest, around 70 degrees, around dawn.

Bright star **Capella** is almost directly overhead at 11:30 p.m. Read about Capella on page 14.

Orion is visible most of the night and is at its highest, around 65 degrees, at 12:15 a.m. It sets in the west around 5:15 a.m. Look for the bright star **Aldebaran** and the **Pleiades** near Orion. See the illustration on page 17.

Bright **Venus** rises in the east around 3:15 a.m.

2026

Monday, December 14

TONIGHT

sunset: **4:36 p.m.**
twilight begins: **5:36 p.m.**

IN THE MORNING

twilight begins: **5:45 a.m.**
sunrise: **6:44 a.m.**

MOON PHASE

moonrise: **10:37 a.m.**
moonset: **9:50 p.m.**

IN THE SKY

The **Geminids meteor shower** happens tonight. A exceptional shower with up to 100 meteors per hour, the best time to watch is after midnight. Tonight moonlight interferes before the moon sets, around 10 p.m., but afterward the viewing conditions are good.

The **moon** is prominent in the sky early in the evening. Using a telescope or binoculars, take a look. You should be able to see rough mountains and craters along the boundary between light and dark, in astronomy called the terminator.

Cassiopeia, shaped like an *M,* is at its highest, around 65 degrees, at 7:20 p.m. and is in the northwestern sky before dawn. Shaped like a *Σ,* it sets in the northwest around 3:50 a.m. See the illustration on page 11.

The brightest star in the sky, **Sirius** is visible most of the night and rises in the east around 7:50 p.m. and at its highest in the south, around 40 degrees, at 1 a.m.

Jupiter rises in the east around 9:40 p.m. and at its highest, around 70 degrees, at 4:10 a.m.

Saturn is at its highest, around 55 degrees, at 6:45 p.m. and sets in the west around 12:40 a.m. Unlike bright Jupiter and Venus and ruddy Mars, Saturn is harder to distinguish from the stars. Stars twinkle and shimmer, while the planets, including Saturn, show a steady flat light.

Castor and Pollux, the twin stars in the constellation Gemini, rise in the northeast around 7:10 p.m. and are almost directly overhead at 2 a.m. Read how to find Castor and Pollux on page 14.

One of the brightest stars in the sky, **Vega** sets in the northwest around 8:55 p.m.

Explore the **dark sky** after the moon sets, from around 10:10 p.m.

After the moon sets, from around 10:10 p.m., look for the **Andromeda Galaxy** with binoculars or a telescope. It is a good target until around 3:40 a.m. See the guide on page 11.

After the moon sets, from around 10:10 p.m., use binoculars or a telescope to look for the **moons of Jupiter.** See the guide on page 18.

Rust-red and bright, **Mars** rises in the east around 10:20 p.m. and at its highest, around 65 degrees, around dawn.

Bright star **Capella** is almost directly overhead at 11:30 p.m. Read about Capella on page 14.

Orion is visible most of the night and is at its highest, around 65 degrees, at 12:10 a.m. It sets in the west around 5:10 a.m. Look for the bright star **Aldebaran** and the **Pleiades** near Orion. See the illustration on page 17.

Bright **Venus** rises in the east around 3:15 a.m.

2026

Tuesday, December 15

TONIGHT

sunset: **4:37 p.m.**
twilight begins: **5:37 p.m.**

IN THE MORNING

twilight begins: **5:46 a.m.**
sunrise: **6:44 a.m.**

MOON PHASE

moonrise: **11:03 a.m.**
moonset: **10:48 p.m.**

IN THE SKY

The **moon** is prominent in the sky early in the evening. Using a telescope or binoculars, take a look. You should be able to see rough mountains and craters along the boundary between light and dark, in astronomy called the terminator.

Saturn is at its highest, around 55 degrees, at 6:40 p.m. and sets in the west around 12:40 a.m. Unlike bright Jupiter and Venus and ruddy Mars, Saturn is harder to distinguish from the stars. Stars twinkle and shimmer, while the planets, including Saturn, show a steady flat light.

Castor and Pollux, the twin stars in the constellation Gemini, rise in the northeast around 7:05 p.m. and are almost directly overhead at 1:55 a.m. Read how to find Castor and Pollux on page 14.

Cassiopeia, shaped like an *M,* is at its highest, around 65 degrees, at 7:15 p.m. and is in the northwestern sky before dawn. Shaped like a *Σ,* it sets in the northwest around 3:50 a.m. See the illustration on page 11.

The brightest star in the sky, **Sirius** is visible most of the night and rises in the east around 7:45 p.m. and at its highest in the south, around 40 degrees, at 12:55 a.m.

The bright star **Altair** sets in the west around 8:15 p.m. Altair can be identified by its less bright neighboring stars, one on either side, together called the shaft of Aquila. See page 18.

One of the brightest stars in the sky, **Vega** sets in the northwest around 8:50 p.m.

The **Great Bear (Big Dipper)** rises in the northeast around 9:15 p.m. and at its highest, around 65 degrees, around dawn. See the illustration on page 10.

Jupiter rises in the east around 9:35 p.m. and at its highest, around 70 degrees, at 4:05 a.m.

Rust-red and bright, Mars rises in the east around 10:20 p.m. and at its highest, around 65 degrees, at 4:45 a.m.

Explore the **dark sky** after the moon sets, from around 11:10 p.m.

After the moon sets, from around 11:10 p.m., use binoculars or a telescope to look for the **moons of Jupiter.** See the guide on page 18.

After the moon sets, from around 11:10 p.m., look for the **Andromeda Galaxy** with binoculars or a telescope. It is a good target until around 3:40 a.m. See the guide on page 11.

Bright star **Capella** is almost directly overhead at 11:25 p.m. Read about Capella on page 14.

Orion is visible most of the night and is at its highest, around 65 degrees, at 12:05 a.m. It sets in the west around 5:10 a.m. Look for the bright star **Aldebaran** and the **Pleiades** near Orion. See the illustration on page 17.

Bright Venus rises in the east around 3:15 a.m.

2026

Wednesday, December 16

TONIGHT

sunset: **4:37 p.m.**
twilight begins: **5:37 p.m.**

IN THE MORNING

twilight begins: **5:46 a.m.**
sunrise: **6:45 a.m.**

MOON PHASE

moonrise: **11:28 a.m.**
moonset: **11:47 p.m.**

IN THE SKY

Saturn is at its highest, around 55 degrees, around dusk and sets in the west around 12:35 a.m. Unlike bright Jupiter and Venus and ruddy Mars, Saturn is harder to distinguish from the stars. Stars twinkle and shimmer, while the planets, including Saturn, show a steady flat light.

The **moon** is prominent in the sky early in the evening. Using a telescope or binoculars, take a look. You should be able to see rough mountains and craters along the boundary between light and dark, in astronomy called the terminator.

Castor and Pollux, the twin stars in the constellation Gemini, rise in the northeast around 7 p.m. and are almost directly overhead at 1:50 a.m. Read how to find Castor and Pollux on page 14.

The brightest star in the sky, **Sirius** is visible most of the night and rises in the east around 7:45 p.m. and at its highest in the south, around 40 degrees, at 12:50 a.m.

The bright star **Altair** sets in the west around 8:10 p.m. Altair can be identified by its less bright neighboring stars, one on either side, together called the shaft of Aquila. See page 18.

One of the brightest stars in the sky, **Vega** sets in the northwest around 8:45 p.m.

The **Great Bear (Big Dipper)** rises in the northeast around 9:10 p.m. and at its highest, around 65 degrees, around dawn. See the illustration on page 10.

Jupiter rises in the east around 9:30 p.m. and at its highest, around 70 degrees, at 4:05 a.m.

Rust-red and bright, **Mars** rises in the east around 10:15 p.m. and at its highest, around 65 degrees, at 4:45 a.m.

Bright star **Capella** is almost directly overhead at 11:20 p.m. Read about Capella on page 14.

Orion is visible most of the night and is at its highest, around 65 degrees, at midnight. It sets in the west around 5:05 a.m. Look for the bright star **Aldebaran** and the **Pleiades** near Orion. See the illustration on page 17.

Explore the **dark sky** after the moon sets, from around 12:05 a.m.

After the moon sets, from around 12:05 a.m., use binoculars or a telescope to look for the **moons of Jupiter.** See the guide on page 18.

After the moon sets, from around 12:05 a.m., look for the **Andromeda Galaxy** with binoculars or a telescope. It is a good target until around 3:35 a.m. See the guide on page 11.

The bright star **Arcturus** rises in the northeast around 1:35 a.m. and is in the eastern sky before dawn. See page 13 for an easy way to locate the star.

Bright **Venus** rises in the east around 3:15 a.m.

2026

Thursday, December 17

TONIGHT

sunset: **4:38 p.m.**
twilight begins: **5:37 p.m.**

IN THE MORNING

twilight begins: **5:47 a.m.**
sunrise: **6:46 a.m.**

MOON PHASE

moonrise: **11:54 a.m.**
moonset: **12:47 a.m.**

IN THE SKY

Saturn is at its highest, around 55 degrees, around dusk and sets in the west around 12:30 a.m.

The **moon** is prominent in the sky early in the evening. Using a telescope or binoculars, take a look. You should be able to see rough mountains and craters along the boundary between light and dark, in astronomy called the terminator.

The **waxing gibbous moon** and **Saturn** appear close together in the evening sky. Look south at about 60 degrees elevation around 5:40 p.m.

Castor and Pollux, the twin stars in the constellation Gemini, rise in the northeast around 7 p.m. and are almost directly overhead at 1:50 a.m. Read how to find Castor and Pollux on page 14.

Cassiopeia, shaped like an *M,* is at its highest, around 65 degrees, at 7:10 p.m. Shaped like a *Σ,* it sets in the northwest around 3:40 a.m. See the illustration on page 11.

The brightest star in the sky, **Sirius** is visible most of the night and rises in the east around 7:40 p.m. and at its highest in the south, around 40 degrees, at 12:50 a.m.

The bright star **Altair** sets in the west around 8:10 p.m. Altair can be identified by its less bright neighboring stars, one on either side, together called the shaft of Aquila. See page 18.

One of the brightest stars in the sky, **Vega** sets in the northwest around 8:45 p.m.

The **Great Bear (Big Dipper)** rises in the northeast around 9:05 p.m. and at its highest, around 65 degrees, around dawn. See the illustration on page 10.

Jupiter rises in the east around 9:25 p.m. and at its highest, around 70 degrees, at 4 a.m.

Rust-red and bright, **Mars** rises in the east around 10:15 p.m. and at its highest, around 65 degrees, at 4:40 a.m.

Bright star **Capella** is almost directly overhead at 11:15 p.m. Read about Capella on page 14.

Orion is visible most of the night and is at its highest, around 65 degrees, at 11:55 p.m. and sets in the west around 5 a.m. Look for the bright star **Aldebaran** and the **Pleiades** near Orion. See the illustration on page 17.

The bright star **Arcturus** rises in the northeast around 1:30 a.m. and is in the eastern sky before dawn. See page 13 for an easy way to locate the star.

Bright **Venus** rises in the east around 3:15 a.m.

2026

Friday, December 18

TONIGHT

sunset: **4:38 p.m.**
twilight begins: **5:38 p.m.**

IN THE MORNING

twilight begins: **5:47 a.m.**
sunrise: **6:46 a.m.**

MOON PHASE

moonrise: **12:22 p.m.**
moonset: **1:52 a.m.**

IN THE SKY

Saturn is at its highest, around 55 degrees, around dusk and sets in the west around 12:25 a.m. Unlike bright Jupiter and Venus and ruddy Mars, Saturn is harder to distinguish from the stars. Stars twinkle and shimmer, while the planets, including Saturn, show a steady flat light.

Castor and Pollux, the twin stars in the constellation Gemini, are visible most of the night. They rise in the northeast around 6:55 p.m. and are almost directly overhead at 1:45 a.m. Read how to find Castor and Pollux on page 14.

Cassiopeia, shaped like an *M,* is at its highest, around 65 degrees, at 7:05 p.m. Shaped like a *Σ,* it sets in the northwest around 3:35 a.m. See the illustration on page 11.

The brightest star in the sky, **Sirius** is visible most of the night and rises in the east around 7:35 p.m. and at its highest in the south, around 40 degrees, at 12:45 a.m.

The bright star **Altair** sets in the west around 8:05 p.m. Altair can be identified by its less bright neighboring stars, one on either side, together called the shaft of Aquila. See page 18.

One of the brightest stars in the sky, **Vega** sets in the northwest around 8:40 p.m.

The **Great Bear (Big Dipper)** rises in the northeast around 9 p.m. and at its highest, around 65 degrees, around dawn. See the illustration on page 10.

Jupiter rises in the east around 9:20 p.m. and at its highest, around 70 degrees, at 3:55 a.m.

Rust-red and bright, **Mars** rises in the east around 10:10 p.m. and at its highest, around 65 degrees, at 4:40 a.m.

Bright star **Capella** is almost directly overhead at 11:15 p.m. Read about Capella on page 14.

Orion is visible most of the night and is at its highest, around 65 degrees, at 11:50 p.m. It sets in the west around 4:55 a.m. Look for the bright star **Aldebaran** and the **Pleiades** near Orion. See the illustration on page 17.

The bright star **Arcturus** rises in the northeast around 1:25 a.m. and is in the eastern sky before dawn. See page 13 for an easy way to locate the star.

Bright **Venus** rises in the east around 3:15 a.m.

Bright star **Vega** rises in the northeast around 4:30 a.m. and is in the eastern sky before dawn.

2026

Saturday, December 19

TONIGHT

sunset: **4:38 p.m.**
twilight begins: **5:38 p.m.**

IN THE MORNING

twilight begins: **5:48 a.m.**
sunrise: **6:47 a.m.**

MOON PHASE

moonrise: **12:53 p.m.**
moonset: **3 a.m.**

IN THE SKY

Saturn is at its highest, around 55 degrees, around dusk and sets in the west around 12:20 a.m. Unlike bright Jupiter and Venus and ruddy Mars, Saturn is harder to distinguish from the stars. Stars twinkle and shimmer, while the planets, including Saturn, show a steady flat light.

Castor and Pollux, the twin stars in the constellation Gemini, are visible most of the night. They rise in the northeast around 6:50 p.m. and are almost directly overhead at 1:40 a.m. Read how to find Castor and Pollux on page 14.

Cassiopeia, shaped like an *M,* is at its highest, around 65 degrees, at 7 p.m. Shaped like a *Σ,* it sets in the northwest around 3:30 a.m. See the illustration on page 11.

The brightest star in the sky, **Sirius** is visible most of the night and rises in the east around 7:30 p.m. and at its highest in the south, around 40 degrees, at 12:40 a.m.

The bright star **Altair** sets in the west around 8 p.m. Altair can be identified by its less bright neighboring stars, one on either side, together called the shaft of Aquila. See page 18.

One of the brightest stars in the sky, **Vega** sets in the northwest around 8:35 p.m.

The **Great Bear (Big Dipper)** rises in the northeast around 9 p.m. and at its highest, around 65 degrees, around dawn. See the illustration on page 10.

Jupiter rises in the east around 9:20 p.m. and at its highest, around 70 degrees, at 3:50 a.m.

Rust-red and bright, **Mars** rises in the east around 10:10 p.m. and at its highest, around 65 degrees, at 4:35 a.m.

Bright star **Capella** is almost directly overhead at 11:10 p.m. Read about Capella on page 14.

Orion is visible most of the night and is at its highest, around 65 degrees, at 11:45 p.m. It sets in the west around 4:50 a.m. Look for the bright star **Aldebaran** and the **Pleiades** near Orion. See the illustration on page 17.

The bright star **Arcturus** rises in the northeast around 1:25 a.m. and is in the eastern sky before dawn. See page 13 for an easy way to locate the star.

Bright **Venus** rises in the east around 3:15 a.m.

Bright star **Vega** rises in the northeast around 4:25 a.m. and is in the eastern sky before dawn.

2026

Sunday, December 20

TONIGHT

sunset: **4:39 p.m.**
twilight begins: **5:39 p.m.**

IN THE MORNING

twilight begins: **5:48 a.m.**
sunrise: **6:47 a.m.**

MOON PHASE

moonrise: **1:30 p.m.**
moonset: **4:13 a.m.**

IN THE SKY

The nearly **full moon** tonight acts like a spotlight. Look around at the desert landscape at how much illumination the moon provides.

Saturn is at its highest, around 55 degrees, around dusk and sets in the west around 12:20 a.m. Unlike bright Jupiter and Venus and ruddy Mars, Saturn is harder to distinguish from the stars. Stars twinkle and shimmer, while the planets, including Saturn, show a steady flat light.

Castor and Pollux, the twin stars in the constellation Gemini, are visible most of the night. They rise in the northeast around 6:45 p.m. and are almost directly overhead at 1:35 a.m. Read how to find Castor and Pollux on page 14.

Cassiopeia, shaped like an *M,* is at its highest, around 65 degrees, at 6:55 p.m. Shaped like a *Σ,* it sets in the northwest around 3:30 a.m. See the illustration on page 11.

The brightest star in the sky, **Sirius** is visible most of the night and rises in the east around 7:25 p.m. and at its highest in the south, around 40 degrees, at 12:35 a.m.

The bright star **Altair** sets in the west around 7:55 p.m. Altair can be identified by its less bright neighboring stars, one on either side, together called the shaft of Aquila. See page 18.

One of the brightest stars in the sky, **Vega** sets in the northwest around 8:30 p.m.

The **Great Bear (Big Dipper)** rises in the northeast around 8:55 p.m. and at its highest, around 65 degrees, around dawn. See the illustration on page 10.

Jupiter rises in the east around 9:15 p.m. and at its highest, around 70 degrees, at 3:45 a.m.

Rust-red and bright, **Mars** rises in the east around 10:05 p.m. and at its highest, around 65 degrees, at 4:30 a.m.

Bright star **Capella** is almost directly overhead at 11:05 p.m. Read about Capella on page 14.

Orion is visible most of the night and is at its highest, around 65 degrees, at 11:45 p.m. It sets in the west around 4:50 a.m. Look for the bright star **Aldebaran** and the **Pleiades** near Orion. See the illustration on page 17.

The bright star **Arcturus** rises in the northeast around 1:20 a.m. and is in the eastern sky before dawn. See page 13 for an easy way to locate the star.

Bright **Venus** rises in the east around 3:15 a.m.

Bright star **Vega** rises in the northeast around 4:20 a.m. and is in the eastern sky before dawn.

2026

Monday, December 21

MOON PHASE

moonrise: **2:16 p.m.**
moonset: **5:27 a.m.**

TONIGHT

sunset: **4:39 p.m.**
twilight begins: **5:39 p.m.**

IN THE MORNING

twilight begins: **5:49 a.m.**
sunrise: **6:48 a.m.**

IN THE SKY

The **Ursids meteor shower** happens tonight. A weak shower with up to 10 meteors per hour, the best time to watch is after midnight. Tonight unfortunately moonlight washes out the shower.

The **full moon** tonight acts like a spotlight. Look around at the desert landscape at how much illumination the moon provides.

Saturn is at its highest, around 55 degrees, around dusk and sets in the west around 12:15 a.m. Unlike bright Jupiter and Venus and ruddy Mars, Saturn is harder to distinguish from the stars. Stars twinkle and shimmer, while the planets, including Saturn, show a steady flat light.

Castor and Pollux, the twin stars in the constellation Gemini, are visible most of the night. They rise in the northeast around 6:40 p.m. and are almost directly overhead at 1:35 a.m. Read how to find Castor and Pollux on page 14.

Cassiopeia, shaped like an *M,* is at its highest, around 65 degrees, at 6:55 p.m. Shaped like a *Σ,* it sets in the northwest around 3:25 a.m. See the illustration on page 11.

The brightest star in the sky, **Sirius** is visible most of the night and rises in the east around 7:25 p.m. and at its highest in the south, around 40 degrees, at 12:30 a.m.

The bright star **Altair** sets in the west around 7:50 p.m. Altair can be identified by its less bright neighboring stars, one on either side, together called the shaft of Aquila. See page 18.

One of the brightest stars in the sky, **Vega** sets in the northwest around 8:25 p.m.

Jupiter rises in the east around 9:10 p.m. and at its highest, around 70 degrees, at 3:45 a.m.

Rust-red and bright, **Mars** rises in the east around 10:05 p.m. and at its highest, around 65 degrees, at 4:30 a.m.

Bright star **Capella** is almost directly overhead at 11 p.m. Read about Capella on page 14.

Orion is visible most of the night and is at its highest, around 65 degrees, at 11:40 p.m. It sets in the west around 4:45 a.m. Look for the bright star **Aldebaran** and the **Pleiades** near Orion. See the illustration on page 17.

The winter solstice means tonight is the longest of the year.

The bright star **Arcturus** rises in the northeast around 1:15 a.m. and is in the eastern sky before dawn. See page 13 for an easy way to locate the star.

Bright **Venus** rises in the east around 3:15 a.m.

Bright star **Vega** rises in the northeast around 4:15 a.m. and is in the eastern sky before dawn.

2026

Tuesday, December 22

TONIGHT

sunset: **4:40 p.m.**
twilight begins: **5:40 p.m.**

IN THE MORNING

twilight begins: **5:49 a.m.**
sunrise: **6:48 a.m.**

MOON PHASE

moonrise: **3:13 p.m.**
moonset: **6:38 a.m.**

IN THE SKY

The **full moon** tonight acts like a spotlight. Look around at the desert landscape at how much illumination the moon provides.

Saturn is at its highest, around 55 degrees, around dusk and sets in the west around 12:10 a.m. Unlike bright Jupiter and Venus and ruddy Mars, Saturn is harder to distinguish from the stars. Stars twinkle and shimmer, while the planets, including Saturn, show a steady flat light.

Castor and Pollux, the twin stars in the constellation Gemini, are visible most of the night. They rise in the northeast around 6:40 p.m. and are almost directly overhead at 1:30 a.m. Read how to find Castor and Pollux on page 14.

Cassiopeia, shaped like an *M,* is at its highest, around 65 degrees, at 6:50 p.m. Shaped like a *Σ,* it sets in the northwest around 3:20 a.m. See the illustration on page 11.

The brightest star in the sky, **Sirius** is visible most of the night and rises in the east around 7:20 p.m. and at its highest in the south, around 40 degrees, at 12:30 a.m.

The bright star **Altair** sets in the west around 7:50 p.m. Altair can be identified by its less bright neighboring stars, one on either side, together called the shaft of Aquila. See page 18.

One of the brightest stars in the sky, **Vega** sets in the northwest around 8:25 p.m.

The **Great Bear (Big Dipper)** rises in the northeast around 8:45 p.m. and at its highest, around 65 degrees, around dawn. See the illustration on page 10.

Jupiter rises in the east around 9:05 p.m. and at its highest, around 70 degrees, at 3:40 a.m.

Rust-red and bright, **Mars** rises in the east around 10 p.m. and at its highest, around 65 degrees, at 4:25 a.m.

Bright star **Capella** is almost directly overhead at 10:55 p.m. Read about Capella on page 14.

Orion is visible most of the night and is at its highest, around 65 degrees, at 11:35 p.m. It sets in the west around 4:40 a.m. Look for the bright star **Aldebaran** and the **Pleiades** near Orion. See the illustration on page 17.

The bright star **Arcturus** rises in the northeast around 1:10 a.m. and is in the eastern sky before dawn. See page 13 for an easy way to locate the star.

Bright **Venus** rises in the east around 3:15 a.m.

Bright star **Vega** rises in the northeast around 4:10 a.m. and is in the eastern sky before dawn.

2026

Wednesday, December 23

TONIGHT

sunset: **4:40 p.m.**
twilight begins: **5:40 p.m.**

IN THE MORNING

twilight begins: **5:50 a.m.**
sunrise: **6:49 a.m.**

MOON PHASE

moonrise: **4:21 p.m.**
moonset: **6:38 a.m.**

IN THE SKY

The **full moon** tonight acts like a spotlight. Look around at the desert landscape at how much illumination the moon provides.

Saturn is at its highest, around 55 degrees, around dusk and sets in the west around 12:05 a.m. Unlike bright Jupiter and Venus and ruddy Mars, Saturn is harder to distinguish from the stars. Stars twinkle and shimmer, while the planets, including Saturn, show a steady flat light.

Castor and Pollux, the twin stars in the constellation Gemini, are visible most of the night. They rise in the northeast around 6:35 p.m. and are almost directly overhead at 1:25 a.m. Read how to find Castor and Pollux on page 14.

Cassiopeia, shaped like an *M,* is at its highest, around 65 degrees, at 6:45 p.m. Shaped like a *Σ,* it sets in the northwest around 3:15 a.m. See the illustration on page 11.

The brightest star in the sky, **Sirius** is visible most of the night and rises in the east around 7:15 p.m. and at its highest in the south, around 40 degrees, at 12:25 a.m.

The bright star **Altair** sets in the west around 7:45 p.m. Altair can be identified by its less bright neighboring stars, one on either side, together called the shaft of Aquila. See page 18.

One of the brightest stars in the sky, **Vega** sets in the northwest around 8:20 p.m.

The **Great Bear (Big Dipper)** rises in the northeast around 8:40 p.m. and at its highest, around 65 degrees, at 4:45 a.m. See the illustration on page 10.

Jupiter rises in the east around 9 p.m. and at its highest, around 70 degrees, at 3:35 a.m.

Rust-red and bright, **Mars** rises in the east around 9:55 p.m. and at its highest, around 65 degrees, at 4:20 a.m.

Bright star **Capella** is almost directly overhead at 10:55 p.m. Read about Capella on page 14.

Orion is visible most of the night and is at its highest, around 65 degrees, at 11:30 p.m. It sets in the west around 4:35 a.m. Look for the bright star **Aldebaran** and the **Pleiades** near Orion. See the illustration on page 17.

The bright star **Arcturus** rises in the northeast around 1:05 a.m. and is in the eastern sky before dawn. See page 13 for an easy way to locate the star.

Bright **Venus** rises in the east around 3:15 a.m.

Bright star **Vega** rises in the northeast around 4:10 a.m. and is in the eastern sky before dawn.

2026

Thursday, December 24

TONIGHT

sunset: **4:41 p.m.**
twilight begins: **5:41 p.m.**

IN THE MORNING

twilight begins: **5:50 a.m.**
sunrise: **6:49 a.m.**

MOON PHASE

moonrise: **5:36 p.m.**
moonset: **7:41 a.m.**

IN THE SKY

The **full moon** tonight acts like a spotlight. Look around at the desert landscape at how much illumination the moon provides.

Saturn is at its highest, around 55 degrees, around dusk and sets in the west around 12:05 a.m. Unlike bright Jupiter and Venus and ruddy Mars, Saturn is harder to distinguish from the stars. Stars twinkle and shimmer, while the planets, including Saturn, show a steady flat light.

Castor and Pollux, the twin stars in the constellation Gemini, are visible most of the night. They rise in the northeast around 6:30 p.m. and are almost directly overhead at 1:20 a.m. Read how to find Castor and Pollux on page 14.

Cassiopeia, shaped like an *M,* is at its highest, around 65 degrees, at 6:40 p.m. Shaped like a *Σ,* it sets in the northwest around 3:10 a.m. See the illustration on page 11.

The brightest star in the sky, **Sirius** is visible most of the night and rises in the east around 7:10 p.m. and at its highest in the south, around 40 degrees, at 12:20 a.m.

The bright star **Altair** sets in the west around 7:40 p.m. Altair can be identified by its less bright neighboring stars, one on either side, together called the shaft of Aquila. See page 18.

One of the brightest stars in the sky, **Vega** sets in the northwest around 8:15 p.m.

The **Great Bear (Big Dipper)** rises in the northeast around 8:40 p.m. and at its highest, around 65 degrees, at 4:45 a.m. See the illustration on page 10.

Jupiter rises in the east around 9 p.m. and at its highest, around 70 degrees, at 3:30 a.m.

Rust-red and bright, **Mars** rises in the east around 9:55 p.m. and at its highest, around 65 degrees, at 4:20 a.m.

Bright star **Capella** is almost directly overhead at 10:50 p.m. Read about Capella on page 14.

Orion is visible most of the night and is at its highest, around 65 degrees, at 11:25 p.m. It sets in the west around 4:35 a.m. Look for the bright star **Aldebaran** and the **Pleiades** near Orion. See the illustration on page 17.

The bright star **Arcturus** rises in the northeast around 1:05 a.m. and is in the eastern sky before dawn. See page 13 for an easy way to locate the star.

Bright **Venus** rises in the east around 3:15 a.m.

Bright star **Vega** rises in the northeast around 4:05 a.m. and is in the eastern sky before dawn.

2026
Friday, December 25

TONIGHT

sunset: **4:41 p.m.**
twilight begins: **5:41 p.m.**

IN THE MORNING

twilight begins: **5:51 a.m.**
sunrise: **6:49 a.m.**

MOON PHASE

moonrise: **6:53 p.m.**
moonset: **8:32 a.m.**

IN THE SKY

The **full moon** tonight acts like a spotlight. Look around at the desert landscape at how much illumination the moon provides.

Cassiopeia, shaped like an *M,* is at its highest, around 65 degrees, around dusk. Shaped like a Σ, it sets in the northwest around 3:10 a.m. See the illustration on page 11.

Castor and Pollux, the twin stars in the constellation Gemini, are visible most of the night. They rise in the northeast around 6:25 p.m. and are almost directly overhead at 1:15 a.m. Read how to find Castor and Pollux on page 14.

The brightest star in the sky, **Sirius** is visible most of the night and rises in the east around 7:10 p.m. and at its highest in the south, around 40 degrees, at 12:15 a.m.

The bright star **Altair** sets in the west around 7:35 p.m. Altair can be identified by its less bright neighboring stars, one on either side, together called the shaft of Aquila. See page 18.

One of the brightest stars in the sky, **Vega** sets in the northwest around 8:10 p.m.

The **Great Bear (Big Dipper)** rises in the northeast around 8:35 p.m. and at its highest, around 65 degrees, at 4:40 a.m. See the illustration on page 10.

Jupiter rises in the east around 8:55 p.m. and at its highest, around 70 degrees, at 3:25 a.m.

Rust-red and bright, **Mars** rises in the east around 9:50 p.m. and at its highest, around 65 degrees, at 4:15 a.m.

Bright star **Capella** is almost directly overhead at 10:45 p.m. Read about Capella on page 14.

Orion is visible most of the night and is at its highest, around 65 degrees, at 11:25 p.m. It sets in the west around 4:30 a.m. Look for the bright star **Aldebaran** and the **Pleiades** near Orion. See the illustration on page 17.

Saturn sets in the west around midnight. Unlike bright Jupiter and Venus and ruddy Mars, Saturn is harder to distinguish from the stars. Stars twinkle and shimmer, while the planets, including Saturn, show a steady flat light.

The bright star **Arcturus** rises in the northeast around 1 a.m. and is in the eastern sky before dawn. See page 13 for an easy way to locate the star.

Bright **Venus** rises in the east around 3:20 a.m.

Bright star **Vega** rises in the northeast around 4 a.m. and is in the eastern sky before dawn.

2026

Saturday, December 26

TONIGHT

sunset: **4:42** p.m.
twilight begins: **5:42** p.m.

IN THE MORNING

twilight begins: **5:51** a.m.
sunrise: **6:50** a.m.

MOON PHASE

moonrise: **8:07** p.m.
moonset: **9:15** a.m.

IN THE SKY

Cassiopeia, shaped like an *M,* is at its highest, around 65 degrees, around dusk. Shaped like a Σ, it sets in the northwest around 3:05 a.m. See the illustration on page 11.

Castor and Pollux, the twin stars in the constellation Gemini, are visible most of the night and rise in the northeast around 6:20 p.m. They are almost directly overhead at 1:15 a.m. Read how to find Castor and Pollux on page 14.

The brightest star in the sky, **Sirius** is visible most of the night and rises in the east around 7:05 p.m. and at its highest in the south, around 40 degrees, at 12:10 a.m. and sets in the west around 5:15 a.m.

The bright star **Altair** sets in the west around 7:30 p.m. Altair can be identified by its less bright neighboring stars, one on either side, together called the shaft of Aquila. See page 18.

One of the brightest stars in the sky, **Vega** sets in the northwest around 8:05 p.m.

Jupiter rises in the east around 8:50 p.m. and at its highest, around 70 degrees, at 3:25 a.m.

The **Great Bear (Big Dipper)** rises in the northeast around 8:30 p.m. and at its highest, around 65 degrees, at 4:35 a.m. See the illustration on page 10.

Rust-red and bright, **Mars** rises in the east around 9:45 p.m. and at its highest, around 65 degrees, at 4:15 a.m.

Bright star **Capella** is almost directly overhead at 10:40 p.m. Read about Capella on page 14.

Orion is visible most of the night and is at its highest, around 65 degrees, at 11:20 p.m. It sets in the west around 4:25 a.m. Look for the bright star **Aldebaran** and the **Pleiades** near Orion. See the illustration on page 17.

Saturn sets in the west around midnight. Unlike bright Jupiter and Venus and ruddy Mars, Saturn is harder to distinguish from the stars. Stars twinkle and shimmer, while the planets, including Saturn, show a steady flat light.

The bright star **Arcturus** rises in the northeast around 12:55 a.m. and is in the eastern sky before dawn. See page 13 for an easy way to locate the star.

Bright **Venus** rises in the east around 3:20 a.m.

Bright star **Vega** rises in the northeast around 3:55 a.m. and is in the eastern sky before dawn.

2026
Sunday, December 27

TONIGHT

sunset: 4:43 p.m.
twilight begins: 5:42 p.m.

IN THE MORNING

twilight begins: 5:51 a.m.
sunrise: 6:50 a.m.

MOON PHASE

moonrise: 9:18 p.m.
moonset: 9:50 a.m.

IN THE SKY

Cassiopeia, shaped like an *M,* is at its highest, around 65 degrees, around dusk. Shaped like a *Σ,* it sets in the northwest around 3 a.m. See the illustration on page 11.

Castor and Pollux, the twin stars in the constellation Gemini, are visible most of the night and rise in the northeast around 6:20 p.m. They are almost directly overhead at 1:10 a.m. Read how to find Castor and Pollux on page 14.

The brightest star in the sky, **Sirius** is visible most of the night and rises in the east around 7 p.m. and at its highest in the south, around 40 degrees, at 12:10 a.m. and sets in the west around 5:15 a.m.

One of the brightest stars in the sky, **Vega** sets in the northwest around 8:05 p.m.

The **Great Bear (Big Dipper)** rises in the northeast around 8:25 p.m. and at its highest, around 65 degrees, at 4:30 a.m. See the illustration on page 10.

Jupiter rises in the east around 8:45 p.m. and at its highest, around 70 degrees, at 3:20 a.m.

Explore the **dark sky** after twilight and before the moon rises, around 9 p.m.

Before the moon rises, until around 9 p.m., look for the **Andromeda Galaxy** with binoculars or a telescope. See the guide on page 11.

Rust-red and bright, **Mars** rises in the east around 9:45 p.m. and at its highest, around 65 degrees, at 4:10 a.m.

The **waning gibbous moon, Mars,** and **Jupiter** appear close together in the evening sky. Look east around 10:30 p.m.

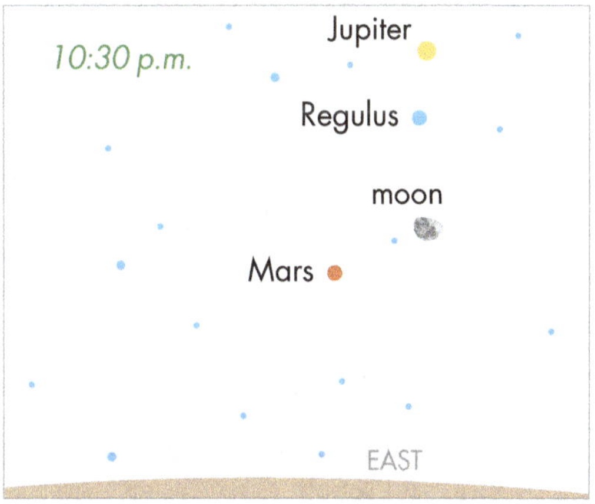

Bright star **Capella** is almost directly overhead at 10:35 p.m. Read about Capella on page 14.

Orion is visible most of the night and is at its highest, around 65 degrees, at 11:15 p.m. It sets in the west around 4:20 a.m. Look for the bright star **Aldebaran** and the **Pleiades** near Orion. See the illustration on page 17.

Saturn sets in the west around midnight. Unlike bright Jupiter and Venus and ruddy Mars, Saturn is harder to distinguish from the stars. Stars twinkle and shimmer, while the planets, including Saturn, show a steady flat light.

Bright **Venus** rises in the east around 3:20 a.m.

2026

Monday, December 28

TONIGHT

sunset: **4:43 p.m.**
twilight begins: **5:43 p.m.**

MOON PHASE

IN THE MORNING

twilight begins: **5:52 a.m.**
sunrise: **6:50 a.m.**

moonrise: **10:24 p.m.**
moonset: **10:21 a.m.**

IN THE SKY

Cassiopeia, shaped like an *M,* is at its highest, around 65 degrees, around dusk. Shaped like a *Σ,* it sets in the northwest around 2:55 a.m. See the illustration on page 11.

Castor and Pollux, the twin stars in the constellation Gemini, are visible most of the night and rise in the northeast around 6:15 p.m. They are almost directly overhead at 1:05 a.m. Read how to find Castor and Pollux on page 14.

The brightest star in the sky, **Sirius** is visible most of the night and rises in the east around 6:55 p.m. and at its highest in the south, around 40 degrees, at 12:05 a.m. and sets in the west around 5:10 a.m.

The bright star **Altair** is in the western sky at dusk and sets in the west around 7:25 p.m. Altair can be identified by its less bright neighboring stars, one on either side, together called the shaft of Aquila. See page 18.

One of the brightest stars in the sky, **Vega** sets in the northwest around 8 p.m.

The **Great Bear (Big Dipper)** rises in the northeast around 8:25 p.m. and at its highest, around 65 degrees, at 4:30 a.m. See the illustration on page 10.

Jupiter rises in the east around 8:40 p.m. and at its highest, around 70 degrees, at 3:15 a.m.

Rust-red and bright, **Mars** rises in the east around 9:40 p.m. and at its highest, around 65 degrees, at 4:05 a.m.

Explore the **dark sky** after twilight and before the moon rises, around 10:05 p.m.

Before the moon rises, until around 10:05 p.m., use binoculars or a telescope to look for the **moons of Jupiter.** See the guide on page 18.

Before the moon rises, until around 10:05 p.m., look for the **Andromeda Galaxy** with binoculars or a telescope. See the guide on page 11.

Bright star **Capella** is almost directly overhead at 10:35 p.m. Read about Capella on page 14.

Orion is visible most of the night and is at its highest, around 65 degrees, at 11:10 p.m. It sets in the west around 4:15 a.m. Look for the bright star **Aldebaran** and the **Pleiades** near Orion. See the illustration on page 17.

Saturn sets in the west around 11:45 p.m. Unlike bright Jupiter and Venus and ruddy Mars, Saturn is harder to distinguish from the stars. Stars twinkle and shimmer, while the planets, including Saturn, show a steady flat light.

The bright star **Arcturus** rises in the northeast around 12:50 a.m. and is in the eastern sky before dawn. See page 13 for an easy way to locate the star.

Bright **Venus** rises in the east around 3:20 a.m.

Bright star **Vega** rises in the northeast around 3:50 a.m. and is in the eastern sky before dawn.

2026

Tuesday, December 29

TONIGHT

sunset: **4:44 p.m.**
twilight begins: **5:44 p.m.**

MOON PHASE

IN THE MORNING

twilight begins: **5:52 a.m.**
sunrise: **6:51 a.m.**

moonrise: **11:27 p.m.**
moonset: **10:49 a.m.**

IN THE SKY

Cassiopeia, shaped like an *M,* is at its highest, around 65 degrees, around dusk. Shaped like a *Σ,* it sets in the northwest around 2:55 a.m. See the illustration on page 11.

Castor and Pollux, the twin stars in the constellation Gemini, are visible most of the night and rise in the northeast around 6:10 p.m. They are almost directly overhead at 1 a.m. Read how to find Castor and Pollux on page 14.

The brightest star in the sky, **Sirius** is visible most of the night and rises in the east around 6:50 p.m. and at its highest in the south, around 40 degrees, at midnight and sets in the west around 5:05 a.m.

The bright star **Altair** is in the western sky at dusk and sets in the west around 7:20 p.m. Altair can be identified by its less bright neighboring stars, one on either side, together called the shaft of Aquila. See page 18.

One of the brightest stars in the sky, **Vega** sets in the northwest around 7:55 p.m.

The **Great Bear (Big Dipper)** rises in the northeast around 8:20 p.m. and at its highest, around 65 degrees, at 4:25 a.m. See the illustration on page 10.

Jupiter rises in the east around 8:35 p.m. and at its highest, around 70 degrees, at 3:10 a.m.

Rust-red and bright, **Mars** rises in the east around 9:40 p.m. and at its highest, around 65 degrees, at 4:05 a.m.

Bright star **Capella** is almost directly overhead at 10:30 p.m. Read about Capella on page 14.

Explore the **dark sky** after twilight and before the moon rises, around 11:05 p.m.

Before the moon rises, until around 11:05 p.m., use binoculars or a telescope to look for the **moons of Jupiter.** See the guide on page 18.

Orion is visible most of the night and is at its highest, around 65 degrees, at 11:05 p.m. It sets in the west around 4:15 a.m. Look for the bright star **Aldebaran** and the **Pleiades** near Orion. See the illustration on page 17.

Before the moon rises, until around 11:05 p.m., look for the **Andromeda Galaxy** with binoculars or a telescope. See the guide on page 11.

Saturn sets in the west around 11:40 p.m. Unlike bright Jupiter and Venus and ruddy Mars, Saturn is harder to distinguish from the stars. Stars twinkle and shimmer, while the planets, including Saturn, show a steady flat light.

The bright star **Arcturus** rises in the northeast around 12:45 a.m. and is in the eastern sky before dawn. See page 13 for an easy way to locate the star.

Bright **Venus** rises in the east around 3:20 a.m.

Bright star **Vega** rises in the northeast around 3:45 a.m. and is in the eastern sky before dawn.

2026

Wednesday, December 30

TONIGHT

sunset: **4:45 p.m.**
twilight begins: **5:44 p.m.**

IN THE MORNING

twilight begins: **5:52 a.m.**
sunrise: **6:51 a.m.**

MOON PHASE

moonrise: **12:29 a.m.**
moonset: **11:16 a.m.**

IN THE SKY

Cassiopeia, shaped like an *M,* is at its highest, around 65 degrees, around dusk. Shaped like a *Σ,* it sets in the northwest around 2:50 a.m. See the illustration on page 11.

Castor and Pollux, the twin stars in the constellation Gemini, are visible most of the night and rise in the northeast around 6:05 p.m. They are almost directly overhead at 12:55 a.m. Read how to find Castor and Pollux on page 14.

The brightest star in the sky, **Sirius** is visible most of the night and rises in the east around 6:50 p.m. and at its highest in the south, around 40 degrees, at 11:55 p.m. and sets in the west around 5 a.m.

The bright star **Altair** is in the western sky at dusk and sets in the west around 7:15 p.m. Altair can be identified by its less bright neighboring stars, one on either side, together called the shaft of Aquila. See page 18.

One of the brightest stars in the sky, **Vega** sets in the northwest around 7:50 p.m.

The **Great Bear (Big Dipper)** rises in the northeast around 8:15 p.m. and at its highest, around 65 degrees, at 4:20 a.m. See the illustration on page 10.

Jupiter rises in the east around 8:35 p.m. and at its highest, around 70 degrees, at 3:05 a.m.

Rust-red and bright, **Mars** rises in the east around 9:35 p.m. and at its highest, around 65 degrees, at 4 a.m.

Bright star **Capella** is almost directly overhead at 10:25 p.m. Read about Capella on page 14.

Orion is visible most of the night and is at its highest, around 65 degrees, at 11:05 p.m. It sets in the west around 4:10 a.m. Look for the bright star **Aldebaran** and the **Pleiades** near Orion. See the illustration on page 17.

Saturn sets in the west around 11:35 p.m. Unlike bright Jupiter and Venus and ruddy Mars, Saturn is harder to distinguish from the stars. Stars twinkle and shimmer, while the planets, including Saturn, show a steady flat light.

Until the moon rises, around 12:20 a.m., use binoculars or a telescope to look for the **moons of Jupiter.** See the guide on page 18.

Until the moon rises, around 12:20 a.m., look for the **Andromeda Galaxy** with binoculars or a telescope. See the guide on page 11.

The bright star **Arcturus** rises in the northeast around 12:40 a.m. and is in the eastern sky before dawn. See page 13 for an easy way to locate the star.

Bright **Venus** rises in the east around 3:20 a.m.

Bright star **Vega** rises in the northeast around 3:40 a.m. and is in the eastern sky before dawn.

2026

Thursday, December 31

TONIGHT

sunset: **4:45 p.m.**
twilight begins: **5:45 p.m.**

IN THE MORNING

twilight begins: **5:53 a.m.**
sunrise: **6:51 a.m.**

MOON PHASE

moonrise: **1:30 a.m.**
moonset: **11:44 a.m.**

IN THE SKY

Cassiopeia, shaped like an *M,* is at its highest, around 65 degrees, around dusk. Shaped like a *Σ,* it sets in the northwest around 2:45 a.m. See the illustration on page 11.

The brightest star in the sky, **Sirius** is visible most of the night and rises in the east around 6:45 p.m. and at its highest in the south, around 40 degrees, at 11:50 p.m. and sets in the west around 5 a.m.

The bright star **Altair** is in the western sky at dusk and sets in the west around 7:15 p.m. Altair can be identified by its less bright neighboring stars, one on either side, together called the shaft of Aquila. See page 18.

One of the brightest stars in the sky, **Vega** sets in the northwest around 7:50 p.m.

The **Great Bear (Big Dipper)** rises in the northeast around 8:10 p.m. and at its highest, around 65 degrees, at 4:15 a.m. See the illustration on page 10.

Jupiter rises in the east around 8:30 p.m. and at its highest, around 70 degrees, at 3 a.m.

Rust-red and bright, **Mars** rises in the east around 9:30 p.m. and at its highest, around 65 degrees, at 3:55 a.m.

Orion is visible most of the night and is at its highest, around 65 degrees, at 11 p.m. It sets in the west around 4:05 a.m. Look for the bright star **Aldebaran** and the **Pleiades** near Orion. See the illustration on page 17.

Saturn sets in the west around 11:35 p.m. Unlike bright Jupiter and Venus and ruddy Mars, Saturn is harder to distinguish from the stars. Stars twinkle and shimmer, while the planets, including Saturn, show a steady flat light.

The bright star **Arcturus** rises in the northeast around 12:35 a.m. and is in the eastern sky before dawn. See page 13 for an easy way to locate the star.

Until the moon rises, around 1:20 a.m., look for the **Andromeda Galaxy** with binoculars or a telescope. See the guide on page 11.

Bright star **Capella** is almost directly overhead at 10:20 p.m. Read about Capella on page 14.

Castor and Pollux, the twin stars in the constellation Gemini, are visible most of the night. They are almost directly overhead at 12:55 a.m. Read how to find Castor and Pollux on page 14.

Until the moon rises, around 1:20 a.m., use binoculars or a telescope to look for the **moons of Jupiter.** See the guide on page 18.

Bright **Venus** rises in the east around 3:20 a.m.

Bright star **Vega** rises in the northeast around 3:35 a.m. and is in the eastern sky before dawn.

Scorpius rises in the southeast around 5:30 a.m. and is in the eastern sky before dawn. See the illustration on page 19.

2027

Friday, January 1

TONIGHT

sunset: **4:46 p.m.**
twilight begins: **5:46 p.m.**

IN THE MORNING

twilight begins: **5:53 a.m.**
sunrise: **6:51 a.m.**

MOON PHASE

moonrise: **2:31 a.m.**
moonset: **12:14 p.m.**

IN THE SKY

Cassiopeia, shaped like an *M,* is at its highest, around 65 degrees, around dusk. Shaped like a *Σ,* sets in the northwest around 2:40 a.m. See the illustration on page 11.

The brightest star in the sky, **Sirius** is visible most of the night. It rises in the east around 6:40 p.m. and is at its highest in the south, around 40 degrees, at 11:45 p.m. It sets in the west around 4:55 a.m.

The **Great Bear (Big Dipper)** rises in the northeast around 8:05 p.m. and is at its highest, around 65 degrees, at 4:10 a.m. See the illustration on page 10.

Jupiter rises in the east around 8:25 p.m. and is at its highest, around 70 degrees, at 3 a.m.

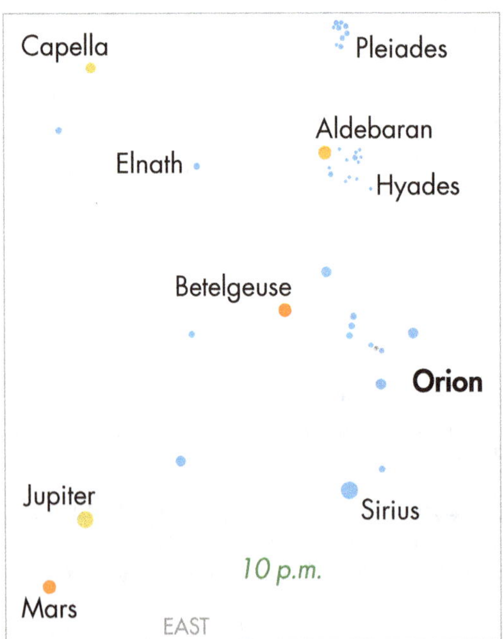

With the dark sky, use binoculars or a telescope to look for the **moons of Jupiter.** See the guide on page 18.

Rust-red and bright, **Mars** rises in the east around 9:25 p.m. and is at its highest, around 65 degrees, at 3:55 a.m.

Bright star **Capella** is almost directly overhead at 10:15 p.m. Read about Capella on page 14.

Orion is visible most of the night and is at its highest, around 65 degrees, at 10:55 p.m. and sets in the west around 4 a.m. Look for the bright star **Aldebaran** and the **Pleiades** near Orion. See the illustration on page 17.

Saturn sets in the west around 11:30 p.m. Unlike bright Jupiter and Venus and ruddy Mars, Saturn is harder to distinguish from the stars. Stars twinkle and shimmer, while the planets, including Saturn, show a steady flat light.

With the dark sky, look for the **Andromeda Galaxy** with binoculars or a telescope. It is a good target until around 2:30 a.m. See the guide on page 11.

Castor and Pollux, the twin stars in the constellation Gemini, are visible most of the night and are almost directly overhead at 12:50 a.m. Read how to find Castor and Pollux on page 14.

At its brightest, **Venus** rises in the east around 3:20 a.m. Venus is at its highest in the sky for the year.

One of the brightest stars in the sky, **Vega** rises in the northeast around 3:30 a.m.

2027

Saturday, January 2

TONIGHT

sunset: **4:47 p.m.**
twilight begins: **5:47 p.m.**

IN THE MORNING

twilight begins: **5:53 a.m.**
sunrise: **6:51 a.m.**

MOON PHASE

moonrise: **3:31 a.m.**
moonset: **12:47 p.m.**

IN THE SKY

The **Quadrantids meteor shower** happens tonight. A good shower with up to 40 meteors per hour, the best time to watch is after midnight. Tonight the viewing is excellent until the moon rises, around 3:20 a.m.

The bright star **Altair** is in the western sky at dusk and sets in the west around 7:05 p.m. Altair can be identified by its less bright neighboring stars, one on either side, together called the shaft of Aquila. See page 18.

The brightest star in the sky, **Sirius** is visible most of the night. It rises in the east around 6:35 p.m. and is at its highest in the south, around 40 degrees, at 11:40 p.m. It sets in the west around 4:50 a.m.

Bright star **Vega** sets in the northwest around 7:40 p.m.

The **Great Bear (Big Dipper)** rises in the northeast around 8 p.m. and is at its highest, around 65 degrees, at 4:05 a.m. See the illustration on page 10.

Jupiter rises in the east around 8:20 p.m. and is at its highest, around 70 degrees, at 2:55 a.m.

With the dark sky, use binoculars or a telescope to look for the **moons of Jupiter.** See the guide on page 18.

Rust-red and bright, **Mars** rises in the east around 9:25 p.m. and is at its highest, around 65 degrees, at 3:50 a.m.

Orion is visible most of the night and is at its highest, around 65 degrees, at 10:50 p.m. It sets in the west around 3:55 a.m. Look for the bright star **Aldebaran** and the **Pleiades** near Orion. See the illustration on page 17.

Saturn sets in the west around 11:25 p.m. Unlike bright Jupiter and Venus and ruddy Mars, Saturn is harder to distinguish from the stars. Stars twinkle and shimmer, while the planets, including Saturn, show a steady flat light.

Castor and Pollux, the twin stars in the constellation Gemini, are visible most of the night and are almost directly overhead at 12:45 a.m. Read how to find Castor and Pollux on page 14.

The absence of the moon tonight provides excellent **dark-sky** viewing conditions.

The bright star **Arcturus** rises in the northeast around 12:25 a.m. and is in the eastern sky before dawn. See page 13 for an easy way to locate the star.

With the dark sky, look for the **Andromeda Galaxy** with binoculars or a telescope. It is a good target until around 2:25 a.m. See the guide on page 11.

Cassiopeia, shaped like a Σ, sets in the northwest around 2:35 a.m. See the illustration on page 11.

At its brightest, **Venus** rises in the east around 3:25 a.m. Venus is at its highest in the sky for the year.

2027

Sunday, January 3

TONIGHT

sunset: **4:48 p.m.**
twilight begins: **5:47 p.m.**

MOON PHASE

IN THE MORNING

twilight begins: **5:53 a.m.**
sunrise: **6:51 a.m.**

moonrise: **4:30 a.m.**
moonset: **1:25 p.m.**

IN THE SKY

The **Quadrantids meteor shower** happens tonight. A good shower with up to 40 meteors per hour, the best time to watch is after midnight. Tonight has moonless dark skies all night.

The absence of the moon tonight provides excellent **dark-sky** viewing conditions.

The brightest star in the sky, **Sirius** is visible most of the night. It rises in the east around 6:30 p.m. and is at its highest in the south, around 40 degrees, at 11:35 p.m. It sets in the west around 4:45 a.m.

The bright star **Altair** is in the western sky at dusk and sets in the west around 7 p.m. Altair can be identified by its less bright neighboring stars, one on either side, together called the shaft of Aquila. See page 18.

Bright star **Vega** is in the western sky at dusk and sets in the northwest around 7:35 p.m.

The **Great Bear (Big Dipper)** rises in the northeast around 8 p.m. and is at its highest, around 65 degrees, at 4:05 a.m. See the illustration on page 10.

Jupiter rises in the east around 8:15 p.m. and is at its highest, around 70 degrees, at 2:50 a.m.

With the dark sky, use binoculars or a telescope to look for the **moons of Jupiter.** See the guide on page 18.

Rust-red and bright, **Mars** rises in the east around 9:20 p.m. and is at its highest, around 65 degrees, at 3:45 a.m.

Bright star **Capella** is almost directly overhead at 10:10 p.m. Read about Capella on page 14.

With the dark sky, look for the **Andromeda Galaxy** with binoculars or a telescope. It is a good target until around 2:20 a.m. See the guide on page 11.

Orion is visible most of the night and is at its highest, around 65 degrees, at 10:45 p.m. It sets in the west around 3:50 a.m. Look for the bright star **Aldebaran** and the **Pleiades** near Orion. See the illustration on page 17.

Saturn sets in the west around 11:20 p.m. Unlike bright Jupiter and Venus and ruddy Mars, Saturn is harder to distinguish from the stars. Stars twinkle and shimmer, while the planets, including Saturn, show a steady flat light.

The bright star **Arcturus** rises in the northeast around 12:25 a.m. and is in the eastern sky before dawn. See page 13 for an easy way to locate the star.

Castor and Pollux, the twin stars in the constellation Gemini, are visible most of the night and are almost directly overhead at 12:40 a.m. Read how to find Castor and Pollux on page 14.

Cassiopeia, shaped like a *Σ*, sets in the northwest around 2:30 a.m. See the illustration on page 11.

At its brightest, **Venus** rises in the east around 3:25 a.m. Venus is at its highest in the sky for the year.

2027

Monday, January 4

TONIGHT

sunset: **4:49 p.m.**
twilight begins: **5:48 p.m.**

IN THE MORNING

twilight begins: **5:53 a.m.**
sunrise: **6:52 a.m.**

MOON PHASE

moonrise: **5:25 a.m.**
moonset: **2:08 p.m.**

IN THE SKY

The **Quadrantids meteor shower** happens tonight. A good shower with up to 40 meteors per hour, the best time to watch is after midnight. Tonight has moonless dark skies all night.

The brightest star in the sky, **Sirius** is visible most of the night. It rises in the east around 6:25 p.m. and is at its highest in the south, around 40 degrees, at 11:30 p.m. It sets in the west around 4:40 a.m.

The bright star **Altair** is in the western sky at dusk and sets in the west around 6:55 p.m. Altair can be identified by its less bright neighboring stars, one on either side, together called the shaft of Aquila. See page 18.

Bright star **Vega** is in the western sky at dusk and sets in the northwest around 7:30 p.m.

The **Great Bear (Big Dipper)** rises in the northeast around 7:55 p.m. and is at its highest, around 65 degrees, at 4 a.m. See the illustration on page 10.

Rust-red and bright, **Mars** rises in the east around 9:15 p.m. and is at its highest, around 65 degrees, at 3:40 a.m.

The absence of the moon tonight provides excellent **dark-sky** viewing conditions.

Jupiter rises in the east around 8:10 p.m. and is at its highest, around 70 degrees, at 2:45 a.m.

With the dark sky, use binoculars or a telescope to look for the **moons of Jupiter.** See the guide on page 18.

Bright star **Capella** is almost directly overhead at 10:05 p.m. Read about Capella on page 14.

Orion is visible most of the night and is at its highest, around 65 degrees, at 10:45 p.m. It sets in the west around 3:50 a.m. Look for the bright star **Aldebaran** and the **Pleiades** near Orion. See the illustration on page 17.

Saturn sets in the west around 11:20 p.m. Unlike bright Jupiter and Venus and ruddy Mars, Saturn is harder to distinguish from the stars. Stars twinkle and shimmer, while the planets, including Saturn, show a steady flat light.

With the dark sky, look for the **Andromeda Galaxy** with binoculars or a telescope. It is a good target until around 2:20 a.m. See the guide on page 11.

Castor and Pollux, the twin stars in the constellation Gemini, are visible most of the night and are almost directly overhead at 12:35 a.m. Read how to find Castor and Pollux on page 14.

Cassiopeia, shaped like a Σ, sets in the northwest around 2:30 a.m. See the illustration on page 11.

One of the brightest stars in the sky, **Vega** rises in the northeast around 3:20 a.m. and is in the eastern sky before dawn.

At its brightest, **Venus** rises in the east around 3:25 a.m. Venus is at its highest in the sky for the year.

2027

Tuesday, January 5

TONIGHT

sunset: **4:50 p.m.**
twilight begins: **5:49 p.m.**

IN THE MORNING

twilight begins: **5:53 a.m.**
sunrise: **6:52 a.m.**

MOON PHASE

moonrise: **6:15 a.m.**
moonset: **2:57 p.m.**

IN THE SKY

The brightest star in the sky, **Sirius** is visible most of the night. It rises in the east around 6:25 p.m. and is at its highest in the south, around 40 degrees, at 11:30 p.m. It sets in the west around 4:35 a.m.

The bright star **Altair** is in the western sky at dusk and sets in the west around 6:50 p.m. Altair can be identified by its less bright neighboring stars, one on either side, together called the shaft of Aquila. See page 18.

Bright star **Vega** is in the western sky at dusk and sets in the northwest around 7:25 p.m.

The **Great Bear (Big Dipper)** rises in the northeast around 7:50 p.m. and is at its highest, around 65 degrees, at 3:55 a.m. See the illustration on page 10.

Jupiter rises in the east around 8:05 p.m. and is at its highest, around 70 degrees, at 2:40 a.m.

With the dark sky, use binoculars or a telescope to look for the **moons of Jupiter.** See the guide on page 18.

The absence of the moon tonight provides excellent **dark-sky** viewing conditions.

Rust-red and bright, **Mars** rises in the east around 9:15 p.m. and is at its highest, around 65 degrees, at 3:40 a.m.

Bright star **Capella** is almost directly overhead at 10 p.m. Read about Capella on page 14.

Orion is visible most of the night and is at its highest, around 65 degrees, at 10:40 p.m. It sets in the west around 3:45 a.m. Look for the bright star **Aldebaran** and the **Pleiades** near Orion. See the illustration on page 17.

Saturn sets in the west around 11:15 p.m. Unlike bright Jupiter and Venus and ruddy Mars, Saturn is harder to distinguish from the stars. Stars twinkle and shimmer, while the planets, including Saturn, show a steady flat light.

With the dark sky, look for the **Andromeda Galaxy** with binoculars or a telescope. It is a good target until around 2:15 a.m. See the guide on page 11.

The bright star **Arcturus** rises in the northeast around 12:15 a.m. and is in the eastern sky before dawn. See page 13 for an easy way to locate the star.

Castor and Pollux, the twin stars in the constellation Gemini, are visible most of the night and are almost directly overhead at 12:35 a.m. Read how to find Castor and Pollux on page 14.

Cassiopeia, shaped like a Σ, sets in the northwest around 2:25 a.m. See the illustration on page 11.

One of the brightest stars in the sky, **Vega** rises in the northeast around 3:15 a.m. and is in the eastern sky before dawn.

At its brightest, **Venus** rises in the east around 3:25 a.m. Venus is at its highest in the sky for the year.

2027

Wednesday, January 6

TONIGHT

sunset: **4:50 p.m.**
twilight begins: **5:50 p.m.**

MOON PHASE 

IN THE MORNING

twilight begins: **5:53 a.m.**
sunrise: **6:52 a.m.**

moonrise: **6:15 a.m.**
moonset: **3:51 p.m.**

IN THE SKY

The bright star **Altair** is in the western sky at dusk and sets in the west around 6:50 p.m. Altair can be identified by its less bright neighboring stars, one on either side, together called the shaft of Aquila. See page 18.

Bright star **Vega** is in the western sky at dusk and sets in the northwest around 7:25 p.m.

The **Great Bear (Big Dipper)** rises in the northeast around 7:45 p.m. and is at its highest, around 65 degrees, at 3:50 a.m. See the illustration on page 10.

Jupiter rises in the east around 8:05 p.m. and is at its highest, around 70 degrees, at 2:35 a.m.

With the dark sky, use binoculars or a telescope to look for the **moons of Jupiter.** See the guide on page 18.

The absence of the moon tonight provides excellent **dark-sky** viewing conditions.

Rust-red and bright, Mars rises in the east around 9:10 p.m. and is at its highest, around 65 degrees, at 3:35 a.m.

Orion is visible most of the night and is at its highest, around 65 degrees, at 10:35 p.m. Look for the bright star **Aldebaran** and the **Pleiades** near Orion. See the illustration on page 17.

Saturn sets in the west around 11:10 p.m. Unlike bright Jupiter and Venus and ruddy Mars, Saturn is harder to distinguish from the stars. Stars twinkle and shimmer, while the planets, including Saturn, show a steady flat light.

With the dark sky, look for the **Andromeda Galaxy** with binoculars or a telescope. It is a good target until around 2:10 a.m. See the guide on page 11.

The brightest star in the sky, **Sirius** is visible most of the night and is at its highest in the south, around 40 degrees, at 11:25 p.m. It sets in the west around 4:35 a.m.

The bright star **Arcturus** rises in the northeast around 12:10 a.m. and is in the eastern sky before dawn. See page 13 for an easy way to locate the star.

Bright star **Capella** is almost directly overhead at 9:55 p.m. Read about Capella on page 14.

Castor and Pollux, the twin stars in the constellation Gemini, are visible most of the night and are almost directly overhead at 12:30 a.m. Read how to find Castor and Pollux on page 14.

Cassiopeia, shaped like a *Σ,* sets in the northwest around 2:20 a.m. See the illustration on page 11.

One of the brightest stars in the sky, **Vega** rises in the northeast around 3:10 a.m. and is in the eastern sky before dawn.

At its brightest, Venus rises in the east around 3:25 a.m. Venus is at its highest in the sky for the year.

2027

Thursday, January 7

TONIGHT

sunset: **4:51 p.m.**
twilight begins: **5:50 p.m.**

MOON PHASE 

IN THE MORNING

twilight begins: **5:54 a.m.**
sunrise: **6:52 a.m.**

moonrise: **7 a.m.**
moonset: **4:49 p.m.**

IN THE SKY

The bright star **Altair** is in the western sky at dusk and sets in the west around 6:45 p.m. Altair can be identified by its less bright neighboring stars, one on either side, together called the shaft of Aquila. See page 18.

Bright star **Vega** is in the western sky at dusk and sets in the northwest around 7:20 p.m.

The **Great Bear (Big Dipper)** rises in the northeast around 7:40 p.m. and is at its highest, around 65 degrees, at 3:45 a.m. See the illustration on page 10.

Jupiter rises in the east around 8 p.m. and is at its highest, around 70 degrees, at 2:35 a.m.

Use binoculars or a telescope to look for the **moons of Jupiter.** See the guide on page 18.

The absence of the moon tonight provides excellent **dark-sky** viewing conditions.

Rust-red and bright, **Mars** rises in the east around 9:05 p.m. and is at its highest, around 65 degrees, at 3:30 a.m.

Bright star **Capella** is almost directly overhead at 9:55 p.m. Read about Capella on page 14.

Look for the **Andromeda Galaxy** with binoculars or a telescope. It is a good target until around 2:05 a.m. See the guide on page 11.

Orion is visible most of the night and is at its highest, around 65 degrees, at 10:30 p.m. It sets in the west around 3:35 a.m. Look for the bright star **Aldebaran** and the **Pleiades** near Orion. See the illustration on page 17.

Saturn sets in the west around 11:05 p.m. Unlike bright Jupiter and Venus and ruddy Mars, Saturn is harder to distinguish from the stars. Stars twinkle and shimmer, while the planets, including Saturn, show a steady flat light.

The brightest star in the sky, **Sirius** is visible most of the night and is at its highest in the south, around 40 degrees, at 11:20 p.m. It sets in the west around 4:30 a.m.

Castor and Pollux, the twin stars in the constellation Gemini, are visible most of the night and are almost directly overhead at 12:25 a.m. Read how to find Castor and Pollux on page 14.

The bright star **Arcturus** rises in the northeast around 12:05 a.m. See page 13 for an easy way to locate the star.

Cassiopeia, shaped like a *Σ*, sets in the northwest around 2:15 a.m. See the illustration on page 11.

One of the brightest stars in the sky, **Vega** rises in the northeast around 3:10 a.m. and is in the eastern sky before dawn.

Bright **Venus** rises in the east around 3:30 a.m.

2027

Friday, January 8

TONIGHT

sunset: 4:52 p.m.
twilight begins: 5:51 p.m.

IN THE MORNING

twilight begins: 5:54 a.m.
sunrise: 6:52 a.m.

MOON PHASE

moonrise: 7:38 a.m.
moonset: 5:47 p.m.

IN THE SKY

The bright star **Altair** is in the western sky at dusk and sets in the west around 6:40 p.m. Altair can be identified by its less bright neighboring stars, one on either side, together called the shaft of Aquila. See page 18.

Bright star **Vega** is in the western sky at dusk and sets in the northwest around 7:15 p.m.

The **Great Bear (Big Dipper)** rises in the northeast around 7:40 p.m. and is at its highest, around 65 degrees, at 3:45 a.m. See the illustration on page 10.

Jupiter rises in the east around 7:55 p.m. and is at its highest, around 70 degrees, at 2:30 a.m.

Use binoculars or a telescope to look for the **moons of Jupiter.** See the guide on page 18.

Rust-red and bright, **Mars** rises in the east around 9 p.m. and is at its highest, around 65 degrees, at 3:25 a.m.

The absence of the moon tonight provides excellent **dark-sky** viewing conditions.

Look for the **Andromeda Galaxy** with binoculars or a telescope. It is a good target until around 2:05 a.m. See the guide on page 11.

Bright star **Capella** is almost directly overhead at 9:50 p.m. Read about Capella on page 14.

Orion is visible most of the night and is at its highest, around 65 degrees, at 10:25 p.m. Look for the bright star **Aldebaran** and the **Pleiades** near Orion. See the illustration on page 17.

Saturn sets in the west around 11:05 p.m. Unlike bright Jupiter and Venus and ruddy Mars, Saturn is harder to distinguish from the stars. Stars twinkle and shimmer, while the planets, including Saturn, show a steady flat light.

The brightest star in the sky, **Sirius** is visible most of the night and is at its highest in the south, around 40 degrees, at 11:15 p.m. It sets in the west around 4:25 a.m.

The bright star **Arcturus** rises in the northeast around midnight. See page 13 for an easy way to locate the star.

Cassiopeia, shaped like a Σ, sets in the northwest around 2:10 a.m. See the illustration on page 11.

One of the brightest stars in the sky, **Vega** rises in the northeast around 3:05 a.m. and is in the eastern sky before dawn.

Castor and Pollux, the twin stars in the constellation Gemini, are visible most of the night and are almost directly overhead at 12:20 a.m. Read how to find Castor and Pollux on page 14.

Scorpius rises in the southeast around 4:55 a.m. and is in the eastern sky before dawn. See the illustration on page 19.

Bright **Venus** rises in the east around 3:30 a.m.

2027

Saturday, January 9

TONIGHT

sunset: **4:53 p.m.**
twilight begins: **5:52 p.m.**

IN THE MORNING

twilight begins: **5:54 a.m.**
sunrise: **6:51 a.m.**

MOON PHASE

moonrise: **8:11 a.m.**
moonset: **6:46 p.m.**

IN THE SKY

The bright star **Altair** is in the western sky at dusk and sets in the west around 6:35 p.m. Altair can be identified by its less bright neighboring stars, one on either side, together called the shaft of Aquila. See page 18.

Bright star **Vega** is in the western sky at dusk and sets in the northwest around 7:10 p.m.

The **Great Bear (Big Dipper)** rises in the northeast around 7:35 p.m. and is at its highest, around 65 degrees, at 3:40 a.m. See the illustration on page 10.

Jupiter rises in the east around 7:50 p.m. and is at its highest, around 70 degrees, at 2:25 a.m.

Use binoculars or a telescope to look for the **moons of Jupiter.** See the guide on page 18.

The absence of the moon tonight provides excellent **dark-sky** viewing conditions.

Rust-red and bright, **Mars** rises in the east around 9 p.m. and is at its highest, around 65 degrees, at 3:25 a.m.

Bright star **Capella** is almost directly overhead at 9:45 p.m. Read about Capella on page 14.

Look for the **Andromeda Galaxy** with binoculars or a telescope. It is a good target until around 2 a.m. See the guide on page 11.

Orion is at its highest, around 65 degrees, at 10:25 p.m. It sets in the west around 3:30 a.m. Look for the bright star **Aldebaran** and the **Pleiades** near Orion. See the illustration on page 17.

Saturn sets in the west around 11 p.m. Unlike bright Jupiter and Venus and ruddy Mars, Saturn is harder to distinguish from the stars. Stars twinkle and shimmer, while the planets, including Saturn, show a steady flat light.

The brightest star in the sky, **Sirius** is visible most of the night and is at its highest in the south, around 40 degrees, at 11:10 p.m. It sets in the west around 4:20 a.m.

The bright star **Arcturus** rises in the northeast around 11:55 p.m. See page 13 for an easy way to locate the star.

Castor and Pollux, the twin stars in the constellation Gemini, are visible most of the night and are almost directly overhead at 12:15 a.m. Read how to find Castor and Pollux on page 14.

Cassiopeia, shaped like a *Σ,* sets in the northwest around 2:10 a.m. See the illustration on page 11.

One of the brightest stars in the sky, **Vega** rises in the northeast around 3 a.m. and is in the eastern sky before dawn.

Bright **Venus** rises in the east around 3:30 a.m.

Scorpius rises in the southeast around 4:50 a.m. and is in the eastern sky before dawn. See the illustration on page 19.

2027

Sunday, January 10

TONIGHT

sunset: **4:54 p.m.**
twilight begins: **5:53 p.m.**

IN THE MORNING

twilight begins: **5:54 a.m.**
sunrise: **6:51 a.m.**

MOON PHASE

moonrise: **8:40 a.m.**
moonset: **7:44 p.m.**

IN THE SKY

The bright star **Altair** is in the western sky at dusk and sets in the west around 6:30 p.m. Altair can be identified by its less bright neighboring stars, one on either side, together called the shaft of Aquila. See page 18.

Bright star **Vega** is in the western sky at dusk and sets in the northwest around 7:05 p.m.

The **Great Bear (Big Dipper)** rises in the northeast around 7:30 p.m. and is at its highest, around 65 degrees, at 3:35 a.m. See the illustration on page 10.

Jupiter rises in the east around 7:45 p.m. and is at its highest, around 70 degrees, at 2:20 a.m.

After the moon sets, from around 8:05 p.m., use binoculars or a telescope to look for the **moons of Jupiter.** See the guide on page 18.

After the moon sets, from around 8:05 p.m., look for the **Andromeda Galaxy** with binoculars or a telescope. It is a good target until around 1:55 a.m. See the guide on page 11.

The absence of the moon tonight provides excellent **dark-sky** viewing conditions. Explore the **dark sky** after the moon sets, from around 8:05 p.m.

Rust-red and bright, **Mars** rises in the east around 8:55 p.m. and is at its highest, around 65 degrees, at 3:20 a.m.

Bright star **Capella** is almost directly overhead at 9:40 p.m. Read about Capella on page 14.

Orion is at its highest, around 65 degrees, at 10:20 p.m. It sets in the west around 3:25 a.m. Look for the bright star **Aldebaran** and the **Pleiades** near Orion. See the illustration on page 17.

Saturn sets in the west around 10:55 p.m. Unlike bright Jupiter and Venus and ruddy Mars, Saturn is harder to distinguish from the stars. Stars twinkle and shimmer, while the planets, including Saturn, show a steady flat light.

The brightest star in the sky, **Sirius** is visible most of the night and is at its highest in the south, around 40 degrees, at 11:10 p.m. It sets in the west around 4:15 a.m.

The bright star **Arcturus** rises in the northeast around 11:50 p.m. See page 13 for an easy way to locate the star.

Castor and Pollux, the twin stars in the constellation Gemini, are visible most of the night and are almost directly overhead at 12:15 a.m. Read how to find Castor and Pollux on page 14.

Cassiopeia, shaped like a *Σ,* sets in the northwest around 2:05 a.m. See the illustration on page 11.

One of the brightest stars in the sky, **Vega** rises in the northeast around 2:55 a.m. and is in the eastern sky before dawn.

Bright **Venus** rises in the east around 3:30 a.m.

Scorpius rises in the southeast around 4:45 a.m. and is in the eastern sky before dawn. See the illustration on page 19.

2027

Monday, January 11

TONIGHT

sunset: **4:55 p.m.**
twilight begins: **5:54 p.m.**

MOON PHASE

IN THE MORNING

twilight begins: **5:53 a.m.**
sunrise: **6:51 a.m.**

moonrise: **9:07 a.m.**
moonset: **8:41 p.m.**

IN THE SKY

The bright star **Altair** is in the western sky at dusk and sets in the west around 6:30 p.m. Altair can be identified by its less bright neighboring stars, one on either side, together called the shaft of Aquila. See page 18.

Bright star **Vega** is in the western sky at dusk and sets in the northwest around 7:05 p.m.

The **Great Bear (Big Dipper)** rises in the northeast around 7:25 p.m. and is at its highest, around 65 degrees, at 3:30 a.m. See the illustration on page 10.

Jupiter rises in the east around 7:40 p.m. and is at its highest, around 70 degrees, at 2:15 a.m.

After the moon sets, from around 9 p.m., use binoculars or a telescope to look for the **moons of Jupiter.** See the guide on page 18.

Rust-red and bright, Mars rises in the east around 8:50 p.m. and is at its highest, around 65 degrees, at 3:15 a.m.

The absence of the moon tonight provides excellent **dark-sky** viewing conditions. Explore the **dark sky** after the moon sets, from around 9 p.m.

After the moon sets, from around 9 p.m., look for the **Andromeda Galaxy** with binoculars or a telescope. It is a good target until around 1:50 a.m. See the guide on page 11.

Bright star **Capella** is almost directly overhead at 9:35 p.m. Read about Capella on page 14.

Orion is at its highest, around 65 degrees, at 10:15 p.m. It sets in the west around 3:20 a.m. Look for the bright star **Aldebaran** and the **Pleiades** near Orion. See the illustration on page 17.

Saturn sets in the west around 10:50 p.m. Unlike bright Jupiter and Venus and ruddy Mars, Saturn is harder to distinguish from the stars. Stars twinkle and shimmer, while the planets, including Saturn, show a steady flat light.

The brightest star in the sky, **Sirius** is visible most of the night and is at its highest in the south, around 40 degrees, at 11:05 p.m. It sets in the west around 4:15 a.m.

The bright star **Arcturus** rises in the northeast around 11:50 p.m. See page 13 for an easy way to locate the star.

Castor and Pollux, the twin stars in the constellation Gemini, are visible most of the night and are almost directly overhead at 12:10 a.m. Read how to find Castor and Pollux on page 14.

Cassiopeia, shaped like a *Σ*, sets in the northwest around 2 a.m. See the illustration on page 11.

One of the brightest stars in the sky, **Vega** rises in the northeast around 2:55 a.m. and is in the eastern sky before dawn.

Bright Venus rises in the east around 3:30 a.m.

Scorpius rises in the southeast around 4:45 a.m. and is in the eastern sky before dawn. See the illustration on page 19.

2027

Tuesday, January 12

TONIGHT

sunset: **4:56 p.m.**
twilight begins: **5:55 p.m.**

IN THE MORNING

twilight begins: **5:53 a.m.**
sunrise: **6:51 a.m.**

MOON PHASE

moonrise: **9:32 a.m.**
moonset: **9:39 p.m.**

IN THE SKY

The bright star **Altair** sets in the west around 6:25 p.m. Altair can be identified by its less bright neighboring stars, one on either side, together called the shaft of Aquila. See page 18.

Bright star **Vega** is in the western sky at dusk and sets in the northwest around 7 p.m.

The **moon** is prominent in the sky early in the evening. Using a telescope or binoculars, take a look. You should be able to see rough mountains and craters along the boundary between light and dark, in astronomy called the terminator.

The **Great Bear (Big Dipper)** rises in the northeast around 7:25 p.m. and is at its highest, around 65 degrees, at 3:30 a.m. See the illustration on page 10.

Jupiter rises in the east around 7:35 p.m. and is at its highest, around 70 degrees, at 2:10 a.m.

After the moon sets, from around 10 p.m., use binoculars or a telescope to look for the **moons of Jupiter.** See the guide on page 18.

Rust-red and bright, **Mars** rises in the east around 8:45 p.m. and is at its highest, around 65 degrees, at 3:10 a.m.

Bright star **Capella** is almost directly overhead at 9:35 p.m. Read about Capella on page 14.

Explore the **dark sky** after the moon sets, from around 10 p.m.

After the moon sets, from around 10 p.m., look for the **Andromeda Galaxy** with binoculars or a telescope. It is a good target until around 1:45 a.m. See the guide on page 11.

Orion is at its highest, around 65 degrees, at 10:10 p.m. It sets in the west around 3:15 a.m. Look for the bright star **Aldebaran** and the **Pleiades** near Orion. See the illustration on page 17.

Saturn sets in the west around 10:50 p.m. Unlike bright Jupiter and Venus and ruddy Mars, Saturn is harder to distinguish from the stars. Stars twinkle and shimmer, while the planets, including Saturn, show a steady flat light.

The brightest star in the sky, **Sirius** is visible most of the night and is at its highest in the south, around 40 degrees, at 11 p.m. It sets in the west around 4:10 a.m.

Castor and Pollux, the twin stars in the constellation Gemini, are visible most of the night and are almost directly overhead at 12:05 a.m. Read how to find Castor and Pollux on page 14.

Cassiopeia, shaped like a Σ, sets in the northwest around 1:55 a.m. See the illustration on page 11.

One of the brightest stars in the sky, **Vega** rises in the northeast around 2:50 a.m. and is in the eastern sky before dawn.

Bright **Venus** rises in the east around 3:35 a.m.

2027

Wednesday, January 13

TONIGHT

sunset: **4:57 p.m.**
twilight begins: **5:55 p.m.**

IN THE MORNING

twilight begins: **5:53 a.m.**
sunrise: **6:51 a.m.**

MOON PHASE

moonrise: **9:57 a.m.**
moonset: **10:38 p.m.**

IN THE SKY

The bright star **Altair** sets in the west around 6:20 p.m. Altair can be identified by its less bright neighboring stars, one on either side, together called the shaft of Aquila. See page 18.

The waxing crescent moon and **Saturn** appear close together in the evening sky. Look southwest at about 45 degrees elevation around 6:45 p.m.

The **moon** is prominent in the sky early in the evening. Using a telescope or binoculars, take a look. You should be able to see rough mountains and craters along the boundary between light and dark, in astronomy called the terminator.

The **Great Bear (Big Dipper)** rises in the northeast around 7:20 p.m. and is at its highest, around 65 degrees, at 3:25 a.m. See the illustration on page 10.

Jupiter rises in the east around 7:30 p.m. and is at its highest, around 70 degrees, at 2:05 a.m.

After the moon sets, from around 11 p.m., use binoculars or a telescope to look for the **moons of Jupiter.** See the guide on page 18.

Rust-red and bright, **Mars** rises in the east around 8:40 p.m. and is at its highest, around 65 degrees, at 3:10 a.m.

Orion is at its highest, around 65 degrees, at 10:05 p.m. It sets in the west around 3:15 a.m. Look for the bright star **Aldebaran** and the **Pleiades** near Orion. See the illustration on page 17.

Saturn sets in the west around 10:45 p.m. Unlike bright Jupiter and Venus and ruddy Mars, Saturn is harder to distinguish from the stars. Stars twinkle and shimmer, while the planets, including Saturn, show a steady flat light.

The brightest star in the sky, **Sirius** is visible most of the night and is at its highest in the south, around 40 degrees, at 10:55 p.m. It sets in the west around 4:05 a.m.

Explore the **dark sky** after the moon sets, from around 11 p.m.

After the moon sets, from around 11 p.m., look for the **Andromeda Galaxy** with binoculars or a telescope. It is a good target until around 1:45 a.m. See the guide on page 11.

The bright star **Arcturus** rises in the northeast around 11:40 p.m. See page 13 for an easy way to locate the star.

Castor and Pollux, the twin stars in the constellation Gemini, are visible most of the night and are almost directly overhead at midnight. Read how to find Castor and Pollux on page 14.

Cassiopeia, shaped like a Σ, sets in the northwest around 1:50 a.m. See the illustration on page 11.

One of the brightest stars in the sky, **Vega** rises in the northeast around 2:45 a.m. and is in the eastern sky before dawn.

Bright **Venus** rises in the east around 3:35 a.m.

2027
Thursday, January 14

TONIGHT

sunset: **4:58 p.m.**
twilight begins: **5:56 p.m.**

IN THE MORNING

twilight begins: **5:53 a.m.**
sunrise: **6:51 a.m.**

MOON PHASE

moonrise: **10:23 a.m.**
moonset: **11:39 p.m.**

IN THE SKY

The bright star **Altair** sets in the west around 6:15 p.m. Altair can be identified by its less bright neighboring stars, one on either side, together called the shaft of Aquila. See page 18.

The waxing first-quarter moon and **Saturn** appear close together in the evening sky. Look southwest at about 55 degrees elevation around 6:45 p.m.

Bright star **Vega** is in the western sky at dusk and sets in the northwest around 6:50 p.m.

The **Great Bear (Big Dipper)** rises in the northeast around 7:15 p.m. and is at its highest, around 65 degrees, at 3:20 a.m. See the illustration on page 10.

Jupiter rises in the east around 7:30 p.m. and is at its highest, around 70 degrees, at 2:05 a.m.

After the moon sets, from around midnight, use binoculars or a telescope to look for the **moons of Jupiter.** See the guide on page 18.

The **moon** is prominent in the sky early in the evening. Using a telescope or binoculars, take a look. You should be able to see rough mountains and craters along the boundary between light and dark, in astronomy called the terminator.

Rust-red and bright, **Mars** rises in the east around 8:35 p.m. and is at its highest, around 65 degrees, at 3:05 a.m.

Bright star **Capella** is almost directly overhead at 9:25 p.m. Read about Capella on page 14.

Orion is at its highest, around 65 degrees, at 10:05 p.m. It sets in the west around 3:10 a.m. Look for the bright star **Aldebaran** and the **Pleiades** near Orion. See the illustration on page 17.

Saturn sets in the west around 10:40 p.m. Unlike bright Jupiter and Venus and ruddy Mars, Saturn is harder to distinguish from the stars. Stars twinkle and shimmer, while the planets, including Saturn, show a steady flat light.

The brightest star in the sky, **Sirius** is visible most of the night and is at its highest in the south, around 40 degrees, at 10:55 p.m. It sets in the west around 4 a.m.

Castor and Pollux, the twin stars in the constellation Gemini, are visible most of the night and are almost directly overhead at 11:55 p.m. Read how to find Castor and Pollux on page 14.

Explore the **dark sky** after the moon sets, from around midnight.

After the moon sets, from around midnight, look for the **Andromeda Galaxy** with binoculars or a telescope. It is a good target until around 1:40 a.m. See the guide on page 11.

Cassiopeia, shaped like a *Σ,* sets in the northwest around 1:50 a.m. See the illustration on page 11.

Bright **Venus** rises in the east around 3:35 a.m.

2027

Friday, January 15

TONIGHT

sunset: **4:58 p.m.**
twilight begins: **5:57 p.m.**

MOON PHASE

IN THE MORNING

twilight begins: **5:53 a.m.**
sunrise: **6:50 a.m.**

moonrise: **10:51 a.m.**
moonset: **12:43 a.m.**

IN THE SKY

Bright star **Vega** is in the western sky at dusk and sets in the northwest around 6:50 p.m.

The **Great Bear (Big Dipper)** rises in the northeast around 7:10 p.m. and is at its highest, around 65 degrees, at 3:15 a.m. See the illustration on page 10.

Jupiter rises in the east around 7:25 p.m. and is at its highest, around 70 degrees, at 2 a.m.

After the moon sets, from around 1:05 a.m., use binoculars or a telescope to look for the **moons of Jupiter.** See the guide on page 18.

The **moon** is prominent in the sky early in the evening. Using a telescope or binoculars, take a look. You should be able to see rough mountains and craters along the boundary between light and dark, in astronomy called the terminator.

Rust-red and bright, **Mars** rises in the east around 8:35 p.m. and is at its highest, around 65 degrees, at 3 a.m.

Bright star **Capella** is almost directly overhead at 9:20 p.m. Read about Capella on page 14.

Orion is at its highest, around 65 degrees, at 10 p.m. It sets in the west around 3:05 a.m. Look for the bright star **Aldebaran** and the **Pleiades** near Orion. See the illustration on page 17.

Saturn sets in the west around 10:35 p.m. Unlike bright Jupiter and Venus and ruddy Mars, Saturn is harder to distinguish from the stars. Stars twinkle and shimmer, while the planets, including Saturn, show a steady flat light.

The brightest star in the sky, **Sirius** is visible most of the night and is at its highest in the south, around 40 degrees, at 10:50 p.m. It sets in the west around 4 a.m.

The bright star **Arcturus** rises in the northeast around 11:30 p.m. See page 13 for an easy way to locate the star.

Castor and Pollux, the twin stars in the constellation Gemini, are visible most of the night and are almost directly overhead at 11:50 p.m. Read how to find Castor and Pollux on page 14.

Explore the **dark sky** after the moon sets, from around 1:05 a.m.

After the moon sets, from around 1:05 a.m., look for the **Andromeda Galaxy** with binoculars or a telescope. It is a good target until around 1:35 a.m. See the guide on page 11.

Cassiopeia, shaped like a *Σ*, sets in the northwest around 1:45 a.m. See the illustration on page 11.

One of the brightest stars in the sky, **Vega** rises in the northeast around 2:35 a.m. and is in the eastern sky before dawn.

Bright **Venus** rises in the east around 3:35 a.m.

2027

Saturday, January 16

TONIGHT

sunset: **4:59 p.m.**
twilight begins: **5:58 p.m.**

IN THE MORNING

twilight begins: **5:53 a.m.**
sunrise: **6:50 a.m.**

MOON PHASE

moonrise: **11:24 a.m.**
moonset: **1:52 a.m.**

IN THE SKY

Bright star **Vega** is in the western sky at dusk and sets in the northwest around 6:45 p.m.

The **Great Bear (Big Dipper)** rises in the northeast around 7:05 p.m. and is at its highest, around 65 degrees, at 3:10 a.m. See the illustration on page 10.

Jupiter rises in the east around 7:20 p.m. and is at its highest, around 70 degrees, at 1:55 a.m.

Rust-red and bright, **Mars** rises in the east around 8:30 p.m. and is at its highest, around 65 degrees, at 2:55 a.m.

The **moon** is prominent in the sky early in the evening. Using a telescope or binoculars, take a look. You should be able to see rough mountains and craters along the boundary between light and dark, in astronomy called the terminator.

Bright star **Capella** is almost directly overhead at 9:15 p.m. Read about Capella on page 14.

Orion is at its highest, around 65 degrees, at 9:55 p.m. It sets in the west around 3 a.m. Look for the bright star **Aldebaran** and the **Pleiades** near Orion. See the illustration on page 17.

Saturn sets in the west around 10:35 p.m. Unlike bright Jupiter and Venus and ruddy Mars, Saturn is harder to distinguish from the stars. Stars twinkle and shimmer, while the planets, including Saturn, show a steady flat light.

The brightest star in the sky, **Sirius** is visible most of the night and is at its highest in the south, around 40 degrees, at 10:45 p.m. It sets in the west around 3:55 a.m.

The bright star **Arcturus** rises in the northeast around 11:30 p.m. See page 13 for an easy way to locate the star.

Castor and Pollux, the twin stars in the constellation Gemini, are visible most of the night and are almost directly overhead at 11:45 p.m. Read how to find Castor and Pollux on page 14.

Cassiopeia, shaped like a Σ, sets in the northwest around 1:40 a.m. See the illustration on page 11.

One of the brightest stars in the sky, **Vega** rises in the northeast around 2:35 a.m. and is in the eastern sky before dawn.

Bright **Venus** rises in the east around 3:35 a.m.

Scorpius rises in the southeast around 4:25 a.m. and is in the eastern sky before dawn. See the illustration on page 19.

The bright star **Altair** rises in the east around 5:35 a.m. Altair can be identified by its less bright neighboring stars, one on either side, together called the shaft of Aquila. See page 18.

2027

Sunday, January 17

TONIGHT

sunset: **5 p.m.**
twilight begins: **5:59 p.m.**

MOON PHASE

IN THE MORNING

twilight begins: **5:53 a.m.**
sunrise: **6:50 a.m.**

moonrise: **12:04 p.m.**
moonset: **3:03 a.m.**

IN THE SKY

Bright star **Vega** is in the western sky at dusk and sets in the northwest around 6:40 p.m.

The **Great Bear (Big Dipper)** rises in the northeast around 7:05 p.m. and is at its highest, around 65 degrees, at 3:10 a.m. See the illustration on page 10.

Jupiter rises in the east around 7:15 p.m. and is at its highest, around 70 degrees, at 1:50 a.m.

Rust-red and bright, **Mars** rises in the east around 8:25 p.m. and is at its highest, around 65 degrees, at 2:50 a.m.

Bright star **Capella** is almost directly overhead at 9:15 p.m. Read about Capella on page 14.

Orion is at its highest, around 65 degrees, at 9:50 p.m. It sets in the west around 2:55 a.m. Look for the bright star **Aldebaran** and the **Pleiades** near Orion. See the illustration on page 17.

Saturn sets in the west around 10:30 p.m. Unlike bright Jupiter and Venus and ruddy Mars, Saturn is harder to distinguish from the stars. Stars twinkle and shimmer, while the planets, including Saturn, show a steady flat light.

The brightest star in the sky, **Sirius** is visible most of the night and is at its highest in the south, around 40 degrees, at 10:40 p.m. It sets in the west around 3:50 a.m.

The waxing gibbous **moon** and the **Pleiades** appear close together in the evening sky. Look west at about 45 degrees elevation around 11 p.m.

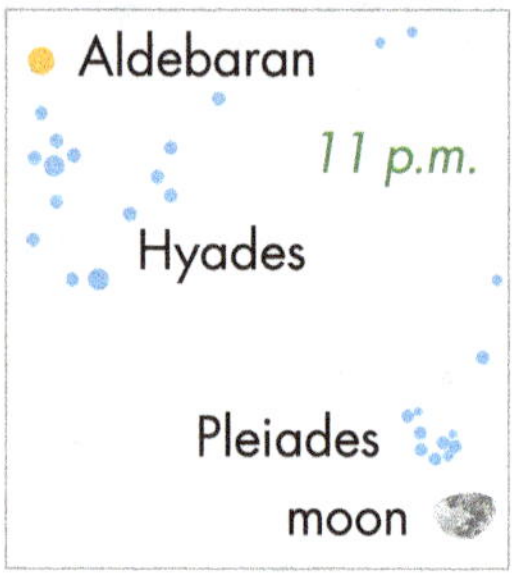

The bright star **Arcturus** rises in the northeast around 11:25 p.m. See page 13 for an easy way to locate the star.

Castor and Pollux, the twin stars in the constellation Gemini, are visible most of the night and are almost directly overhead at 11:40 p.m. Read how to find Castor and Pollux on page 14.

Cassiopeia, shaped like a Σ, sets in the northwest around 1:35 a.m. See the illustration on page 11.

One of the brightest stars in the sky, **Vega** rises in the northeast around 2:30 a.m. and is in the eastern sky before dawn.

Bright **Venus** rises in the southeast around 3:40 a.m.

Scorpius rises in the southeast around 4:20 a.m. and is in the eastern sky before dawn. See the illustration on page 19.

The bright star **Altair** rises in the east around 5:30 a.m. and is in the eastern sky before dawn. Altair can be identified by its less bright neighboring stars, one on either side, together called the shaft of Aquila. See page 18.

2027

Monday, January 18

TONIGHT

sunset: **5:01 p.m.**
twilight begins: **6 p.m.**

IN THE MORNING

twilight begins: **5:52 a.m.**
sunrise: **6:49 a.m.**

MOON PHASE

moonrise: **12:54 p.m.**
moonset: **4:13 a.m.**

IN THE SKY

Bright star **Vega** is in the western sky at dusk and sets in the northwest around 6:35 p.m.

The **Great Bear (Big Dipper)** rises in the northeast around 7 p.m. and is at its highest, around 65 degrees, at 3:05 a.m. See the illustration on page 10.

Jupiter rises in the east around 7:10 p.m. and is at its highest, around 70 degrees, at 1:45 a.m.

Rust-red and bright, **Mars** rises in the east around 8:20 p.m. and is at its highest, around 65 degrees, at 2:45 a.m.

Bright star **Capella** is almost directly overhead at 9:10 p.m. Read about Capella on page 14.

Orion is at its highest, around 65 degrees, at 9:45 p.m. It sets in the west around 2:55 a.m. Look for the bright star **Aldebaran** and the **Pleiades** near Orion. See the illustration on page 17.

Saturn sets in the west around 10:25 p.m. Unlike bright Jupiter and Venus and ruddy Mars, Saturn is harder to distinguish from the stars. Stars twinkle and shimmer, while the planets, including Saturn, show a steady flat light.

The brightest star in the sky, **Sirius** is visible most of the night and is at its highest in the south, around 40 degrees, at 10:35 p.m. It sets in the west around 3:45 a.m.

The bright star **Arcturus** rises in the northeast around 11:20 p.m. See page 13 for an easy way to locate the star.

Castor and Pollux, the twin stars in the constellation Gemini, are visible most of the night and are almost directly overhead at 11:40 p.m. Read how to find Castor and Pollux on page 14.

Cassiopeia, shaped like a Σ, sets in the northwest around 1:35 a.m. See the illustration on page 11.

One of the brightest stars in the sky, **Vega** rises in the northeast around 2:25 a.m. and is in the eastern sky before dawn.

Bright **Venus** rises in the southeast around 3:40 a.m.

Scorpius rises in the southeast around 4:15 a.m. and is in the eastern sky before dawn. See the illustration on page 19.

The bright star **Altair** rises in the east around 5:30 a.m. and is in the eastern sky before dawn. Altair can be identified by its less bright neighboring stars, one on either side, together called the shaft of Aquila. See page 18.

2027

Tuesday, January 19

TONIGHT

sunset: **5:02 p.m.**
twilight begins: **6:01 p.m.**

IN THE MORNING

twilight begins: **5:52 a.m.**
sunrise: **6:49 a.m.**

MOON PHASE

moonrise: **1:54 p.m.**
moonset: **5:19 a.m.**

IN THE SKY

Bright star **Vega** is in the western sky at dusk and sets in the northwest around 6:30 p.m.

The **Great Bear (Big Dipper)** rises in the northeast around 6:55 p.m. and is at its highest, around 65 degrees, at 3 a.m. See the illustration on page 10.

Jupiter rises in the east around 7:05 p.m. and is at its highest, around 70 degrees, at 1:40 a.m.

Rust-red and bright, **Mars** rises in the east around 8:15 p.m. and is at its highest, around 65 degrees, at 2:45 a.m.

The **full moon** tonight acts like a spotlight. Look around at the desert landscape at how much illumination the moon provides.

Bright star **Capella** is almost directly overhead at 9:05 p.m. Read about Capella on page 14.

Orion is at its highest, around 65 degrees, at 9:45 p.m. It sets in the west around 2:50 a.m. Look for the bright star **Aldebaran** and the **Pleiades** near Orion. See the illustration on page 17.

Saturn sets in the west around 10:25 p.m. Unlike bright Jupiter and Venus and ruddy Mars, Saturn is harder to distinguish from the stars. Stars twinkle and shimmer, while the planets, including Saturn, show a steady flat light.

The brightest star in the sky, **Sirius** is visible most of the night and is at its highest in the south, around 40 degrees, at 10:35 p.m. It sets in the west around 3:40 a.m.

The bright star **Arcturus** rises in the northeast around 11:15 p.m. See page 13 for an easy way to locate the star.

Castor and Pollux, the twin stars in the constellation Gemini, are visible most of the night and are almost directly overhead at 11:35 p.m. Read how to find Castor and Pollux on page 14.

Cassiopeia, shaped like a Σ, sets in the northwest around 1:30 a.m. See the illustration on page 11.

One of the brightest stars in the sky, **Vega** rises in the northeast around 2:20 a.m. and is in the eastern sky before dawn.

Bright **Venus** rises in the southeast around 3:40 a.m.

Scorpius rises in the southeast around 4:10 a.m. and is in the eastern sky before dawn. See the illustration on page 19.

The bright star **Altair** rises in the east around 5:25 a.m. and is in the eastern sky before dawn. Altair can be identified by its less bright neighboring stars, one on either side, together called the shaft of Aquila. See page 18.

2027

Wednesday, January 20

TONIGHT

sunset: **5:03 p.m.**
twilight begins: **6:01 p.m.**

MOON PHASE

IN THE MORNING

twilight begins: **5:52 a.m.**
sunrise: **6:49 a.m.**

moonrise: **3:05 p.m.**
moonset: **6:16 a.m.**

IN THE SKY

Bright star **Vega** sets in the northwest around 6:30 p.m.

The **Great Bear (Big Dipper)** rises in the northeast around 6:50 p.m. and is at its highest, around 65 degrees, at 2:55 a.m. See the illustration on page 10.

Jupiter rises in the east around 7 p.m. and is at its highest, around 70 degrees, at 1:35 a.m.

Rust-red and bright, **Mars** rises in the east around 8:10 p.m. and is at its highest, around 65 degrees, at 2:40 a.m.

The **full moon** tonight acts like a spotlight. Look around at the desert landscape at how much illumination the moon provides.

Bright star **Capella** is almost directly overhead at 9 p.m. Read about Capella on page 14.

Orion is at its highest, around 65 degrees, at 9:40 p.m. It sets in the west around 2:45 a.m. Look for the bright star **Aldebaran** and the **Pleiades** near Orion. See the illustration on page 17.

Saturn sets in the west around 10:20 p.m. Unlike bright Jupiter and Venus and ruddy Mars, Saturn is harder to distinguish from the stars. Stars twinkle and shimmer, while the planets, including Saturn, show a steady flat light.

The brightest star in the sky, **Sirius** is visible most of the night and is at its highest in the south, around 40 degrees, at 10:30 p.m. It sets in the west around 3:40 a.m.

The bright star **Arcturus** rises in the northeast around 11:10 p.m. See page 13 for an easy way to locate the star.

Castor and Pollux, the twin stars in the constellation Gemini, are visible most of the night and are almost directly overhead at 11:30 p.m. Read how to find Castor and Pollux on page 14.

Cassiopeia, shaped like a Σ, sets in the northwest around 1:25 a.m. See the illustration on page 11.

One of the brightest stars in the sky, **Vega** rises in the northeast around 2:15 a.m.

Bright **Venus** rises in the southeast around 3:40 a.m.

Scorpius rises in the southeast around 4:10 a.m. and is in the eastern sky before dawn. See the illustration on page 19.

The bright star **Altair** rises in the east around 5:20 a.m. and is in the eastern sky before dawn. Altair can be identified by its less bright neighboring stars, one on either side, together called the shaft of Aquila. See page 18.

2027

Thursday, January 21

TONIGHT

sunset: **5:04 p.m.**
twilight begins: **6:02 p.m.**

IN THE MORNING

twilight begins: **5:51 a.m.**
sunrise: **6:48 a.m.**

MOON PHASE

moonrise: **4:21 p.m.**
moonset: **6:16 a.m.**

IN THE SKY

The **full moon** and the bright stars Castor and Pollux appear close together in the evening sky. Look east near the horizon around 6:15 p.m.

Bright star **Vega** sets in the northwest around 6:25 p.m.

The **Great Bear (Big Dipper)** rises in the northeast around 6:45 p.m. and is at its highest, around 65 degrees, at 2:50 a.m. See the illustration on page 10.

Jupiter is visible most of the night. It rises in the east around 6:55 p.m. and is at its highest, around 70 degrees, at 1:30 a.m.

Rust-red and bright, **Mars** rises in the east around 8:05 p.m. and is at its highest, around 65 degrees, at 2:35 a.m.

The **full moon** tonight acts like a spotlight. Look around at the desert landscape at how much illumination the moon provides.

Bright star **Capella** is almost directly overhead at 9 p.m. Read about Capella on page 14.

Orion is at its highest, around 65 degrees, at 9:35 p.m. It sets in the west around 2:40 a.m. Look for the bright star **Aldebaran** and the **Pleiades** near Orion. See the illustration on page 17.

Saturn sets in the west around 10:15 p.m. Unlike bright Jupiter and Venus and ruddy Mars, Saturn is harder to distinguish from the stars. Stars twinkle and shimmer, while the planets, including Saturn, show a steady flat light.

The brightest star in the sky, **Sirius** is visible most of the night and is at its highest in the south, around 40 degrees, at 10:25 p.m. It sets in the west around 3:35 a.m.

The bright star **Arcturus** rises in the northeast around 11:10 p.m. See page 13 for an easy way to locate the star.

Castor and Pollux, the twin stars in the constellation Gemini, are visible most of the night and are almost directly overhead at 11:25 p.m. Read how to find Castor and Pollux on page 14.

Cassiopeia, shaped like a *Σ,* sets in the northwest around 1:20 a.m. See the illustration on page 11.

One of the brightest stars in the sky, **Vega** rises in the northeast around 2:15 a.m.

Bright **Venus** rises in the southeast around 3:45 a.m.

Scorpius rises in the southeast around 4:05 a.m. and is in the eastern sky before dawn. See the illustration on page 19.

The bright star **Altair** rises in the east around 5:15 a.m. and is in the eastern sky before dawn. Altair can be identified by its less bright neighboring stars, one on either side, together called the shaft of Aquila. See page 18.

2027

Friday, January 22

TONIGHT

sunset: **5:05 p.m.**
twilight begins: **6:03 p.m.**

IN THE MORNING

twilight begins: **5:51 a.m.**
sunrise: **6:48 a.m.**

MOON PHASE

moonrise: **5:39 p.m.**
moonset: **7:04 a.m.**

IN THE SKY

Bright star **Vega** sets in the northwest around 6:20 p.m.

The **Great Bear (Big Dipper)** rises in the northeast around 6:45 p.m. and is at its highest, around 65 degrees, at 2:50 a.m. See the illustration on page 10.

Jupiter is visible most of the night. It rises in the east around 6:50 p.m. and is at its highest, around 70 degrees, at 1:30 a.m.

The **full moon** tonight acts like a spotlight. Look around at the desert landscape at how much illumination the moon provides.

Rust-red and bright, **Mars** rises in the east around 8 p.m. and is at its highest, around 70 degrees, at 2:30 a.m.

Bright star **Capella** is almost directly overhead at 8:55 p.m. Read about Capella on page 14.

Orion is at its highest, around 65 degrees, at 9:30 p.m. It sets in the west around 2:40 a.m. Look for the bright star **Aldebaran** and the **Pleiades** near Orion. See the illustration on page 17.

Saturn sets in the west around 10:10 p.m. Unlike bright Jupiter and Venus and ruddy Mars, Saturn is harder to distinguish from the stars. Stars twinkle and shimmer, while the planets, including Saturn, show a steady flat light.

The brightest star in the sky, **Sirius** is visible most of the night and is at its highest in the south, around 40 degrees, at 10:20 p.m. It sets in the west around 3:30 a.m.

The bright star **Arcturus** rises in the northeast around 11:05 p.m. See page 13 for an easy way to locate the star.

Castor and Pollux, the twin stars in the constellation Gemini, are visible most of the night and are almost directly overhead at 11:20 p.m. Read how to find Castor and Pollux on page 14.

Cassiopeia, shaped like a *Σ*, sets in the northwest around 1:15 a.m. See the illustration on page 11.

One of the brightest stars in the sky, **Vega** rises in the northeast around 2:10 a.m.

Bright **Venus** rises in the southeast around 3:45 a.m.

Scorpius rises in the southeast around 4 a.m. and is in the eastern sky before dawn. See the illustration on page 19.

The bright star **Altair** rises in the east around 5:15 a.m. and is in the eastern sky before dawn. Altair can be identified by its less bright neighboring stars, one on either side, together called the shaft of Aquila. See page 18.

2027

Saturday, January 23

TONIGHT

sunset: **5:06 p.m.**
twilight begins: **6:04 p.m.**

MOON PHASE

IN THE MORNING

twilight begins: **5:51 a.m.**
sunrise: **6:47 a.m.**

moonrise: **6:53 p.m.**
moonset: **7:43 a.m.**

IN THE SKY

Bright star **Vega** sets in the northwest around 6:15 p.m.

The **Great Bear (Big Dipper)** rises in the northeast around 6:40 p.m. It is in the northeastern sky at dusk and is at its highest, around 65 degrees, at 2:45 a.m. See the illustration on page 10.

Jupiter is visible most of the night. It rises in the east around 6:50 p.m. and is at its highest, around 70 degrees, at 1:25 a.m.

Rust-red and bright, **Mars** rises in the east around 7:55 p.m. and is at its highest, around 70 degrees, at 2:25 a.m.

The **full moon** tonight acts like a spotlight. Look around at the desert landscape at how much illumination the moon provides.

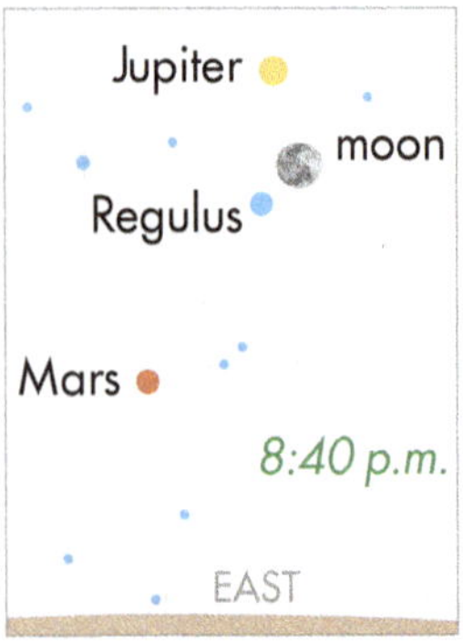

The **full moon, Jupiter,** and **Mars** appear close together in the evening sky. Look east near the horizon around 8:40 p.m.

Bright star **Capella** is almost directly overhead at 8:50 p.m. Read about Capella on page 14.

Orion is at its highest, around 65 degrees, at 9:30 p.m. It sets in the west around 2:35 a.m. Look for the bright star **Aldebaran** and the **Pleiades** near Orion. See the illustration on page 17.

Saturn sets in the west around 10:10 p.m. Unlike bright Jupiter and Venus and ruddy Mars, Saturn is harder to distinguish from the stars. Stars twinkle and shimmer, while the planets, including Saturn, show a steady flat light.

The brightest star in the sky, **Sirius** is visible most of the night and is at its highest in the south, around 40 degrees, at 10:15 p.m. It sets in the west around 3:25 a.m.

The bright star **Arcturus** rises in the northeast around 11 p.m. See page 13 for an easy way to locate the star.

Castor and Pollux, the twin stars in the constellation Gemini, are visible most of the night and are almost directly overhead at 11:20 p.m. Read how to find Castor and Pollux on page 14.

Cassiopeia, shaped like a Σ, sets in the northwest around 1:15 a.m. See the illustration on page 11.

One of the brightest stars in the sky, **Vega** rises in the northeast around 2:05 a.m.

Bright **Venus** rises in the southeast around 3:45 a.m.

Scorpius rises in the southeast around 3:55 a.m. and is in the eastern sky before dawn. See the illustration on page 19.

The bright star **Altair** rises in the east around 5:10 a.m. and is in the eastern sky before dawn. Altair can be identified by its less bright neighboring stars, one on either side, together called the shaft of Aquila. See page 18.

2027

Sunday, January 24

TONIGHT

sunset: **5:07 p.m.**
twilight begins: **6:05 p.m.**

MOON PHASE

IN THE MORNING

twilight begins: **5:50 a.m.**
sunrise: **6:47 a.m.**

moonrise: **8:04 p.m.**
moonset: **8:17 a.m.**

IN THE SKY

The **Great Bear (Big Dipper)** rises in the northeast around 6:35 p.m. It is in the northeastern sky at dusk and is at its highest, around 65 degrees, at 2:40 a.m. See the illustration on page 10.

Jupiter is visible most of the night. It rises in the east around 6:45 p.m. and is at its highest, around 70 degrees, at 1:20 a.m.

Rust-red and bright, **Mars** rises in the east around 7:50 p.m. and is at its highest, around 70 degrees, at 2:20 a.m.

The nearly **full moon** tonight acts like a spotlight. Look around at the desert landscape at how much illumination the moon provides.

Bright star **Capella** is almost directly overhead at 8:45 p.m. Read about Capella on page 14.

Orion is at its highest, around 65 degrees, at 9:25 p.m. It sets in the west around 2:30 a.m. Look for the bright star **Aldebaran** and the **Pleiades** near Orion. See the illustration on page 17.

Saturn sets in the west around 10:05 p.m. Unlike bright Jupiter and Venus and ruddy Mars, Saturn is harder to distinguish from the stars. Stars twinkle and shimmer, while the planets, including Saturn, show a steady flat light.

The brightest star in the sky, **Sirius** is visible most of the night and is at its highest in the south, around 40 degrees, at 10:15 p.m. It sets in the west around 3:20 a.m.

The bright star **Arcturus** rises in the northeast around 10:55 p.m. See page 13 for an easy way to locate the star.

Castor and Pollux, the twin stars in the constellation Gemini, are visible most of the night and are almost directly overhead at 11:15 p.m. Read how to find Castor and Pollux on page 14.

Cassiopeia, shaped like a Σ, sets in the northwest around 1:10 a.m. See the illustration on page 11.

One of the brightest stars in the sky, **Vega** rises in the northeast around 2 a.m. and is in the eastern sky before dawn.

Bright **Venus** rises in the southeast around 3:45 a.m.

Scorpius rises in the southeast around 3:50 a.m. and is in the eastern sky before dawn. See the illustration on page 19.

The bright star **Altair** rises in the east around 5:05 a.m. and is in the eastern sky before dawn. Altair can be identified by its less bright neighboring stars, one on either side, together called the shaft of Aquila. See page 18.

2027

Monday, January 25

TONIGHT

sunset: **5:08 p.m.**
twilight begins: **6:06 p.m.**

IN THE MORNING

twilight begins: **5:50 a.m.**
sunrise: **6:46 a.m.**

MOON PHASE

moonrise: **9:11 p.m.**
moonset: **8:47 a.m.**

IN THE SKY

The **Great Bear (Big Dipper)** rises in the northeast around 6:30 p.m. It is in the northeastern sky at dusk and is at its highest, around 65 degrees, at 2:35 a.m. See the illustration on page 10.

Jupiter is visible most of the night. It rises in the east around 6:40 p.m. and is at its highest, around 70 degrees, at 1:15 a.m.

Rust-red and bright, **Mars** rises in the east around 7:45 p.m. and is at its highest, around 70 degrees, at 2:15 a.m.

Bright star **Capella** is almost directly overhead at 8:40 p.m. Read about Capella on page 14.

Orion is at its highest, around 65 degrees, at 9:20 p.m. It sets in the west around 2:25 a.m. Look for the bright star **Aldebaran** and the **Pleiades** near Orion. See the illustration on page 17.

Saturn sets in the west around 10 p.m. Unlike bright Jupiter and Venus and ruddy Mars, Saturn is harder to distinguish from the stars. Stars twinkle and shimmer, while the planets, including Saturn, show a steady flat light.

The brightest star in the sky, **Sirius** is visible most of the night and is at its highest in the south, around 40 degrees, at 10:10 p.m. It sets in the west around 3:20 a.m.

The bright star **Arcturus** rises in the northeast around 10:55 p.m. See page 13 for an easy way to locate the star.

Castor and Pollux, the twin stars in the constellation Gemini, are visible most of the night and are almost directly overhead at 11:10 p.m. Read how to find Castor and Pollux on page 14.

Cassiopeia, shaped like a Σ, sets in the northwest around 1:05 a.m. See the illustration on page 11.

One of the brightest stars in the sky, **Vega** rises in the northeast around 2 a.m. and is in the eastern sky before dawn.

Bright **Venus** rises in the southeast around 3:50 a.m.

Scorpius rises in the southeast around 3:50 a.m. and is in the eastern sky before dawn. See the illustration on page 19.

The bright star **Altair** rises in the east around 5 a.m. and is in the eastern sky before dawn. Altair can be identified by its less bright neighboring stars, one on either side, together called the shaft of Aquila. See page 18.

2027
Tuesday, January 26

TONIGHT

sunset: **5:09 p.m.**
twilight begins: **6:07 p.m.**

IN THE MORNING

twilight begins: **5:49 a.m.**
sunrise: **6:45 a.m.**

MOON PHASE

moonrise: **10:16 p.m.**
moonset: **9:16 a.m.**

IN THE SKY

Jupiter is visible most of the night. It is in the eastern sky at dusk and is at its highest, around 70 degrees, at 1:10 a.m.

Before the moon rises, until around 9:55 p.m., use binoculars or a telescope to look for the **moons of Jupiter.** See the guide on page 18.

The **Great Bear (Big Dipper)** rises in the northeast around 6:30 p.m. It is in the northeastern sky at dusk and is at its highest, around 65 degrees, at 2:35 a.m. See the illustration on page 10.

Rust-red and bright, Mars rises in the east around 7:40 p.m. and is at its highest, around 70 degrees, at 2:10 a.m.

Bright star **Capella** is almost directly overhead at 8:40 p.m. Read about Capella on page 14.

Orion is at its highest, around 65 degrees, at 9:15 p.m. It sets in the west around 2:20 a.m. Look for the bright star **Aldebaran** and the **Pleiades** near Orion. See the illustration on page 17.

Explore the **dark sky** after twilight and before the moon rises, around 9:55 p.m.

Before the moon rises, until around 9:55 p.m., look for the **Andromeda Galaxy** with binoculars or a telescope. It is a good target until around 12:50 a.m. See the guide on page 11.

Saturn sets in the west around 10 p.m. Unlike bright Jupiter and Venus and ruddy Mars, Saturn is harder to distinguish from the stars. Stars twinkle and shimmer, while the planets, including Saturn, show a steady flat light.

The brightest star in the sky, **Sirius** is at its highest in the south, around 40 degrees, at 10:05 p.m. It sets in the west around 3:15 a.m.

The bright star **Arcturus** rises in the northeast around 10:50 p.m. See page 13 for an easy way to locate the star.

Castor and Pollux, the twin stars in the constellation Gemini, are visible most of the night and are almost directly overhead at 11:05 p.m. Read how to find Castor and Pollux on page 14.

Cassiopeia, shaped like a *Σ*, sets in the northwest around 1 a.m. See the illustration on page 11.

One of the brightest stars in the sky, **Vega** rises in the northeast around 1:55 a.m. and is in the eastern sky before dawn.

Scorpius rises in the southeast around 3:45 a.m. and is in the eastern sky before dawn. See the illustration on page 19.

Bright Venus rises in the southeast around 3:50 a.m.

The bright star **Altair** rises in the east around 4:55 a.m. and is in the eastern sky before dawn. Altair can be identified by its less bright neighboring stars, one on either side, together called the shaft of Aquila. See page 18.

2027

Wednesday, January 27

TONIGHT

sunset: **5:10 p.m.**
twilight begins: **6:08 p.m.**

IN THE MORNING

twilight begins: **5:49 a.m.**
sunrise: **6:45 a.m.**

MOON PHASE

moonrise: **11:20 p.m.**
moonset: **9:44 a.m.**

IN THE SKY

Jupiter is visible most of the night. It is in the eastern sky at dusk and is at its highest, around 70 degrees, at 1:05 a.m.

Before the moon rises, until around 11 p.m., use binoculars or a telescope to look for the **moons of Jupiter.** See the guide on page 18.

Rust-red and bright, Mars rises in the east around 7:35 p.m. and is at its highest, around 70 degrees, at 2:05 a.m.

Bright star **Capella** is almost directly overhead at 8:35 p.m. Read about Capella on page 14.

Orion is at its highest, around 65 degrees, at 9:10 p.m. It sets in the west around 2:20 a.m. Look for the bright star **Aldebaran** and the **Pleiades** near Orion. See the illustration on page 17.

Saturn sets in the west around 9:55 p.m. Unlike bright Jupiter and Venus and ruddy Mars, Saturn is harder to distinguish from the stars. Stars twinkle and shimmer, while the planets, including Saturn, show a steady flat light.

The brightest star in the sky, **Sirius** is at its highest in the south, around 40 degrees, at 10 p.m. It sets in the west around 3:10 a.m.

The bright star **Arcturus** rises in the northeast around 10:45 p.m. See page 13 for an easy way to locate the star.

Explore the **dark sky** after twilight and before the moon rises, around 11 p.m.

Castor and Pollux, the twin stars in the constellation Gemini, are visible most of the night and are almost directly overhead at 11 p.m. Read how to find Castor and Pollux on page 14.

Before the moon rises, until around 11 p.m., look for the **Andromeda Galaxy** with binoculars or a telescope. It is a good target until around 12:50 a.m. See the guide on page 11.

Cassiopeia, shaped like a *Σ,* sets in the northwest around 12:55 a.m. See the illustration on page 11.

One of the brightest stars in the sky, **Vega** rises in the northeast around 1:50 a.m. and is in the eastern sky before dawn.

The **Great Bear (Big Dipper)** is in the northeastern sky at dusk and is at its highest, around 65 degrees, at 2:30 a.m. See the illustration on page 10.

Scorpius rises in the southeast around 3:40 a.m. and is in the eastern sky before dawn. See the illustration on page 19.

Bright Venus rises in the southeast around 3:50 a.m.

The bright star **Altair** rises in the east around 4:55 a.m. and is in the eastern sky before dawn. Altair can be identified by its less bright neighboring stars, one on either side, together called the shaft of Aquila. See page 18.

2027
Thursday, January 28

TONIGHT

sunset: **5:11 p.m.**
twilight begins: **6:09 p.m.**

MOON PHASE

IN THE MORNING

twilight begins: **5:48 a.m.**
sunrise: **6:44 a.m.**

moonrise: **12:22 a.m.**
moonset: **10:14 a.m.**

IN THE SKY

Jupiter is visible most of the night. It is in the eastern sky at dusk and is at its highest, around 70 degrees, at 1 a.m.

Before the moon rises, until around midnight, use binoculars or a telescope to look for the **moons of Jupiter.** See the guide on page 18.

Rust-red and bright, Mars rises in the east around 7:30 p.m. and is at its highest, around 70 degrees, at 2 a.m.

Bright star **Capella** is almost directly overhead at 8:30 p.m. Read about Capella on page 14.

Orion is at its highest, around 65 degrees, at 9:10 p.m. It sets in the west around 2:15 a.m. Look for the bright star **Aldebaran** and the **Pleiades** near Orion. See the illustration on page 17.

Saturn sets in the west around 9:50 p.m. Unlike bright Jupiter and Venus and ruddy Mars, Saturn is harder to distinguish from the stars. Stars twinkle and shimmer, while the planets, including Saturn, show a steady flat light.

The brightest star in the sky, **Sirius** is at its highest in the south, around 40 degrees, at 10 p.m. It sets in the west around 3:05 a.m.

The bright star **Arcturus** rises in the northeast around 10:40 p.m. See page 13 for an easy way to locate the star.

Castor and Pollux, the twin stars in the constellation Gemini, are visible most of the night and are almost directly overhead at 11 p.m. Read how to find Castor and Pollux on page 14.

Explore the **dark sky** after twilight and before the moon rises, around midnight.

Cassiopeia, shaped like a Σ, sets in the northwest around 12:55 a.m. See the illustration on page 11.

One of the brightest stars in the sky, **Vega** rises in the northeast around 1:45 a.m. and is in the eastern sky before dawn.

The **Great Bear (Big Dipper)** is in the northeastern sky at dusk and is at its highest, around 65 degrees, at 2:25 a.m. See the illustration on page 10.

Scorpius rises in the southeast around 3:35 a.m. and is in the eastern sky before dawn. See the illustration on page 19.

Bright Venus rises in the southeast around 3:50 a.m.

The bright star **Altair** rises in the east around 4:50 a.m. and is in the eastern sky before dawn. Altair can be identified by its less bright neighboring stars, one on either side, together called the shaft of Aquila. See page 18.

2027

Friday, January 29

TONIGHT

sunset: **5:12 p.m.**
twilight begins: **6:09 p.m.**

MOON PHASE

IN THE MORNING

twilight begins: **5:48 a.m.**
sunrise: **6:43 a.m.**

moonrise: **1:24 a.m.**
moonset: **10:47 a.m.**

IN THE SKY

Jupiter is visible most of the night. It is in the eastern sky at dusk and is at its highest, around 70 degrees, at 12:55 a.m.

Use binoculars or a telescope to look for the **moons of Jupiter.** See the guide on page 18.

The **Great Bear (Big Dipper)** is in the northeastern sky at dusk and is at its highest, around 65 degrees, at 2:20 a.m. See the illustration on page 10.

Rust-red and bright, **Mars** rises in the east around 7:25 p.m. and is at its highest, around 70 degrees, at 1:55 a.m.

Bright star **Capella** is almost directly overhead at 8:25 p.m. Read about Capella on page 14.

Orion is at its highest, around 65 degrees, at 9:05 p.m. It sets in the west around 2:10 a.m. Look for the bright star **Aldebaran** and the **Pleiades** near Orion. See the illustration on page 17.

Saturn sets in the west around 9:45 p.m. Unlike bright Jupiter and Venus and ruddy Mars, Saturn is harder to distinguish from the stars. Stars twinkle and shimmer, while the planets, including Saturn, show a steady flat light.

The brightest star in the sky, **Sirius** is at its highest in the south, around 40 degrees, at 9:55 p.m. It sets in the west around 3:05 a.m.

The bright star **Arcturus** rises in the northeast around 10:35 p.m. and is at its highest, around 75 degrees, around dawn. See page 13 for an easy way to locate the star.

Castor and Pollux, the twin stars in the constellation Gemini, are visible most of the night and are almost directly overhead at 10:55 p.m. and set in the northwest around 5:25 a.m. Read how to find Castor and Pollux on page 14.

Cassiopeia, shaped like a Σ, sets in the northwest around 12:50 a.m. See the illustration on page 11.

One of the brightest stars in the sky, **Vega** rises in the northeast around 1:40 a.m. and is in the eastern sky before dawn.

Scorpius rises in the southeast around 3:35 a.m. and is in the eastern sky before dawn. See the illustration on page 19.

Bright **Venus** rises in the southeast around 3:55 a.m.

The bright star **Altair** rises in the east around 4:45 a.m. and is in the eastern sky before dawn. Altair can be identified by its less bright neighboring stars, one on either side, together called the shaft of Aquila. See page 18.

2027
Saturday, January 30

TONIGHT

sunset: **5:13 p.m.**
twilight begins: **6:10 p.m.**

IN THE MORNING

twilight begins: **5:47 a.m.**
sunrise: **6:43 a.m.**

MOON PHASE

moonrise: **2:24 a.m.**
moonset: **11:23 a.m.**

IN THE SKY

Jupiter is visible most of the night. It is in the eastern sky at dusk and is at its highest, around 70 degrees, at 12:55 a.m.

Use binoculars or a telescope to look for the **moons of Jupiter.** See the guide on page 18.

The **Great Bear (Big Dipper)** is in the northeastern sky at dusk and is at its highest, around 65 degrees, at 2:15 a.m. See the illustration on page 10.

Rust-red and bright, **Mars** rises in the east around 7:20 p.m. and is at its highest, around 70 degrees, at 1:50 a.m.

Bright star **Capella** is almost directly overhead at 8:20 p.m. Read about Capella on page 14.

Orion is at its highest, around 65 degrees, at 9 p.m. It sets in the west around 2:05 a.m. Look for the bright star **Aldebaran** and the **Pleiades** near Orion. See the illustration on page 17.

Saturn sets in the west around 9:45 p.m. Unlike bright Jupiter and Venus and ruddy Mars, Saturn is harder to distinguish from the stars. Stars twinkle and shimmer, while the planets, including Saturn, show a steady flat light.

The brightest star in the sky, **Sirius** is at its highest in the south, around 40 degrees, at 9:50 p.m. It sets in the west around 3 a.m.

The bright star **Arcturus** rises in the northeast around 10:35 p.m. and is at its highest, around 75 degrees, around dawn. See page 13 for an easy way to locate the star.

Castor and Pollux, the twin stars in the constellation Gemini, are visible most of the night and are almost directly overhead at 10:50 p.m. They set in the northwest around 5:20 a.m. Read how to find Castor and Pollux on page 14.

Look for the **Andromeda Galaxy** with binoculars or a telescope. It is a good target until around 12:35 a.m. See the guide on page 11.

Cassiopeia, shaped like a *Σ,* sets in the northwest around 12:45 a.m. See the illustration on page 11.

One of the brightest stars in the sky, **Vega** rises in the northeast around 1:40 a.m. and is in the eastern sky before dawn.

Scorpius rises in the southeast around 3:30 a.m. and is in the eastern sky before dawn. See the illustration on page 19.

Bright **Venus** rises in the southeast around 3:55 a.m.

The bright star **Altair** rises in the east around 4:40 a.m. and is in the eastern sky before dawn. Altair can be identified by its less bright neighboring stars, one on either side, together called the shaft of Aquila. See page 18.

2027

Sunday, January 31

TONIGHT

sunset: **5:14 p.m.**
twilight begins: **6:11 p.m.**

IN THE MORNING

twilight begins: **5:46 a.m.**
sunrise: **6:42 a.m.**

MOON PHASE

moonrise: **3:20 a.m.**
moonset: **12:05 p.m.**

IN THE SKY

Jupiter is visible most of the night. It is in the eastern sky at dusk and is at its highest, around 70 degrees, at 12:50 a.m.

With the dark sky, use binoculars or a telescope to look for the **moons of Jupiter.** See the guide on page 18.

Rust-red and bright, Mars rises in the east around 7:15 p.m. and is at its highest, around 70 degrees, at 1:45 a.m.

Bright star **Capella** is almost directly overhead at 8:20 p.m. Read about Capella on page 14.

Orion is at its highest, around 65 degrees, at 8:55 p.m. It sets in the west around 2 a.m. Look for the bright star **Aldebaran** and the **Pleiades** near Orion. See the illustration on page 17.

Saturn sets in the west around 9:40 p.m. Unlike bright Jupiter and Venus and ruddy Mars, Saturn is harder to distinguish from the stars. Stars twinkle and shimmer, while the planets, including Saturn, show a steady flat light.

The brightest star in the sky, **Sirius** is at its highest in the south, around 40 degrees, at 9:45 p.m. It sets in the west around 2:55 a.m.

Castor and Pollux, the twin stars in the constellation Gemini, are almost directly overhead at 10:45 p.m. and set in the northwest around 5:15 a.m. Read how to find Castor and Pollux on page 14.

Look for the **Andromeda Galaxy** with binoculars or a telescope. It is a good target until around 12:30 a.m. See the guide on page 11.

Cassiopeia, shaped like a *Σ*, sets in the northwest around 12:40 a.m. See the illustration on page 11.

The **Great Bear (Big Dipper)** is in the northeastern sky at dusk and is at its highest, around 65 degrees, at 2:15 a.m. See the illustration on page 10.

Scorpius rises in the southeast around 3:25 a.m. and is in the eastern sky before dawn. See the illustration on page 19.

Bright Venus rises in the southeast around 3:55 a.m.

The waning **crescent moon,** bright reddish star **Antares,** and Venus appear close together in the morning sky. Look southeast near the horizon around 4:30 a.m.

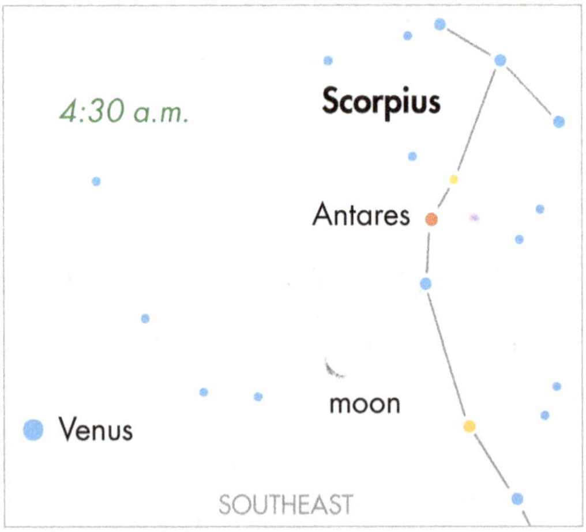

2027

Monday, February 1

TONIGHT

sunset: **5:15 p.m.**
twilight begins: **6:12 p.m.**

IN THE MORNING

twilight begins: **5:46 a.m.**
sunrise: **6:41 a.m.**

MOON PHASE

moonrise: **4:12 a.m.**
moonset: **12:53 p.m.**

IN THE SKY

Jupiter is visible most of the night. It is in the eastern sky at dusk and is at its highest, around 70 degrees, at 12:45 a.m.

With the dark sky, use binoculars or a telescope to look for the **moons of Jupiter.** See the guide on page 18.

The absence of the moon tonight provides excellent **dark-sky** viewing conditions.

The **Great Bear (Big Dipper)** is in the northeastern sky at dusk and is at its highest, around 65 degrees, at 2:10 a.m. See the illustration on page 10.

Rust-red and bright, Mars rises in the east around 7:10 p.m. and is at its highest, around 70 degrees, at 1:45 a.m.

Bright star **Capella** is almost directly overhead at 8:15 p.m. and is in the northwestern sky before dawn. It sets in the northwest around 5:10 a.m. Read about Capella on page 14.

Orion is at its highest, around 65 degrees, at 8:50 p.m. It sets in the west around 2 a.m. Look for the bright star **Aldebaran** and the **Pleiades** near Orion. See the illustration on page 17.

Saturn sets in the west around 9:35 p.m. Unlike bright Jupiter and Venus and ruddy Mars, Saturn is harder to distinguish from the stars. Stars twinkle and shimmer, while the planets, including Saturn, show a steady flat light.

The brightest star in the sky, **Sirius** is at its highest in the south, around 40 degrees, at 9:40 p.m. It sets in the west around 2:50 a.m.

The bright star **Arcturus** rises in the northeast around 10:25 p.m. and is at its highest, around 75 degrees, around dawn. See page 13 for an easy way to locate the star.

Castor and Pollux, the twin stars in the constellation Gemini, are visible most of the night and are almost directly overhead at 10:40 p.m. and set in the northwest around 5:15 a.m. Read how to find Castor and Pollux on page 14.

Look for the **Andromeda Galaxy** with binoculars or a telescope. It is a good target until around 12:30 a.m. See the guide on page 11.

Cassiopeia, shaped like a Σ, sets in the northwest around 12:40 a.m. See the illustration on page 11.

One of the brightest stars in the sky, **Vega** rises in the northeast around 1:30 a.m. and is in the eastern sky before dawn.

Scorpius rises in the southeast around 3:20 a.m. and is in the eastern sky before dawn. See the illustration on page 19.

Bright Venus rises in the southeast around 3:55 a.m. and is in the eastern sky before dawn.

The bright star **Altair** rises in the east around 4:35 a.m. and is in the eastern sky before dawn. Altair can be identified by its less bright neighboring stars, one on either side, together called the shaft of Aquila. See page 18.

2027

Tuesday, February 2

TONIGHT

sunset: **5:16 p.m.**
twilight begins: **6:13 p.m.**

MOON PHASE

IN THE MORNING

twilight begins: **5:45 a.m.**
sunrise: **6:41 a.m.**

moonrise: **4:58 a.m.**
moonset: **1:45 p.m.**

IN THE SKY

Look for **Mercury** in the evening sky, just after sunset in the west. It's difficult but possible to see the tiny dot in the glare of the sun. It's easier to find with the naked eye, not using binoculars or a telescope. Never look directly at the sun or point binoculars or a telescope at it.

Jupiter is visible most of the night. It is in the eastern sky at dusk and is at its highest, around 70 degrees, at 12:40 a.m.

With the dark sky, use binoculars or a telescope to look for the **moons of Jupiter.** See the guide on page 18.

The **Great Bear (Big Dipper)** is in the northeastern sky at dusk and is at its highest, around 65 degrees, at 2:05 a.m. See the illustration on page 10.

The absence of the moon tonight provides excellent **dark-sky** viewing conditions.

Rust-red and bright, **Mars** rises in the east around 7:05 p.m. and is at its highest, around 70 degrees, at 1:40 a.m.

Bright star **Capella** is almost directly overhead at 8:10 p.m. and is in the northwestern sky before dawn. It sets in the northwest around 5:10 a.m. Read about Capella on page 14.

Orion is at its highest, around 65 degrees, at 8:50 p.m. It sets in the west around 1:55 a.m. Look for the bright star **Aldebaran** and the **Pleiades** near Orion. See the illustration on page 17.

Saturn sets in the west around 9:35 p.m. Unlike bright Jupiter and Venus and ruddy Mars, Saturn is harder to distinguish from the stars. Stars twinkle and shimmer, while the planets, including Saturn, show a steady flat light.

The brightest star in the sky, **Sirius** is at its highest in the south, around 40 degrees, at 9:40 p.m. It sets in the west around 2:45 a.m.

The bright star **Arcturus** rises in the northeast around 10:20 p.m. and is at its highest, around 75 degrees, around dawn. See page 13 for an easy way to locate the star.

Castor and Pollux, the twin stars in the constellation Gemini, are visible most of the night and are almost directly overhead at 10:40 p.m. and set in the northwest around 5:10 a.m. Read how to find Castor and Pollux on page 14.

With the dark sky, look for the **Andromeda Galaxy** with binoculars or a telescope. It is a good target until around 12:25 a.m. See the guide on page 11.

Cassiopeia, shaped like a *Σ,* sets in the northwest around 12:35 a.m. See the illustration on page 11.

Scorpius rises in the southeast around 3:15 a.m. and is in the eastern sky before dawn. See the illustration on page 19.

Bright **Venus** rises in the southeast around 4 a.m. and is in the eastern sky before dawn.

2027

Wednesday, February 3

TONIGHT

sunset: **5:17 p.m.**
twilight begins: **6:14 p.m.**

IN THE MORNING

twilight begins: **5:44 a.m.**
sunrise: **6:40 a.m.**

MOON PHASE

moonrise: **5:38 a.m.**
moonset: **2:42 p.m.**

IN THE SKY

Look for **Mercury** in the evening sky, just after sunset in the west. It's difficult but possible to see the tiny dot in the glare of the sun. It's easier to find with the naked eye, not using binoculars or a telescope. Never look directly at the sun or point binoculars or a telescope at it.

Jupiter is visible most of the night. It is in the eastern sky at dusk and is at its highest, around 70 degrees, at 12:35 a.m.

With the dark sky, use binoculars or a telescope to look for the **moons of Jupiter.** See the guide on page 18.

The **Great Bear (Big Dipper)** is in the northeastern sky at dusk and is at its highest, around 65 degrees, at 2 a.m. See the illustration on page 10.

The absence of the moon tonight provides excellent **dark-sky** viewing conditions.

Rust-red and bright, **Mars** rises in the east around 7 p.m. and is at its highest, around 70 degrees, at 1:35 a.m.

Bright star **Capella** is almost directly overhead at 8:05 p.m. and is in the northwestern sky before dawn. It sets in the northwest around 5:05 a.m. Read about Capella on page 14.

Orion is at its highest, around 65 degrees, at 8:45 p.m. It sets in the west around 1:50 a.m. Look for the bright star **Aldebaran** and the **Pleiades** near Orion. See the illustration on page 17.

Saturn sets in the west around 9:30 p.m. Unlike bright Jupiter and Venus and ruddy Mars, Saturn is harder to distinguish from the stars. Stars twinkle and shimmer, while the planets, including Saturn, show a steady flat light.

The brightest star in the sky, **Sirius** is at its highest in the south, around 40 degrees, at 9:35 p.m. It sets in the west around 2:45 a.m.

The bright star **Arcturus** rises in the northeast around 10:15 p.m. and is at its highest, around 75 degrees, around dawn. See page 13 for an easy way to locate the star.

Castor and Pollux, the twin stars in the constellation Gemini, are almost directly overhead at 10:35 p.m. and set in the northwest around 5:05 a.m. Read how to find Castor and Pollux on page 14.

With the dark sky, look for the **Andromeda Galaxy** with binoculars or a telescope. It is a good target until around 12:20 a.m. See the guide on page 11.

Cassiopeia, shaped like a Σ, sets in the northwest around 12:30 a.m. See the illustration on page 11.

One of the brightest stars in the sky, **Vega** rises in the northeast around 1:20 a.m. and is in the eastern sky before dawn.

Bright **Venus** rises in the southeast around 4 a.m. and is in the eastern sky before dawn.

2027

Thursday, February 4

TONIGHT

sunset: **5:18 p.m.**
twilight begins: **6:15 p.m.**

IN THE MORNING

twilight begins: **5:44 a.m.**
sunrise: **6:39 a.m.**

MOON PHASE

moonrise: **6:13 a.m.**
moonset: **3:40 p.m.**

IN THE SKY

Look for **Mercury** in the evening sky, just after sunset in the west. It's difficult but possible to see the tiny dot in the glare of the sun. It's easier to find with the naked eye, not using binoculars or a telescope. Never look directly at the sun or point binoculars or a telescope at it.

Jupiter is visible most of the night. It is in the eastern sky at dusk and is at its highest, around 70 degrees, at 12:30 a.m.

With the dark sky, use binoculars or a telescope to look for the **moons of Jupiter.** See the guide on page 18.

The **Great Bear (Big Dipper)** is in the northeastern sky at dusk and is at its highest, around 65 degrees, at 1:55 a.m. See the illustration on page 10.

The absence of the moon tonight provides excellent **dark-sky** viewing conditions.

Rust-red and bright, **Mars** is visible most of the night. It rises in the east around 6:55 p.m. and is at its highest, around 70 degrees, at 1:25 a.m.

Bright star **Capella** is almost directly overhead at 8:05 p.m. and is in the northwestern sky before dawn. It sets in the northwest around 5 a.m. Read about Capella on page 14.

Orion is at its highest, around 65 degrees, at 8:40 p.m. It sets in the west around 1:45 a.m. Look for the bright star **Aldebaran** and the **Pleiades** near Orion. See the illustration on page 17.

Saturn sets in the west around 9:25 p.m. Unlike bright Jupiter and Venus and ruddy Mars, Saturn is harder to distinguish from the stars. Stars twinkle and shimmer, while the planets, including Saturn, show a steady flat light.

The brightest star in the sky, **Sirius** is at its highest in the south, around 40 degrees, at 9:30 p.m. It sets in the west around 2:40 a.m.

The bright star **Arcturus** rises in the northeast around 10:15 p.m. and is at its highest, around 75 degrees, around dawn. See page 13 for an easy way to locate the star.

Castor and Pollux, the twin stars in the constellation Gemini, are almost directly overhead at 10:30 p.m. and set in the northwest around 5 a.m. Read how to find Castor and Pollux on page 14.

With the dark sky, look for the **Andromeda Galaxy** with binoculars or a telescope. It is a good target until around 12:15 a.m. See the guide on page 11.

Cassiopeia, shaped like a Σ, sets in the northwest around 12:25 a.m. See the illustration on page 11.

One of the brightest stars in the sky, **Vega** rises in the northeast around 1:20 a.m. and is in the eastern sky before dawn.

Bright **Venus** rises in the southeast around 4 a.m. and is in the eastern sky before dawn.

2027

Friday, February 5

TONIGHT

sunset: **5:19 p.m.**
twilight begins: **6:16 p.m.**

IN THE MORNING

twilight begins: **5:43 a.m.**
sunrise: **6:38 a.m.**

MOON PHASE

moonrise: **6:13 a.m.**
moonset: **4:39 p.m.**

IN THE SKY

At its closest and brightest, **Jupiter** is visible most of the night. It is in the eastern sky at dusk and is at its highest, around 70 degrees, at 12:25 a.m.

With the dark sky, use binoculars or a telescope to look for the **moons of Jupiter.** See the guide on page 18.

The **Great Bear (Big Dipper)** is in the northeastern sky at dusk and is at its highest, around 65 degrees, at 1:55 a.m. See the illustration on page 10.

The absence of the moon tonight provides excellent **dark-sky** viewing conditions.

Rust-red and bright, **Mars** is visible most of the night. It rises in the east around 6:50 p.m. and is at its highest, around 70 degrees, at 1:20 a.m.

Bright star **Capella** is almost directly overhead at 8 p.m. and is in the northwestern sky before dawn. It sets in the northwest around 4:55 a.m. Read about Capella on page 14.

Orion is at its highest, around 65 degrees, at 8:35 p.m. It sets in the west around 1:45 a.m. Look for the bright star **Aldebaran** and the **Pleiades** near Orion. See the illustration on page 17.

Saturn sets in the west around 9:25 p.m. Unlike bright Jupiter and Venus and ruddy Mars, Saturn is harder to distinguish from the stars. Stars twinkle and shimmer, while the planets, including Saturn, show a steady flat light.

The brightest star in the sky, **Sirius** is at its highest in the south, around 40 degrees, at 9:25 p.m. It sets in the west around 2:35 a.m.

The bright star **Arcturus** rises in the northeast around 10:10 p.m. and is at its highest, around 75 degrees, around dawn. See page 13 for an easy way to locate the star.

Castor and Pollux, the twin stars in the constellation Gemini, are almost directly overhead at 10:25 p.m. and set in the northwest around 4:55 a.m. Read how to find Castor and Pollux on page 14.

With the dark sky, look for the **Andromeda Galaxy** with binoculars or a telescope. It is a good target until around 12:15 a.m. See the guide on page 11.

Cassiopeia, shaped like a *Σ*, sets in the northwest around 12:20 a.m. See the illustration on page 11.

One of the brightest stars in the sky, **Vega** rises in the northeast around 1:15 a.m. and is in the eastern sky before dawn.

Scorpius rises in the southeast around 3:05 a.m. and is in the eastern sky before dawn. See the illustration on page 19.

Bright **Venus** rises in the southeast around 4 a.m. and is in the eastern sky before dawn.

2027

Saturday, February 6

TONIGHT

sunset: **5:20 p.m.**
twilight begins: **6:17 p.m.**

IN THE MORNING

twilight begins: **5:42 a.m.**
sunrise: **6:37 a.m.**

MOON PHASE

moonrise: **6:43 a.m.**
moonset: **5:38 p.m.**

IN THE SKY

Rust-red and bright, **Mars** is visible most of the night. It is in the eastern sky at dusk and is at its highest, around 70 degrees, at 1:15 a.m.

At its closest and brightest, **Jupiter** is visible most of the night and is at its highest, around 70 degrees, at 12:20 a.m.

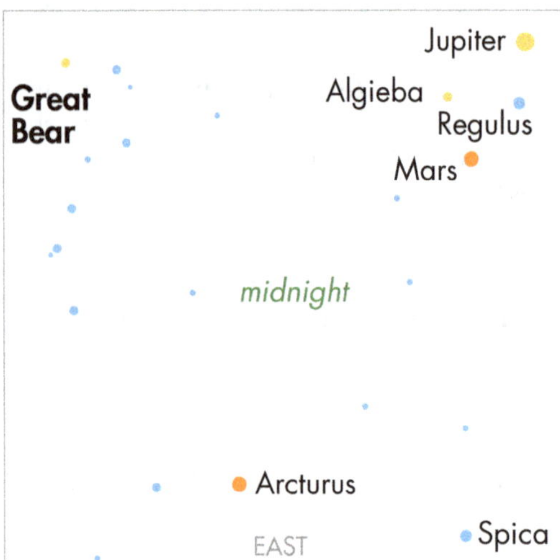

After the moon sets, from around 6 p.m., use binoculars or a telescope to look for the **moons of Jupiter.** See the guide on page 18.

The absence of the moon tonight provides excellent **dark-sky** viewing conditions.

The **Great Bear (Big Dipper)** is at its highest, around 65 degrees, at 1:50 a.m. See the illustration on page 10.

After the moon sets, from around 6 p.m., look for the **Andromeda Galaxy** with binoculars or a telescope. It is a good target until around 12:10 a.m. See the guide on page 11.

Orion is at its highest, around 65 degrees, at 8:35 p.m. It sets in the west around 1:40 a.m. Look for the bright star **Aldebaran** and the **Pleiades** near Orion. See the illustration on page 17.

Saturn sets in the west around 9:20 p.m.

The brightest star in the sky, **Sirius** is at its highest in the south, around 40 degrees, at 9:20 p.m. It sets in the west around 2:30 a.m.

The bright star **Arcturus** rises in the northeast around 10:05 p.m. and is at its highest, around 75 degrees, around dawn. See page 13 for an easy way to locate the star.

Castor and Pollux, the twin stars in the constellation Gemini, are almost directly overhead at 10:25 p.m. and set in the northwest around 4:55 a.m. Read how to find Castor and Pollux on page 14.

Cassiopeia, shaped like a Σ, sets in the northwest around 12:20 a.m. See the illustration on page 11.

One of the brightest stars in the sky, **Vega** rises in the northeast around 1:10 a.m. and is in the eastern sky before dawn.

Scorpius rises in the southeast around 3 a.m. and is in the eastern sky before dawn. See the illustration on page 19.

Bright **Venus** rises in the southeast around 4 a.m. and is in the eastern sky before dawn.

2027
Sunday, February 7

TONIGHT

sunset: **5:21 p.m.**
twilight begins: **6:17 p.m.**

IN THE MORNING

twilight begins: **5:41 a.m.**
sunrise: **6:36 a.m.**

MOON PHASE

moonrise: **7:11 a.m.**
moonset: **6:36 p.m.**

IN THE SKY

Rust-red and bright, **Mars** is visible most of the night. It is in the eastern sky at dusk and is at its highest, around 70 degrees, at 1:10 a.m.

At its closest and brightest, **Jupiter** is visible most of the night and is at its highest, around 70 degrees, at 12:15 a.m.

After the moon sets, from around 6:55 p.m., use binoculars or a telescope to look for the **moons of Jupiter.** See the guide on page 18.

The **Great Bear (Big Dipper)** is in the northeastern sky at dusk and is at its highest, around 65 degrees, at 1:45 a.m. See the illustration on page 10.

Bright star **Capella** is almost directly overhead at 7:50 p.m. and is in the northwestern sky before dawn. It sets in the northwest around 4:50 a.m. Read about Capella on page 14.

Orion is at its highest, around 65 degrees, at 8:30 p.m. It sets in the west around 1:35 a.m. Look for the bright star **Aldebaran** and the **Pleiades** near Orion. See the illustration on page 17.

Saturn sets in the west around 9:15 p.m. Unlike bright Jupiter and Venus and ruddy Mars, Saturn is harder to distinguish from the stars. Stars twinkle and shimmer, while the planets, including Saturn, show a steady flat light.

The absence of the moon tonight provides excellent **dark-sky** viewing conditions.

After the moon sets, from around 6:55 p.m., look for the **Andromeda Galaxy** with binoculars or a telescope. It is a good target until around 12:05 a.m. See the guide on page 11.

The brightest star in the sky, **Sirius** is at its highest in the south, around 40 degrees, at 9:20 p.m. It sets in the west around 2:25 a.m.

The bright star **Arcturus** rises in the northeast around 10 p.m. and is at its highest, around 75 degrees, around dawn. See page 13 for an easy way to locate the star.

Castor and Pollux, the twin stars in the constellation Gemini, are almost directly overhead at 10:20 p.m. and set in the northwest around 4:50 a.m. Read how to find Castor and Pollux on page 14.

Cassiopeia, shaped like a Σ, sets in the northwest around 12:15 a.m. See the illustration on page 11.

One of the brightest stars in the sky, **Vega** rises in the northeast around 1:05 a.m. and is in the eastern sky before dawn.

Scorpius rises in the southeast around 2:55 a.m. and is in the eastern sky before dawn. See the illustration on page 19.

Bright **Venus** rises in the southeast around 4:05 a.m. and is in the eastern sky before dawn.

2027

Monday, February 8

TONIGHT

sunset: **5:22 p.m.**
twilight begins: **6:18 p.m.**

IN THE MORNING

twilight begins: **5:41 a.m.**
sunrise: **6:35 a.m.**

MOON PHASE

moonrise: **7:36 a.m.**
moonset: **7:34 p.m.**

IN THE SKY

Rust-red and bright, Mars is visible most of the night. It is in the eastern sky at dusk and is at its highest, around 70 degrees, at 1:05 a.m.

Bright star **Capella** is almost directly overhead at 7:45 p.m. and is in the northwestern sky before dawn. It sets in the northwest around 4:45 a.m. Read about Capella on page 14.

At its closest and brightest, Jupiter is visible most of the night and is at its highest, around 70 degrees, at 12:15 a.m.

After the moon sets, from around 7:55 p.m., use binoculars or a telescope to look for the **moons of Jupiter.** See the guide on page 18.

The absence of the moon tonight provides excellent **dark-sky** viewing conditions.

After the moon sets, from around 7:55 p.m., look for the **Andromeda Galaxy** with binoculars or a telescope. It is a good target until around midnight. See the guide on page 11.

Saturn sets in the west around 9:10 p.m. Unlike bright Jupiter and Venus and ruddy Mars, Saturn is harder to distinguish from the stars. Stars twinkle and shimmer, while the planets, including Saturn, show a steady flat light.

Orion is at its highest, around 65 degrees, at 8:25 p.m. It sets in the west around 1:30 a.m. Look for the bright star **Aldebaran** and the **Pleiades** near Orion. See the illustration on page 17.

Castor and Pollux, the twin stars in the constellation Gemini, are almost directly overhead at 10:15 p.m. and set in the northwest around 4:45 a.m. Read how to find Castor and Pollux on page 14.

The brightest star in the sky, **Sirius** is at its highest in the south, around 40 degrees, at 9:15 p.m. It sets in the west around 2:25 a.m.

The bright star **Arcturus** rises in the northeast around 10 p.m. and is at its highest, around 75 degrees, around dawn. See page 13 for an easy way to locate the star.

One of the brightest stars in the sky, **Vega** rises in the northeast around 1:05 a.m. and is in the eastern sky before dawn.

Cassiopeia, shaped like a Σ, sets in the northwest around 12:10 a.m. See the illustration on page 11.

The **Great Bear (Big Dipper)** is at its highest, around 65 degrees, at 1:40 a.m. See the illustration on page 10.

Scorpius rises in the southeast around 2:55 a.m. and is in the eastern sky before dawn. See the illustration on page 19.

Bright Venus rises in the southeast around 4:05 a.m. and is in the eastern sky before dawn.

The bright star **Altair** rises in the east around 4:05 a.m. and is in the eastern sky before dawn. Altair can be identified by its less bright neighboring stars, one on either side, together called the shaft of Aquila. See page 18.

2027
Tuesday, February 9

TONIGHT

sunset: **5:23 p.m.**
twilight begins: **6:19 p.m.**

IN THE MORNING

twilight begins: **5:40 a.m.**
sunrise: **6:34 a.m.**

MOON PHASE

moonrise: **8:01 a.m.**
moonset: **8:32 p.m.**

IN THE SKY

Rust-red and bright, **Mars** is visible most of the night. It is in the eastern sky at dusk and is at its highest, around 70 degrees, at 1 a.m.

Bright star **Capella** is almost directly overhead at 7:45 p.m. and is in the northwestern sky before dawn. It sets in the northwest around 4:40 a.m. Read about Capella on page 14.

At its closest and brightest, **Jupiter** is visible most of the night and is at its highest, around 70 degrees, at 12:10 a.m.

After the moon sets, from around 8:50 p.m., use binoculars or a telescope to look for the **moons of Jupiter.** See the guide on page 18.

Saturn sets in the west around 9:10 p.m. Unlike bright Jupiter and Venus and ruddy Mars, Saturn is harder to distinguish from the stars. Stars twinkle and shimmer, while the planets, including Saturn, show a steady flat light.

Orion is at its highest, around 65 degrees, at 8:20 p.m. It sets in the west around 1:25 a.m. Look for the bright star **Aldebaran** and the **Pleiades** near Orion. See the illustration on page 17.

The absence of the moon tonight provides excellent **dark-sky** viewing conditions. Explore the **dark sky** after the moon sets, from around 8:50 p.m.

After the moon sets, from around 8:50 p.m., look for the **Andromeda Galaxy** with binoculars or a telescope. It is a good target until around 12:55 a.m. See the guide on page 11.

The brightest star in the sky, **Sirius** is at its highest in the south, around 40 degrees, at 9:10 p.m. It sets in the west around 2:20 a.m.

The bright star **Arcturus** rises in the northeast around 9:55 p.m. and is at its highest, around 75 degrees, around dawn. See page 13 for an easy way to locate the star.

Castor and Pollux, the twin stars in the constellation Gemini, are almost directly overhead at 10:10 p.m. and set in the northwest around 4:40 a.m. Read how to find Castor and Pollux on page 14.

Cassiopeia, shaped like a Σ, sets in the northwest around 12:05 a.m. See the illustration on page 11.

One of the brightest stars in the sky, **Vega** rises in the northeast around 1 a.m. and is in the eastern sky before dawn.

The **Great Bear (Big Dipper)** is at its highest, around 65 degrees, at 1:40 a.m. See the illustration on page 10.

Scorpius rises in the southeast around 2:50 a.m. and is in the eastern sky before dawn. See the illustration on page 19.

The bright star **Altair** rises in the east around 4 a.m. and is in the eastern sky before dawn. Altair can be identified by its less bright neighboring stars, one on either side, together called the shaft of Aquila. See page 18.

Bright **Venus** rises in the southeast around 4:05 a.m. and is in the eastern sky before dawn.

2027

Wednesday, February 10

TONIGHT

sunset: **5:24 p.m.**
twilight begins: **6:20 p.m.**

MOON PHASE

IN THE MORNING

twilight begins: **5:39 a.m.**
sunrise: **6:33 a.m.**

moonrise: **8:27 a.m.**
moonset: **9:32 p.m.**

IN THE SKY

Rust-red and bright, Mars is visible most of the night. It is in the eastern sky at dusk and is at its highest, around 70 degrees, at 12:55 a.m.

Bright star **Capella** is almost directly overhead at 7:40 p.m. and is in the northwestern sky before dawn. It sets in the northwest around 4:35 a.m. Read about Capella on page 14.

Orion is at its highest, around 65 degrees, at 8:15 p.m. It sets in the west around 1:25 a.m. Look for the bright star **Aldebaran** and the **Pleiades** near Orion. See the illustration on page 17.

Saturn sets in the west around 9:05 p.m. Unlike bright Jupiter and Venus and ruddy Mars, Saturn is harder to distinguish from the stars. Stars twinkle and shimmer, while the planets, including Saturn, show a steady flat light.

The brightest star in the sky, **Sirius** is at its highest in the south, around 40 degrees, at 9:05 p.m. It sets in the west around 2:15 a.m.

Explore the **dark sky** after the moon sets, from around 9:50 p.m.

At its closest and brightest, Jupiter is visible most of the night and is at its highest, around 70 degrees, at 12:05 a.m.

After the moon sets, from around 9:50 p.m., use binoculars or a telescope to look for the **moons of Jupiter.** See the guide on page 18.

After the moon sets, from around 9:50 p.m., look for the **Andromeda Galaxy** with binoculars or a telescope. It is a good target until around 12:55 a.m. See the guide on page 11.

The bright star **Arcturus** rises in the northeast around 9:50 p.m. and is at its highest, around 75 degrees, around dawn. See page 13 for an easy way to locate the star.

Castor and Pollux, the twin stars in the constellation Gemini, are almost directly overhead at 10:05 p.m. and set in the northwest around 4:40 a.m. Read how to find Castor and Pollux on page 14.

Cassiopeia, shaped like a Σ, sets in the northwest around midnight. See the illustration on page 11.

One of the brightest stars in the sky, **Vega** rises in the northeast around 12:55 a.m. and is in the eastern sky before dawn.

The **Great Bear (Big Dipper)** is at its highest, around 65 degrees, at 1:35 a.m. See the illustration on page 10.

Scorpius rises in the southeast around 2:45 a.m. and is in the eastern sky before dawn. See the illustration on page 19.

Bright Venus rises in the southeast around 4:05 a.m. and is in the eastern sky before dawn.

Thursday, February 11

TONIGHT

sunset: **5:25 p.m.**
twilight begins: **6:21 p.m.**

IN THE MORNING

twilight begins: **5:38 a.m.**
sunrise: **6:33 a.m.**

MOON PHASE

moonrise: **8:54 a.m.**
moonset: **10:35 p.m.**

IN THE SKY

The **moon** is prominent in the sky early in the evening. Using a telescope or binoculars, take a look. You should be able to see rough mountains and craters along the boundary between light and dark, in astronomy called the terminator.

Rust-red and bright, **Mars** is visible most of the night. It is in the eastern sky at dusk and is at its highest, around 70 degrees, at 12:50 a.m.

Bright star **Capella** is almost directly overhead at 7:35 p.m. and is in the northwestern sky before dawn. It sets in the northwest around 4:30 a.m. Read about Capella on page 14.

Orion is at its highest, around 65 degrees, at 8:15 p.m. It sets in the west around 1:20 a.m. Look for the bright star **Aldebaran** and the **Pleiades** near Orion. See the illustration on page 17.

Saturn sets in the west around 9 p.m. Unlike bright Jupiter and Venus and ruddy Mars, Saturn is harder to distinguish from the stars. Stars twinkle and shimmer, while the planets, including Saturn, show a steady flat light.

The brightest star in the sky, **Sirius** is at its highest in the south, around 40 degrees, at 9:05 p.m. It sets in the west around 2:10 a.m.

The bright star **Arcturus** rises in the northeast around 9:45 p.m. and is at its highest, around 75 degrees, at 4:35 a.m. See page 13 for an easy way to locate the star.

Castor and Pollux, the twin stars in the constellation Gemini, are almost directly overhead at 10:05 p.m. and set in the northwest around 4:35 a.m. Read how to find Castor and Pollux on page 14.

Explore the **dark sky** after the moon sets, from around 10:55 p.m.

After the moon sets, from around 10:55 p.m., look for the **Andromeda Galaxy** with binoculars or a telescope. It is a good target until around 12:50 a.m. See the guide on page 11.

At its closest and brightest, **Jupiter** is visible most of the night and is at its highest, around 70 degrees, at 11:55 p.m.

After the moon sets, from around 10:55 p.m., use binoculars or a telescope to look for the **moons of Jupiter.** See the guide on page 18.

The **Great Bear (Big Dipper)** is at its highest, around 65 degrees, at 1:30 a.m. See the illustration on page 10.

One of the brightest stars in the sky, **Vega** rises in the northeast around 12:50 a.m. and is in the eastern sky before dawn.

Cassiopeia, shaped like a Σ, sets in the northwest around 1 a.m. See the illustration on page 11.

Scorpius rises in the southeast around 2:40 a.m. and is in the eastern sky before dawn. See the illustration on page 19.

Bright **Venus** rises in the southeast around 4:10 a.m. and is in the eastern sky before dawn.

2027

Friday, February 12

TONIGHT

sunset: **5:26 p.m.**
twilight begins: **6:22 p.m.**

IN THE MORNING

twilight begins: **5:37 a.m.**
sunrise: **6:32 a.m.**

MOON PHASE

moonrise: **9:25 a.m.**
moonset: **11:40 p.m.**

IN THE SKY

Rust-red and bright, **Mars** is visible most of the night. It is in the eastern sky at dusk and is at its highest, around 70 degrees, at 12:45 a.m.

At its closest and brightest, **Jupiter** is visible most of the night and is at its highest, around 70 degrees, at 11:50 p.m.

After the moon sets, from around midnight, use binoculars or a telescope to look for the **moons of Jupiter.** See the guide on page 18.

The **moon** is prominent in the sky early in the evening. Using a telescope or binoculars, take a look. You should be able to see rough mountains and craters along the boundary between light and dark, in astronomy called the terminator.

Bright star **Capella** is almost directly overhead at 7:30 p.m. and is in the northwestern sky before dawn. It sets in the northwest around 4:30 a.m. Read about Capella on page 14.

Orion is at its highest, around 65 degrees, at 8:10 p.m. It sets in the west around 1:15 a.m. Look for the bright star **Aldebaran** and the **Pleiades** near Orion. See the illustration on page 17.

Saturn sets in the west around 9 p.m. Unlike bright Jupiter and Venus and ruddy Mars, Saturn is harder to distinguish from the stars. Stars twinkle and shimmer, while the planets, including Saturn, show a steady flat light.

The brightest star in the sky, **Sirius** is at its highest in the south, around 40 degrees, at 9 p.m. It sets in the west around 2:10 a.m.

The bright star **Arcturus** rises in the northeast around 9:40 p.m. and is at its highest, around 75 degrees, at 4:30 a.m. See page 13 for an easy way to locate the star.

Castor and Pollux, the twin stars in the constellation Gemini, are almost directly overhead at 10 p.m. and set in the northwest around 4:30 a.m. Read how to find Castor and Pollux on page 14.

Explore the **dark sky** after the moon sets, from around midnight.

After the moon sets, from around midnight, look for the **Andromeda Galaxy** with binoculars or a telescope. It is a good target until around 12:45 a.m. See the guide on page 11.

One of the brightest stars in the sky, **Vega** rises in the northeast around 12:45 a.m. and is in the eastern sky before dawn.

Cassiopeia, shaped like a *Σ*, sets in the northwest around 12:55 a.m. See the illustration on page 11.

The **Great Bear (Big Dipper)** is at its highest, around 65 degrees, at 1:25 a.m. See the illustration on page 10.

Bright **Venus** rises in the southeast around 4:10 a.m. and is in the eastern sky before dawn.

2027

Saturday, February 13

TONIGHT

sunset: **5:27 p.m.**
twilight begins: **6:23 p.m.**

IN THE MORNING

twilight begins: **5:36 a.m.**
sunrise: **6:31 a.m.**

MOON PHASE

moonrise: **10:01 a.m.**
moonset: **12:48 a.m.**

IN THE SKY

Rust-red and bright, **Mars** is visible most of the night. It is in the eastern sky at dusk and is at its highest, around 70 degrees, at 12:40 a.m.

At its closest and brightest, **Jupiter** is visible most of the night and is at its highest, around 70 degrees, at 11:45 p.m.

After the moon sets, from around 1:10 a.m., use binoculars or a telescope to look for the **moons of Jupiter.** See the guide on page 18.

Orion is at its highest, around 65 degrees, at 8:05 p.m. It sets in the west around 1:10 a.m. Look for the bright star **Aldebaran** and the **Pleiades** near Orion. See the illustration on page 17.

The waxing first-quarter **moon** and the **Pleiades** appear close together in the evening sky. Look west at about 40 degrees elevation around 9 p.m.

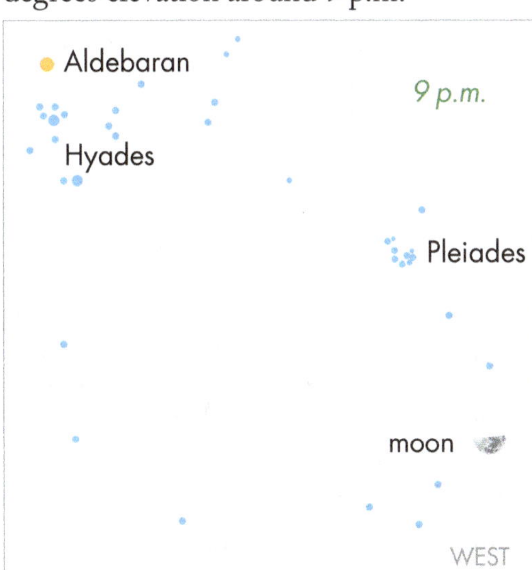

Saturn sets in the west around 8:55 p.m. Unlike bright Jupiter and Venus and ruddy Mars, Saturn is harder to distinguish from the stars. Stars twinkle and shimmer, while the planets, including Saturn, show a steady flat light.

The brightest star in the sky, **Sirius** is at its highest in the south, around 40 degrees, at 8:55 p.m. It sets in the west around 2:05 a.m.

Castor and Pollux, the twin stars in the constellation Gemini, are almost directly overhead at 9:55 p.m. and set in the northwest around 4:25 a.m. Read how to find Castor and Pollux on page 14.

One of the brightest stars in the sky, **Vega** rises in the northeast around 12:45 a.m. and is in the eastern sky before dawn.

Cassiopeia, shaped like a *Σ,* sets in the northwest around 12:50 a.m. See the illustration on page 11.

Explore the **dark sky** after the moon sets, from around 1:10 a.m.

The **Great Bear (Big Dipper)** is at its highest, around 65 degrees, at 1:20 a.m. See the illustration on page 10.

Scorpius rises in the southeast around 2:35 a.m. and is in the eastern sky before dawn. See the illustration on page 19.

Bright **Venus** rises in the southeast around 4:10 a.m. and is in the eastern sky before dawn.

2027

Sunday, February 14

TONIGHT

sunset: **5:28 p.m.**
twilight begins: **6:24 p.m.**

IN THE MORNING

twilight begins: **5:35 a.m.**
sunrise: **6:29 a.m.**

MOON PHASE

moonrise: **10:46 a.m.**
moonset: **1:57 a.m.**

IN THE SKY

Rust-red **Mars** is at its closest and brightest and is visible most of the night. It is in the eastern sky at dusk and is at its highest, around 70 degrees, at 12:35 a.m.

Bright star **Capella** is almost directly overhead around dusk and is in the northwestern sky before dawn. It sets in the northwest around 4:20 a.m. Read about Capella on page 14.

At its closest and brightest, **Jupiter** is visible most of the night and is at its highest, around 70 degrees, at 11:40 p.m.

The **moon** is prominent in the sky early in the evening. Using a telescope or binoculars, take a look. You should be able to see rough mountains and craters along the boundary between light and dark, in astronomy called the terminator.

Orion is at its highest, around 65 degrees, at 8 p.m. It sets in the west around 1:05 a.m. Look for the bright star **Aldebaran** and the **Pleiades** near Orion. See the illustration on page 17.

Saturn sets in the west around 8:50 p.m. Unlike bright Jupiter and Venus and ruddy Mars, Saturn is harder to distinguish from the stars. Stars twinkle and shimmer, while the planets, including Saturn, show a steady flat light.

The brightest star in the sky, **Sirius** is at its highest in the south, around 40 degrees, at 8:50 p.m. It sets in the west around 2 a.m.

The waxing gibbous moon and the **Pleiades** appear close together in the evening sky. Look west at about 55 degrees elevation around 9 p.m.

The bright star **Arcturus** rises in the northeast around 9:35 p.m. and is at its highest, around 75 degrees, at 4:25 a.m. See page 13 for an easy way to locate the star.

Castor and Pollux, the twin stars in the constellation Gemini, are almost directly overhead at 9:50 p.m. and set in the northwest around 4:20 a.m. Read how to find Castor and Pollux on page 14.

One of the brightest stars in the sky, **Vega** rises in the northeast around 12:40 a.m. and is in the eastern sky before dawn.

Cassiopeia, shaped like a Σ, sets in the northwest around 12:45 a.m. See the illustration on page 11.

The **Great Bear (Big Dipper)** is at its highest, around 65 degrees, at 1:20 a.m. See the illustration on page 10.

Scorpius rises in the southeast around 2:30 a.m. and is in the eastern sky before dawn. See the illustration on page 19.

The bright star **Altair** rises in the east around 3:40 a.m. and is in the eastern sky before dawn. Altair can be identified by its less bright neighboring stars, one on either side, together called the shaft of Aquila. See page 18.

Bright **Venus** rises in the southeast around 4:10 a.m. and is in the eastern sky before dawn.

2027

Monday, February 15

MOON PHASE

moonrise: **11:39 a.m.**
moonset: **3:03 a.m.**

TONIGHT

sunset: **5:29 p.m.**
twilight begins: **6:24 p.m.**

IN THE MORNING

twilight begins: **5:34 a.m.**
sunrise: **6:28 a.m.**

IN THE SKY

Bright star **Capella** is almost directly overhead around dusk and is in the northwestern sky before dawn. It sets in the northwest around 4:15 a.m. Read about Capella on page 14.

At its closest and brightest, **Jupiter** is visible most of the night and is at its highest, around 70 degrees, at 11:35 p.m.

Orion is at its highest, around 65 degrees, at 7:55 p.m. It sets in the west around 1:05 a.m. Look for the bright star **Aldebaran** and the **Pleiades** near Orion. See the illustration on page 17.

The brightest star in the sky, **Sirius** is at its highest in the south, around 40 degrees, at 8:45 p.m. It sets in the west around 1:55 a.m.

Saturn sets in the west around 8:50 p.m. Unlike bright Jupiter and Venus and ruddy Mars, Saturn is harder to distinguish from the stars. Stars twinkle and shimmer, while the planets, including Saturn, show a steady flat light.

The bright star **Arcturus** rises in the northeast around 9:30 p.m. and is at its highest, around 75 degrees, at 4:20 a.m. See page 13 for an easy way to locate the star.

Castor and Pollux, the twin stars in the constellation Gemini, are almost directly overhead at 9:45 p.m. and set in the northwest around 4:20 a.m. Read how to find Castor and Pollux on page 14.

Rust-red **Mars** is at its closest and brightest and is visible most of the night. It is at its highest, around 70 degrees, at 12:30 a.m.

One of the brightest stars in the sky, **Vega** rises in the northeast around 12:35 a.m. and is in the eastern sky before dawn.

Cassiopeia, shaped like a *Σ*, sets in the northwest around 12:45 a.m. See the illustration on page 11.

The **Great Bear (Big Dipper)** is at its highest, around 65 degrees, at 1:15 a.m. See the illustration on page 10.

Scorpius rises in the southeast around 2:25 a.m. and is in the eastern sky before dawn. See the illustration on page 19.

The bright star **Altair** rises in the east around 3:40 a.m. and is in the eastern sky before dawn. Altair can be identified by its less bright neighboring stars, one on either side, together called the shaft of Aquila. See page 18.

Bright **Venus** rises in the southeast around 4:10 a.m. and is in the eastern sky before dawn.

Just east of **Scorpius**, Sagittarius rises in the southeast around 5:15 a.m. and is in the eastern sky before dawn.

2027

Tuesday, February 16

TONIGHT

sunset: **5:30 p.m.**
twilight begins: **6:25 p.m.**

IN THE MORNING

twilight begins: **5:33 a.m.**
sunrise: **6:27 a.m.**

MOON PHASE

moonrise: **12:43 p.m.**
moonset: **4:02 a.m.**

IN THE SKY

Bright star **Capella** is almost directly overhead around dusk and is in the northwestern sky before dawn. It sets in the northwest around 4:15 a.m. Read about Capella on page 14.

Orion is at its highest, around 65 degrees, at 7:55 p.m. It sets in the west around 1 a.m. Look for the bright star **Aldebaran** and the **Pleiades** near Orion. See the illustration on page 17.

Saturn sets in the west around 8:45 p.m. Unlike bright Jupiter and Venus and ruddy Mars, Saturn is harder to distinguish from the stars. Stars twinkle and shimmer, while the planets, including Saturn, show a steady flat light.

The brightest star in the sky, **Sirius** is at its highest in the south, around 40 degrees, at 8:45 p.m. It sets in the west around 1:50 a.m.

The bright star **Arcturus** rises in the northeast around 9:25 p.m. and is at its highest, around 75 degrees, at 4:15 a.m. See page 13 for an easy way to locate the star.

Castor and Pollux, the twin stars in the constellation Gemini, are almost directly overhead at 9:45 p.m. and set in the northwest around 4:15 a.m. Read how to find Castor and Pollux on page 14.

Jupiter is visible most of the night and is at its highest, around 70 degrees, at 11:35 p.m.

Rust-red **Mars** is at its closest and brightest and is visible most of the night. It is at its highest, around 70 degrees, at 12:25 a.m.

One of the brightest stars in the sky, **Vega** rises in the northeast around 12:30 a.m. and is in the eastern sky before dawn.

Cassiopeia, shaped like a *Σ,* sets in the northwest around 12:40 a.m. See the illustration on page 11.

The **Great Bear (Big Dipper)** is at its highest, around 65 degrees, at 1:10 a.m. See the illustration on page 10.

Scorpius rises in the southeast around 2:20 a.m. and is in the eastern sky before dawn. See the illustration on page 19.

The bright star **Altair** rises in the east around 3:35 a.m. and is in the eastern sky before dawn. Altair can be identified by its less bright neighboring stars, one on either side, together called the shaft of Aquila. See page 18.

Bright **Venus** rises in the southeast around 4:10 a.m. and is in the eastern sky before dawn.

Just east of **Scorpius**, Sagittarius rises in the southeast around 5:10 a.m. and is in the eastern sky before dawn.

2027

Wednesday, February 17

TONIGHT

sunset: **5:31 p.m.**
twilight begins: **6:26 p.m.**

IN THE MORNING

twilight begins: **5:32 a.m.**
sunrise: **6:26 a.m.**

MOON PHASE

moonrise: **1:55 p.m.**
moonset: **4:52 a.m.**

IN THE SKY

Bright star **Capella** is almost directly overhead around dusk and is in the northwestern sky before dawn. It sets in the northwest around 4:10 a.m. Read about Capella on page 14.

The nearly **full moon** tonight acts like a spotlight. Look around at the desert landscape at how much illumination the moon provides.

Orion is at its highest, around 65 degrees, at 7:50 p.m. It sets in the west around 12:55 a.m. Look for the bright star **Aldebaran** and the **Pleiades** near Orion. See the illustration on page 17.

The moon and the bright stars Castor and Pollux appear close together in the evening sky. Look east at about 70 degrees elevation around 8 p.m.

Saturn sets in the west around 8:40 p.m. Unlike bright Jupiter and Venus and ruddy Mars, Saturn is harder to distinguish from the stars. Stars twinkle and shimmer, while the planets, including Saturn, show a steady flat light.

The brightest star in the sky, **Sirius** is at its highest in the south, around 40 degrees, at 8:40 p.m. It sets in the west around 1:50 a.m.

The bright star **Arcturus** rises in the northeast around 9:20 p.m. and is at its highest, around 75 degrees, at 4:10 a.m. See page 13 for an easy way to locate the star.

Castor and Pollux, the twin stars in the constellation Gemini, are almost directly overhead at 9:40 p.m. and set in the northwest around 4:10 a.m. Read how to find Castor and Pollux on page 14.

Jupiter is visible most of the night and is at its highest, around 70 degrees, at 11:30 p.m.

Rust-red **Mars** is at its closest and brightest and is visible most of the night. It is at its highest, around 70 degrees, at 12:20 a.m.

One of the brightest stars in the sky, **Vega** rises in the northeast around 12:25 a.m. and is in the eastern sky before dawn.

Cassiopeia, shaped like a Σ, sets in the northwest around 12:35 a.m. See the illustration on page 11.

The **Great Bear (Big Dipper)** is at its highest, around 65 degrees, at 1:05 a.m. See the illustration on page 10.

Scorpius rises in the southeast around 2:20 a.m. and is in the eastern sky before dawn. See the illustration on page 19.

The bright star **Altair** rises in the east around 3:30 a.m. and is in the eastern sky before dawn. Altair can be identified by its less bright neighboring stars, one on either side, together called the shaft of Aquila. See page 18.

Bright **Venus** rises in the southeast around 4:15 a.m. and is in the eastern sky before dawn.

2027

Thursday, February 18

TONIGHT

sunset: **5:31 p.m.**
twilight begins: **6:27 p.m.**

IN THE MORNING

twilight begins: **5:31 a.m.**
sunrise: **6:25 a.m.**

MOON PHASE

moonrise: **3:10 p.m.**
moonset: **5:35 a.m.**

IN THE SKY

Bright star **Capella** is almost directly overhead around dusk and is in the northwestern sky before dawn. It sets in the northwest around 4:05 a.m. Read about Capella on page 14.

The **full moon** tonight acts like a spotlight. Look around at the desert landscape at how much illumination the moon provides.

Orion is at its highest, around 65 degrees, at 7:45 p.m. It sets in the west around 12:50 a.m. Look for the bright star **Aldebaran** and the **Pleiades** near Orion. See the illustration on page 17.

The brightest star in the sky, **Sirius** is at its highest in the south, around 40 degrees, at 8:35 p.m. It sets in the west around 1:45 a.m.

Saturn sets in the west around 8:40 p.m. Unlike bright Jupiter and Venus and ruddy Mars, Saturn is harder to distinguish from the stars. Stars twinkle and shimmer, while the planets, including Saturn, show a steady flat light.

The bright star **Arcturus** rises in the northeast around 9:20 p.m. and is at its highest, around 75 degrees, at 4:10 a.m. See page 13 for an easy way to locate the star.

Castor and Pollux, the twin stars in the constellation Gemini, are almost directly overhead at 9:35 p.m. and set in the northwest around 4:05 a.m. Read how to find Castor and Pollux on page 14.

Jupiter is visible most of the night and is at its highest, around 70 degrees, at 11:25 p.m.

Rust-red **Mars** is at its closest and brightest. It is visible most of the night and is at its highest, around 70 degrees, at 12:15 a.m.

One of the brightest stars in the sky, **Vega** rises in the northeast around 12:25 a.m. and is in the eastern sky before dawn.

Cassiopeia, shaped like a *Σ,* sets in the northwest around 12:30 a.m. See the illustration on page 11.

The **Great Bear (Big Dipper)** is at its highest, around 65 degrees, at 1 a.m. See the illustration on page 10.

Scorpius rises in the southeast around 2:15 a.m. and is in the eastern sky before dawn. See the illustration on page 19.

The bright star **Altair** rises in the east around 3:25 a.m. and is in the eastern sky before dawn. Altair can be identified by its less bright neighboring stars, one on either side, together called the shaft of Aquila. See page 18.

Bright **Venus** rises in the southeast around 4:15 a.m. and is in the eastern sky before dawn.

Just east of **Scorpius,** Sagittarius rises in the southeast around 5 a.m. and is in the eastern sky before dawn.

2027

Friday, February 19

TONIGHT

sunset: **5:32 p.m.**
twilight begins: **6:28 p.m.**

IN THE MORNING

twilight begins: **5:30 a.m.**
sunrise: **6:24 a.m.**

MOON PHASE

moonrise: **4:25 p.m.**
moonset: **6:11 a.m.**

IN THE SKY

The **full moon** tonight acts like a spotlight. Look around at the desert landscape at how much illumination the moon provides.

Bright star **Capella** is almost directly overhead around dusk and is in the northwestern sky before dawn. It sets in the northwest around 4 a.m. Read about Capella on page 14.

Orion is at its highest, around 65 degrees, at 7:40 p.m. It sets in the west around 12:50 a.m. Look for the bright star **Aldebaran** and the **Pleiades** near Orion. See the illustration on page 17.

The brightest star in the sky, **Sirius** is at its highest in the south, around 40 degrees, at 8:30 p.m. It sets in the west around 1:40 a.m.

Saturn sets in the west around 8:35 p.m. Unlike bright Jupiter and Venus and ruddy Mars, Saturn is harder to distinguish from the stars. Stars twinkle and shimmer, while the planets, including Saturn, show a steady flat light.

The bright star **Arcturus** rises in the northeast around 9:15 p.m. and is at its highest, around 75 degrees, at 4:05 a.m. See page 13 for an easy way to locate the star.

Castor and Pollux, the twin stars in the constellation Gemini, are almost directly overhead at 9:30 p.m. and set in the northwest around 4 a.m. Read how to find Castor and Pollux on page 14.

The **full moon** and **Jupiter** appear close together in the evening sky. Look southeast at about 60 degrees elevation around 9:45 p.m.

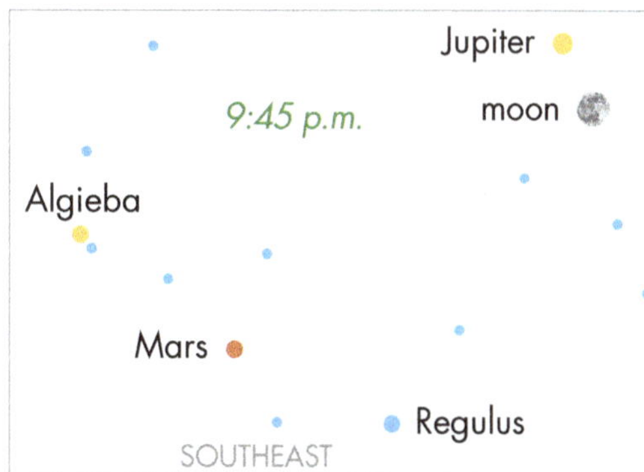

Jupiter is visible most of the night and is at its highest, around 70 degrees, at 11:20 p.m.

Rust-red **Mars** is at its closest and brightest. It is visible most of the night and is at its highest, around 70 degrees, at 12:10 a.m.

Cassiopeia, shaped like a *Σ,* sets in the northwest around 12:25 a.m. See the illustration on page 11.

The **Great Bear (Big Dipper)** is at its highest, around 65 degrees, at 1 a.m. See the illustration on page 10.

Scorpius rises in the southeast around 2:10 a.m. and is in the eastern sky before dawn. See the illustration on page 19.

Bright **Venus** rises in the southeast around 4:15 a.m. and is in the eastern sky before dawn.

2027

Saturday, February 20

TONIGHT

sunset: **5:33 p.m.**
twilight begins: **6:29 p.m.**

IN THE MORNING

twilight begins: **5:29 a.m.**
sunrise: **6:23 a.m.**

MOON PHASE

moonrise: **5:38 p.m.**
moonset: **6:11 a.m.**

IN THE SKY

The **full moon** tonight acts like a spotlight. Look around at the desert landscape at how much illumination the moon provides.

Bright star **Capella** is almost directly overhead around dusk and is in the northwestern sky before dawn. It sets in the northwest around 3:55 a.m. Read about Capella on page 14.

Orion is at its highest, around 65 degrees, at 7:40 p.m. It sets in the west around 12:45 a.m. Look for the bright star **Aldebaran** and the **Pleiades** near Orion. See the illustration on page 17.

The brightest star in the sky, **Sirius** is at its highest in the south, around 40 degrees, at 8:25 p.m. It sets in the west around 1:35 a.m.

Saturn sets in the west around 8:30 p.m. Unlike bright Jupiter and Venus and ruddy Mars, Saturn is harder to distinguish from the stars. Stars twinkle and shimmer, while the planets, including Saturn, show a steady flat light.

The bright star **Arcturus** rises in the northeast around 9:10 p.m. and is at its highest, around 75 degrees, at 4 a.m. See page 13 for an easy way to locate the star.

Castor and Pollux, the twin stars in the constellation Gemini, are almost directly overhead at 9:30 p.m. and set in the northwest around 4 a.m. Read how to find Castor and Pollux on page 14.

Jupiter is visible most of the night and is at its highest, around 70 degrees, at 11:15 p.m.

Rust-red **Mars** is at its closest and brightest. It is visible most of the night and is at its highest, around 70 degrees, at midnight.

One of the brightest stars in the sky, **Vega** rises in the northeast around 12:15 a.m. and is in the eastern sky before dawn.

Cassiopeia, shaped like a Σ, sets in the northwest around 12:25 a.m. See the illustration on page 11.

The **Great Bear (Big Dipper)** is at its highest, around 65 degrees, at 12:55 a.m. See the illustration on page 10.

Scorpius rises in the southeast around 2:05 a.m. and is in the eastern sky before dawn. See the illustration on page 19.

The bright star **Altair** rises in the east around 3:20 a.m. and is in the eastern sky before dawn. Altair can be identified by its less bright neighboring stars, one on either side, together called the shaft of Aquila. See page 18.

Bright **Venus** rises in the southeast around 4:15 a.m. and is in the eastern sky before dawn.

Just east of **Scorpius,** Sagittarius rises in the southeast around 4:55 a.m. and is in the eastern sky before dawn.

2027

Sunday, February 21

TONIGHT

sunset: **5:34 p.m.**
twilight begins: **6:29 p.m.**

IN THE MORNING

twilight begins: **5:28 a.m.**
sunrise: **6:22 a.m.**

MOON PHASE

moonrise: **6:48 p.m.**
moonset: **6:43 a.m.**

IN THE SKY

The **full moon** tonight acts like a spotlight. Look around at the desert landscape at how much illumination the moon provides.

Orion is at its highest, around 65 degrees, at 7:35 p.m. It sets in the west around 12:40 a.m. Look for the bright star **Aldebaran** and the **Pleiades** near Orion. See the illustration on page 17.

Saturn sets in the west around 8:25 p.m. Unlike bright Jupiter and Venus and ruddy Mars, Saturn is harder to distinguish from the stars. Stars twinkle and shimmer, while the planets, including Saturn, show a steady flat light.

The brightest star in the sky, **Sirius** is at its highest in the south, around 40 degrees, at 8:25 p.m. It sets in the west around 1:30 a.m.

The bright star **Arcturus** rises in the northeast around 9:05 p.m. and is at its highest, around 75 degrees, at 3:55 a.m. See page 13 for an easy way to locate the star.

Castor and Pollux, the twin stars in the constellation Gemini, are almost directly overhead at 9:25 p.m. and set in the northwest around 3:55 a.m. Read how to find Castor and Pollux on page 14.

Jupiter is visible most of the night and is at its highest, around 70 degrees, at 11:10 p.m.

Rust-red **Mars** is at its closest and brightest. It is visible most of the night and is at its highest, around 70 degrees, at 11:50 p.m.

One of the brightest stars in the sky, **Vega** rises in the northeast around 12:10 a.m. and is in the eastern sky before dawn.

Cassiopeia, shaped like a Σ, sets in the northwest around 12:20 a.m. See the illustration on page 11.

The **Great Bear (Big Dipper)** is at its highest, around 65 degrees, at 12:50 a.m. See the illustration on page 10.

Scorpius rises in the southeast around 2 a.m. and is in the eastern sky before dawn. See the illustration on page 19.

The bright star **Altair** rises in the east around 3:15 a.m. and is in the eastern sky before dawn. Altair can be identified by its less bright neighboring stars, one on either side, together called the shaft of Aquila. See page 18.

Bright star **Capella** is almost directly overhead around dusk and is in the northwestern sky before dawn. It sets in the northwest around 3:55 a.m. Read about Capella on page 14.

Bright **Venus** rises in the southeast around 4:15 a.m. and is in the eastern sky before dawn.

Just east of **Scorpius**, Sagittarius rises in the southeast around 4:50 a.m. and is in the eastern sky before dawn.

2027

Monday, February 22

TONIGHT

sunset: **5:35 p.m.**
twilight begins: **6:30 p.m.**

MOON PHASE

IN THE MORNING

twilight begins: **5:27 a.m.**
sunrise: **6:21 a.m.**

moonrise: **7:55 p.m.**
moonset: **7:13 a.m.**

IN THE SKY

The **full moon** tonight acts like a spotlight. Look around at the desert landscape at how much illumination the moon provides.

Orion is at its highest, around 65 degrees, at 7:30 p.m. It sets in the west around 12:35 a.m. Look for the bright star **Aldebaran** and the **Pleiades** near Orion. See the illustration on page 17.

Saturn sets in the west around 8:25 p.m. Unlike bright Jupiter and Venus and ruddy Mars, Saturn is harder to distinguish from the stars. Stars twinkle and shimmer, while the planets, including Saturn, show a steady flat light.

The bright star **Arcturus** rises in the northeast around 9:05 p.m. and is at its highest, around 75 degrees, at 3:55 a.m. See page 13 for an easy way to locate the star.

Castor and Pollux, the twin stars in the constellation Gemini, are almost directly overhead at 9:20 p.m. and set in the northwest around 3:50 a.m. Read how to find Castor and Pollux on page 14.

Jupiter is visible most of the night and is at its highest, around 70 degrees, at 11:05 p.m.

Rust-red Mars is at its closest and brightest. It is visible most of the night and is at its highest, around 70 degrees, at 11:45 p.m.

One of the brightest stars in the sky, **Vega** rises in the northeast around 12:05 a.m.

Cassiopeia, shaped like a Σ, sets in the northwest around 12:15 a.m. See the illustration on page 11.

The **Great Bear (Big Dipper)** is at its highest, around 65 degrees, at 12:45 a.m. See the illustration on page 10.

The brightest star in the sky, **Sirius** is at its highest in the south, around 40 degrees, at 8:20 p.m. It sets in the west around 1:30 a.m.

Scorpius rises in the southeast around 2 a.m. and is in the eastern sky before dawn. See the illustration on page 19.

The bright star **Altair** rises in the east around 3:10 a.m. and is in the eastern sky before dawn. Altair can be identified by its less bright neighboring stars, one on either side, together called the shaft of Aquila. See page 18.

Bright star **Capella** is almost directly overhead around dusk and is in the northwestern sky before dawn. It sets in the northwest around 3:50 a.m. Read about Capella on page 14.

Bright Venus rises in the southeast around 4:15 a.m. and is in the eastern sky before dawn.

Just east of **Scorpius**, Sagittarius rises in the southeast around 4:45 a.m. and is in the eastern sky before dawn.

2027

Tuesday, February 23

MOON PHASE

TONIGHT

sunset: **5:36 p.m.**
twilight begins: **6:31 p.m.**

IN THE MORNING

twilight begins: **5:26 a.m.**
sunrise: **6:19 a.m.**

moonrise: **9:01 p.m.**
moonset: **7:42 a.m.**

IN THE SKY

The brightest star in the sky, **Sirius** is at its highest in the south, around 40 degrees, at 8:15 p.m. It sets in the west around 1:25 a.m.

Saturn is in the western sky at dusk and sets in the west around 8:20 p.m. Unlike bright Jupiter and Venus and ruddy Mars, Saturn is harder to distinguish from the stars. Stars twinkle and shimmer, while the planets, including Saturn, show a steady flat light.

The bright star **Arcturus** rises in the northeast around 9 p.m. and is at its highest, around 75 degrees, at 3:50 a.m. See page 13 for an easy way to locate the star.

Castor and Pollux, the twin stars in the constellation Gemini, are almost directly overhead at 9:15 p.m. and set in the northwest around 3:45 a.m. Read how to find Castor and Pollux on page 14.

Jupiter is visible most of the night and is at its highest, around 70 degrees, at 11 p.m.

Rust-red **Mars** is at its closest and brightest. It is visible most of the night and is at its highest, around 70 degrees, at 11:40 p.m.

One of the brightest stars in the sky, **Vega** rises in the northeast around midnight.

Cassiopeia, shaped like a Σ, sets in the northwest around 12:10 a.m. See the illustration on page 11.

Orion is at its highest, around 65 degrees, around dusk and sets in the west around 12:30 a.m. Look for the bright star **Aldebaran** and the **Pleiades** near Orion. See the illustration on page 17.

The **Great Bear (Big Dipper)** is at its highest, around 65 degrees, at 12:45 a.m. See the illustration on page 10.

Scorpius rises in the southeast around 1:55 a.m. and is in the eastern sky before dawn. See the illustration on page 19.

The bright star **Altair** rises in the east around 3:05 a.m. and is in the eastern sky before dawn. Altair can be identified by its less bright neighboring stars, one on either side, together called the shaft of Aquila. See page 18.

Bright star **Capella** is in the northwestern sky before dawn and sets in the northwest around 3:45 a.m. Read about Capella on page 14.

Bright **Venus** rises in the southeast around 4:15 a.m. and is in the eastern sky before dawn.

Just east of **Scorpius**, Sagittarius rises in the southeast around 4:40 a.m. and is in the eastern sky before dawn.

2027

Wednesday, February 24

TONIGHT

sunset: **5:37 p.m.**
twilight begins: **6:32 p.m.**

MOON PHASE

IN THE MORNING

twilight begins: **5:25 a.m.**
sunrise: **6:18 a.m.**

moonrise: **10:06 p.m.**
moonset: **8:12 a.m.**

IN THE SKY

Jupiter is visible most of the night and is at its highest, around 70 degrees, at 11 p.m.

Before the moon rises, until around 9:45 p.m., use binoculars or a telescope to look for the **moons of Jupiter.** See the guide on page 18.

The brightest star in the sky, **Sirius** is at its highest in the south, around 40 degrees, at 8:10 p.m. It sets in the west around 1:20 a.m.

Saturn is in the western sky at dusk and sets in the west around 8:15 p.m. Unlike bright Jupiter and Venus and ruddy Mars, Saturn is harder to distinguish from the stars. Stars twinkle and shimmer, while the planets, including Saturn, show a steady flat light.

The bright star **Arcturus** rises in the northeast around 8:55 p.m. and is at its highest, around 75 degrees, at 3:45 a.m. See page 13 for an easy way to locate the star.

Castor and Pollux, the twin stars in the constellation Gemini, are almost directly overhead at 9:10 p.m. and set in the northwest around 3:45 a.m. Read how to find Castor and Pollux on page 14.

Before the moon rises, until around 9:45 p.m., look for the **Andromeda Galaxy** with binoculars or a telescope. See the guide on page 11.

Explore the **dark sky** after twilight and before the moon rises, around 9:45 p.m.

Rust-red and bright, Mars is at its closest and brightest. It is visible most of the night and is at its highest, around 70 degrees, at 11:35 p.m.

One of the brightest stars in the sky, **Vega** rises in the northeast around 11:55 p.m.

Cassiopeia, shaped like a Σ, sets in the northwest around 12:05 a.m. See the illustration on page 11.

Orion is at its highest, around 65 degrees, around dusk and sets in the west around 12:30 a.m. Look for the bright star **Aldebaran** and the **Pleiades** near Orion. See the illustration on page 17.

The **Great Bear (Big Dipper)** is at its highest, around 65 degrees, at 12:40 a.m. See the illustration on page 10.

Scorpius rises in the southeast around 1:50 a.m. and is in the eastern sky before dawn. See the illustration on page 19.

The bright star **Altair** rises in the east around 3:05 a.m. and is in the eastern sky before dawn. Altair can be identified by its less bright neighboring stars, one on either side, together called the shaft of Aquila. See page 18.

Bright star **Capella** is in the northwestern sky before dawn and sets in the northwest around 3:40 a.m. Read about Capella on page 14.

Bright Venus rises in the southeast around 4:20 a.m. and is in the eastern sky before dawn.

2027

Thursday, February 25

TONIGHT

sunset: **5:38 p.m.**
twilight begins: **6:33 p.m.**

IN THE MORNING

twilight begins: **5:23 a.m.**
sunrise: **6:17 a.m.**

MOON PHASE

moonrise: **11:10 p.m.**
moonset: **8:44 a.m.**

IN THE SKY

The brightest star in the sky, **Sirius** is at its highest in the south, around 40 degrees, at 8:10 p.m. It sets in the west around 1:15 a.m.

Saturn is in the western sky at dusk and sets in the west around 8:15 p.m. Unlike bright Jupiter and Venus and ruddy Mars, Saturn is harder to distinguish from the stars. Stars twinkle and shimmer, while the planets, including Saturn, show a steady flat light.

The bright star **Arcturus** rises in the northeast around 8:50 p.m. and is at its highest, around 75 degrees, at 3:40 a.m. See page 13 for an easy way to locate the star.

Castor and Pollux, the twin stars in the constellation Gemini, are almost directly overhead at 9:10 p.m. and set in the northwest around 3:40 a.m. Read how to find Castor and Pollux on page 14.

Before the moon rises, until around 10:50 p.m., look for the **Andromeda Galaxy** with binoculars or a telescope. See the guide on page 11.

Explore the **dark sky** after twilight and before the moon rises, around 10:50 p.m.

Jupiter is at its highest, around 70 degrees, at 10:55 p.m.

Rust-red and bright, **Mars** is visible most of the night and is at its highest, around 70 degrees, at 11:30 p.m.

One of the brightest stars in the sky, **Vega** rises in the northeast around 11:50 p.m.

Cassiopeia, shaped like a Σ, sets in the northwest around 12:05 a.m. See the illustration on page 11.

Orion is at its highest, around 65 degrees, around dusk and sets in the west around 12:25 a.m. Look for the bright star **Aldebaran** and the **Pleiades** near Orion. See the illustration on page 17.

The **Great Bear (Big Dipper)** is at its highest, around 65 degrees, at 12:35 a.m. See the illustration on page 10.

Scorpius rises in the southeast around 1:45 a.m. and is in the eastern sky before dawn. See the illustration on page 19.

The bright star **Altair** rises in the east around 3 a.m. and is in the eastern sky before dawn. Altair can be identified by its less bright neighboring stars, one on either side, together called the shaft of Aquila. See page 18.

Bright star **Capella** is in the northwestern sky before dawn and sets in the northwest around 3:35 a.m. Read about Capella on page 14.

Bright **Venus** rises in the southeast around 4:20 a.m. and is in the eastern sky before dawn.

Just east of **Scorpius**, Sagittarius rises in the southeast around 4:35 a.m. and is in the eastern sky before dawn.

2027

Friday, February 26

TONIGHT

sunset: **5:39 p.m.**
twilight begins: **6:34 p.m.**

IN THE MORNING

twilight begins: **5:22 a.m.**
sunrise: **6:16 a.m.**

MOON PHASE

moonrise: **12:12 a.m.**
moonset: **9:20 a.m.**

IN THE SKY

Saturn is in the western sky at dusk and sets in the west around 8:10 p.m. Unlike bright Jupiter and Venus and ruddy Mars, Saturn is harder to distinguish from the stars. Stars twinkle and shimmer, while the planets, including Saturn, show a steady flat light.

Jupiter is at its highest, around 70 degrees, at 10:50 p.m.

Before the moon rises, until around 12:10 a.m., use binoculars or a telescope to look for the **moons of Jupiter.** See the guide on page 18.

The brightest star in the sky, **Sirius** is at its highest in the south, around 40 degrees, at 8:05 p.m. It sets in the west around 1:10 a.m.

The bright star **Arcturus** rises in the northeast around 8:45 p.m. and is at its highest, around 75 degrees, at 3:35 a.m. See page 13 for an easy way to locate the star.

Castor and Pollux, the twin stars in the constellation Gemini, are almost directly overhead at 9:05 p.m. and set in the northwest around 3:35 a.m. Read how to find Castor and Pollux on page 14.

Look for the **Andromeda Galaxy** with binoculars or a telescope. It is a good target until around 10:45 p.m. See the guide on page 11.

Cassiopeia, shaped like a Σ, sets in the northwest around 10:55 p.m. See the illustration on page 11.

Rust-red and bright, **Mars** is visible most of the night and is at its highest, around 70 degrees, at 11:25 p.m.

One of the brightest stars in the sky, **Vega** rises in the northeast around 11:50 p.m.

Explore the **dark sky** after twilight and before the moon rises, around 12:10 a.m.

Orion is at its highest, around 65 degrees, around dusk and sets in the west around 12:20 a.m. Look for the bright star **Aldebaran** and the **Pleiades** near Orion. See the illustration on page 17.

The **Great Bear (Big Dipper)** is at its highest, around 65 degrees, at 12:30 a.m. See the illustration on page 10.

Scorpius rises in the southeast around 1:40 a.m. and is in the eastern sky before dawn. See the illustration on page 19.

The bright star **Altair** rises in the east around 2:55 a.m. and is in the eastern sky before dawn. Altair can be identified by its less bright neighboring stars, one on either side, together called the shaft of Aquila. See page 18.

Bright star **Capella** is in the northwestern sky before dawn and sets in the northwest around 3:35 a.m. Read about Capella on page 14.

Bright **Venus** rises in the southeast around 4:20 a.m. and is in the eastern sky before dawn.

Just east of **Scorpius**, Sagittarius rises in the southeast around 4:30 a.m. and is in the eastern sky before dawn.

2027

Saturday, February 27

TONIGHT

sunset: **5:39 p.m.**
twilight begins: **6:34 p.m.**

IN THE MORNING

twilight begins: **5:21 a.m.**
sunrise: **6:15 a.m.**

MOON PHASE

moonrise: **1:11 a.m.**
moonset: **10:01 a.m.**

IN THE SKY

The brightest star in the sky, **Sirius** is at its highest in the south, around 40 degrees, at 8 p.m. It sets in the west around 1:10 a.m.

Saturn is in the western sky at dusk and sets in the west around 8:05 p.m. Unlike bright Jupiter and Venus and ruddy Mars, Saturn is harder to distinguish from the stars. Stars twinkle and shimmer, while the planets, including Saturn, show a steady flat light.

Castor and Pollux, the twin stars in the constellation Gemini, are almost directly overhead at 9 p.m. and set in the northwest around 3:30 a.m. Read how to find Castor and Pollux on page 14.

Look for the **Andromeda Galaxy** with binoculars or a telescope. It is a good target until around 10:40 p.m. See the guide on page 11.

Jupiter is at its highest, around 70 degrees, at 10:45 p.m.

Before the moon rises, until around 12:50 a.m., use binoculars or a telescope to look for the **moons of Jupiter.** See the guide on page 18.

Cassiopeia, shaped like a Σ, sets in the northwest around 10:50 p.m. See the illustration on page 11.

Rust-red and bright, **Mars** is visible most of the night and is at its highest, around 70 degrees, at 11:20 p.m.

One of the brightest stars in the sky, **Vega** rises in the northeast around 11:45 p.m.

Orion is at its highest, around 65 degrees, around dusk and sets in the west around 12:15 a.m. Look for the bright star **Aldebaran** and the **Pleiades** near Orion. See the illustration on page 17.

The **Great Bear (Big Dipper)** is at its highest, around 65 degrees, at 12:25 a.m. See the illustration on page 10.

Explore the **dark sky** after twilight and before the moon rises, around 12:50 a.m.

Scorpius rises in the southeast around 1:40 a.m. and is in the eastern sky before dawn. See the illustration on page 19.

The bright star **Altair** rises in the east around 2:50 a.m. and is in the eastern sky before dawn. Altair can be identified by its less bright neighboring stars, one on either side, together called the shaft of Aquila. See page 18.

The waning third-quarter **moon** and the bright yellowish star **Antares** appear close together in the morning sky. Look southeast near the horizon around 3 a.m.

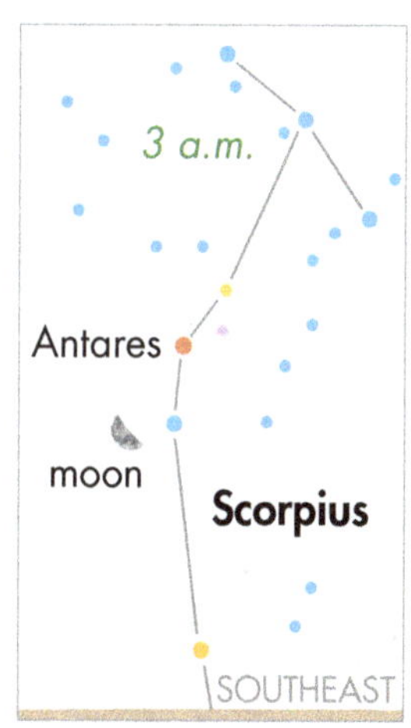

Bright **Venus** rises in the southeast around 4:20 a.m. and is in the eastern sky before dawn.

2027

Sunday, February 28

TONIGHT

sunset: **5:40 p.m.**
twilight begins: **6:35 p.m.**

MOON PHASE

IN THE MORNING

twilight begins: **5:20 a.m.**
sunrise: **6:13 a.m.**

moonrise: **2:06 a.m.**
moonset: **10:47 a.m.**

IN THE SKY

Saturn is in the western sky at dusk and sets in the west around 8:05 p.m. Unlike bright Jupiter and Venus and ruddy Mars, Saturn is harder to distinguish from the stars. Stars twinkle and shimmer, while the planets, including Saturn, show a steady flat light.

The bright star **Arcturus** rises in the northeast around 8:40 p.m. and is at its highest, around 75 degrees, at 3:30 a.m. See page 13 for an easy way to locate the star.

Castor and Pollux, the twin stars in the constellation Gemini, are almost directly overhead at 8:55 p.m. and set in the northwest around 3:25 a.m. Read how to find Castor and Pollux on page 14.

Jupiter is at its highest, around 70 degrees, at 10:40 p.m.

With the dark sky, use binoculars or a telescope to look for the **moons of Jupiter.** See the guide on page 18.

Look for the **Andromeda Galaxy** with binoculars or a telescope. It is a good target until around 10:40 p.m. See the guide on page 11.

Cassiopeia, shaped like a *Σ*, sets in the northwest around 10:50 p.m. See the illustration on page 11.

Rust-red and bright, Mars is visible most of the night and is at its highest, around 70 degrees, at 11:15 p.m.

One of the brightest stars in the sky, **Vega** rises in the northeast around 11:40 p.m.

Orion is at its highest, around 65 degrees, around dusk and sets in the west around 12:10 a.m. Look for the bright star **Aldebaran** and the **Pleiades** near Orion. See the illustration on page 17.

The **Great Bear (Big Dipper)** is at its highest, around 65 degrees, at 12:25 a.m. See the illustration on page 10.

The brightest star in the sky, **Sirius** is at its highest in the south, around 40 degrees, at 7:55 p.m. It sets in the west around 1:05 a.m.

Scorpius rises in the southeast around 1:35 a.m. and is in the eastern sky before dawn. See the illustration on page 19.

The bright star **Altair** rises in the east around 2:45 a.m. and is in the eastern sky before dawn. Altair can be identified by its less bright neighboring stars, one on either side, together called the shaft of Aquila. See page 18.

Bright star **Capella** is in the northwestern sky before dawn and sets in the northwest around 3:25 a.m. Read about Capella on page 14.

Bright Venus rises in the southeast around 4:20 a.m. and is in the eastern sky before dawn.

Just east of **Scorpius**, Sagittarius rises in the southeast around 4:20 a.m. and is in the eastern sky before dawn.

2027

Monday, March 1

TONIGHT

sunset: **5:41 p.m.**
twilight begins: **6:36 p.m.**

IN THE MORNING

twilight begins: **5:19 a.m.**
sunrise: **6:12 a.m.**

MOON PHASE

moonrise: **2:54 a.m.**
moonset: **11:38 a.m.**

IN THE SKY

The brightest star in the sky, **Sirius** is at its highest in the south, around 40 degrees, at 7:50 p.m. It sets in the west around 1 a.m.

Saturn is in the western sky at dusk and sets in the west around 8 p.m. Unlike bright Jupiter and Venus and ruddy Mars, Saturn is harder to distinguish from the stars. Stars twinkle and shimmer, while the planets, including Saturn, show a steady flat light.

The bright star **Arcturus** rises in the northeast around 8:35 p.m. and is at its highest, around 75 degrees, at 3:25 a.m. See page 13 for an easy way to locate the star.

Castor and Pollux, the twin stars in the constellation Gemini, are almost directly overhead at 8:50 p.m. and set in the northwest around 3:25 a.m. Read how to find Castor and Pollux on page 14.

Jupiter is at its highest, around 70 degrees, at 10:35 p.m.

With the dark sky, use binoculars or a telescope to look for the **moons of Jupiter.** See the guide on page 18.

Look for the **Andromeda Galaxy** with binoculars or a telescope. It is a good target until around 10:35 p.m. See the guide on page 11.

Cassiopeia, shaped like a *Σ,* sets in the northwest around 10:45 p.m. See the illustration on page 11.

Rust-red and bright, **Mars** is visible most of the night and is at its highest, around 70 degrees, at 11:10 p.m.

One of the brightest stars in the sky, **Vega** rises in the northeast around 11:35 p.m.

Orion is at its highest, around 65 degrees, around dusk and sets in the west around 12:10 a.m. Look for the bright star **Aldebaran** and the **Pleiades** near Orion. See the illustration on page 17.

The **Great Bear (Big Dipper)** is at its highest, around 65 degrees, at 12:20 a.m. See the illustration on page 10.

Scorpius rises in the southeast around 1:30 a.m. and is in the eastern sky before dawn. See the illustration on page 19.

The bright star **Altair** rises in the east around 2:45 a.m. and is in the eastern sky before dawn. Altair can be identified by its less bright neighboring stars, one on either side, together called the shaft of Aquila. See page 18.

Bright star **Capella** is in the northwestern sky before dawn and sets in the northwest around 3:20 a.m. Read about Capella on page 14.

Bright **Venus** rises in the southeast around 4:20 a.m. and is in the eastern sky before dawn.

Just east of **Scorpius,** Sagittarius rises in the southeast around 4:20 a.m. and is in the eastern sky before dawn.

2027

Tuesday, March 2

TONIGHT

sunset: **5:42 p.m.**
twilight begins: **6:37 p.m.**

MOON PHASE

IN THE MORNING

twilight begins: **5:17 a.m.**
sunrise: **6:11 a.m.**

moonrise: **3:36 a.m.**
moonset: **12:33 p.m.**

IN THE SKY

The brightest star in the sky, **Sirius** is at its highest in the south, around 40 degrees, at 7:50 p.m. It sets in the west around 12:55 a.m.

Saturn is in the western sky at dusk and sets in the west around 7:55 p.m. Unlike bright Jupiter and Venus and ruddy Mars, Saturn is harder to distinguish from the stars. Stars twinkle and shimmer, while the planets, including Saturn, show a steady flat light.

The absence of the moon tonight provides excellent **dark-sky** viewing conditions.

The bright star **Arcturus** rises in the northeast around 8:30 p.m. and is at its highest, around 75 degrees, at 3:20 a.m. See page 13 for an easy way to locate the star.

Castor and Pollux, the twin stars in the constellation Gemini, are almost directly overhead at 8:50 p.m. and set in the northwest around 3:20 a.m. Read how to find Castor and Pollux on page 14.

Jupiter is at its highest, around 70 degrees, at 10:30 p.m.

With the dark sky, use binoculars or a telescope to look for the **moons of Jupiter.** See the guide on page 18.

Look for the **Andromeda Galaxy** with binoculars or a telescope. It is a good target until around 10:30 p.m. See the guide on page 11.

Cassiopeia, shaped like a *Σ*, sets in the northwest around 10:40 p.m. See the illustration on page 11.

Rust-red and bright, **Mars** is visible most of the night and is at its highest, around 75 degrees, at 11 p.m.

One of the brightest stars in the sky, **Vega** rises in the northeast around 11:30 p.m.

Orion is at its highest, around 65 degrees, around dusk and sets in the west around 12:05 a.m. Look for the bright star **Aldebaran** and the **Pleiades** near Orion. See the illustration on page 17.

The **Great Bear (Big Dipper)** is at its highest, around 65 degrees, at 12:15 a.m. See the illustration on page 10.

Scorpius rises in the southeast around 1:25 a.m. and is in the eastern sky before dawn. See the illustration on page 19.

The bright star **Altair** rises in the east around 2:40 a.m. and is in the eastern sky before dawn. Altair can be identified by its less bright neighboring stars, one on either side, together called the shaft of Aquila. See page 18.

Bright star **Capella** is in the northwestern sky before dawn and sets in the northwest around 3:20 a.m. Read about Capella on page 14.

Just east of **Scorpius**, Sagittarius rises in the southeast around 4:15 a.m. and is in the eastern sky before dawn.

Bright **Venus** rises in the southeast around 4:20 a.m. and is in the eastern sky before dawn.

2027
Wednesday, March 3

TONIGHT

sunset: **5:43 p.m.**
twilight begins: **6:38 p.m.**

IN THE MORNING

twilight begins: **5:16 a.m.**
sunrise: **6:09 a.m.**

MOON PHASE

moonrise: **4:13 a.m.**
moonset: **1:31 p.m.**

IN THE SKY

The absence of the moon tonight provides excellent **dark-sky** viewing conditions.

The brightest star in the sky, **Sirius** is at its highest in the south, around 40 degrees, at 7:45 p.m. It sets in the west around 12:55 a.m.

Saturn is in the western sky at dusk and sets in the west around 7:55 p.m. Unlike bright Jupiter and Venus and ruddy Mars, Saturn is harder to distinguish from the stars. Stars twinkle and shimmer, while the planets, including Saturn, show a steady flat light.

The bright star **Arcturus** rises in the northeast around 8:25 p.m. and is at its highest, around 75 degrees, at 3:15 a.m. See page 13 for an easy way to locate the star.

Castor and Pollux, the twin stars in the constellation Gemini, are almost directly overhead at 8:45 p.m. and set in the northwest around 3:15 a.m. Read how to find Castor and Pollux on page 14.

Jupiter is at its highest, around 70 degrees, at 10:25 p.m.

With the dark sky, use binoculars or a telescope to look for the **moons of Jupiter.** See the guide on page 18.

Look for the **Andromeda Galaxy** with binoculars or a telescope. It is a good target until around 10:25 p.m. See the guide on page 11.

Cassiopeia, shaped like a Σ, sets in the northwest around 10:35 p.m. See the illustration on page 11.

Rust-red and bright, **Mars** is at its highest, around 75 degrees, at 10:55 p.m.

One of the brightest stars in the sky, **Vega** rises in the northeast around 11:30 p.m.

Orion sets in the west around midnight. Look for the bright star **Aldebaran** and the **Pleiades** near Orion. See the illustration on page 17.

The **Great Bear (Big Dipper)** is at its highest, around 65 degrees, at 12:10 a.m. See the illustration on page 10.

Scorpius rises in the southeast around 1:25 a.m. and is in the eastern sky before dawn. See the illustration on page 19.

The bright star **Altair** rises in the east around 2:35 a.m. and is in the eastern sky before dawn. Altair can be identified by its less bright neighboring stars, one on either side, together called the shaft of Aquila. See page 18.

Bright star **Capella** is in the northwestern sky before dawn and sets in the northwest around 3:15 a.m. Read about Capella on page 14.

Just east of **Scorpius**, Sagittarius rises in the southeast around 4:10 a.m. and is in the eastern sky before dawn.

Bright **Venus** rises in the southeast around 4:20 a.m. and is in the eastern sky before dawn.

The waning sliver-crescent moon and **Venus** appear close together in the morning sky. Look southeast near the horizon around 5 a.m.

2027

Thursday, March 4

TONIGHT

sunset: **5:44 p.m.**
twilight begins: **6:39 p.m.**

MOON PHASE

IN THE MORNING

twilight begins: **5:15 a.m.**
sunrise: **6:08 a.m.**

moonrise: **4:45 a.m.**
moonset: **2:30 p.m.**

IN THE SKY

The absence of the moon tonight provides excellent **dark-sky** viewing conditions.

The brightest star in the sky, **Sirius** is at its highest in the south, around 40 degrees, at 7:40 p.m. It sets in the west around 12:50 a.m.

Saturn is in the western sky at dusk and sets in the west around 7:50 p.m. Unlike bright Jupiter and Venus and ruddy Mars, Saturn is harder to distinguish from the stars. Stars twinkle and shimmer, while the planets, including Saturn, show a steady flat light.

The bright star **Arcturus** rises in the northeast around 8:25 p.m. and is at its highest, around 75 degrees, at 3:15 a.m. See page 13 for an easy way to locate the star.

Castor and Pollux, the twin stars in the constellation Gemini, are almost directly overhead at 8:40 p.m. and set in the northwest around 3:10 a.m. Read how to find Castor and Pollux on page 14.

Jupiter is at its highest, around 70 degrees, at 10:20 p.m.

With the dark sky, use binoculars or a telescope to look for the **moons of Jupiter.** See the guide on page 18.

Look for the **Andromeda Galaxy** with binoculars or a telescope. It is a good target until around 10:20 p.m. See the guide on page 11.

Cassiopeia, shaped like a Σ, sets in the northwest around 10:30 p.m. See the illustration on page 11.

Rust-red and bright, **Mars** is at its highest, around 75 degrees, at 10:50 p.m.

One of the brightest stars in the sky, **Vega** rises in the northeast around 11:25 p.m.

The **Great Bear (Big Dipper)** is at its highest, around 65 degrees, at 12:05 a.m. See the illustration on page 10.

Orion sets in the west around 12:55 a.m. Look for the bright star **Aldebaran** and the **Pleiades** near Orion. See the illustration on page 17.

Scorpius rises in the southeast around 1:20 a.m. and is in the eastern sky before dawn. See the illustration on page 19.

The bright star **Altair** rises in the east around 2:30 a.m. and is in the eastern sky before dawn. Altair can be identified by its less bright neighboring stars, one on either side, together called the shaft of Aquila. See page 18.

Bright star **Capella** is in the northwestern sky before dawn and sets in the northwest around 3:10 a.m. Read about Capella on page 14.

Just east of **Scorpius**, Sagittarius rises in the southeast around 4:05 a.m. and is in the eastern sky before dawn.

Bright **Venus** rises in the east around 4:20 a.m. and is in the eastern sky before dawn.

2027

Friday, March 5

MOON PHASE

TONIGHT

sunset: **5:45 p.m.**
twilight begins: **6:39 p.m.**

IN THE MORNING

twilight begins: **5:14 a.m.**
sunrise: **6:07 a.m.**

moonrise: **5:13 a.m.**
moonset: **3:29 p.m.**

IN THE SKY

The absence of the moon tonight provides excellent **dark-sky** viewing conditions.

Saturn is in the western sky at dusk and sets in the west around 7:45 p.m. Unlike bright Jupiter and Venus and ruddy Mars, Saturn is harder to distinguish from the stars. Stars twinkle and shimmer, while the planets, including Saturn, show a steady flat light.

The bright star **Arcturus** rises in the northeast around 8:20 p.m. and is at its highest, around 75 degrees, at 3:10 a.m. See page 13 for an easy way to locate the star.

Castor and Pollux, the twin stars in the constellation Gemini, are almost directly overhead at 8:35 p.m. and set in the northwest around 3:05 a.m. Read how to find Castor and Pollux on page 14.

Jupiter is at its highest, around 70 degrees, at 10:20 p.m.

With the dark sky, use binoculars or a telescope to look for the **moons of Jupiter.** See the guide on page 18.

Look for the **Andromeda Galaxy** with binoculars or a telescope. It is a good target until around 10:20 p.m. See the guide on page 11.

Cassiopeia, shaped like a *Σ,* sets in the northwest around 10:30 p.m. See the illustration on page 11.

Rust-red and bright, Mars is at its highest, around 75 degrees, at 10:45 p.m.

One of the brightest stars in the sky, **Vega** rises in the northeast around 11:20 p.m.

The **Great Bear (Big Dipper)** is at its highest, around 65 degrees, at 12:05 a.m. See the illustration on page 10.

The brightest star in the sky, **Sirius** is at its highest in the south, around 40 degrees, around dusk and sets in the west around 12:45 a.m.

Orion sets in the west around 12:55 a.m. Look for the bright star **Aldebaran** and the **Pleiades** near Orion. See the illustration on page 17.

Scorpius rises in the southeast around 1:15 a.m. and is in the eastern sky before dawn. See the illustration on page 19.

The bright star **Altair** rises in the east around 2:25 a.m. and is in the eastern sky before dawn. Altair can be identified by its less bright neighboring stars, one on either side, together called the shaft of Aquila. See page 18.

Bright star **Capella** is in the northwestern sky before dawn and sets in the northwest around 3:05 a.m. Read about Capella on page 14.

Just east of **Scorpius**, Sagittarius rises in the southeast around 4 a.m. and is in the eastern sky before dawn.

Bright Venus rises in the east around 4:20 a.m. and is in the eastern sky before dawn.

2027

Saturday, March 6

TONIGHT

sunset: **5:45 p.m.**
twilight begins: **6:40 p.m.**

IN THE MORNING

twilight begins: **5:12 a.m.**
sunrise: **6:06 a.m.**

MOON PHASE

moonrise: **5:40 a.m.**
moonset: **4:27 p.m.**

IN THE SKY

The absence of the moon tonight provides excellent **dark-sky** viewing conditions.

Saturn is in the western sky at dusk and sets in the west around 7:45 p.m.

Castor and Pollux, the twin stars in the constellation Gemini, are almost directly overhead at 8:35 p.m. and set in the northwest around 3:05 a.m. Read how to find Castor and Pollux on page 14.

Jupiter is at its highest, around 70 degrees, at 10:15 p.m.

With the dark sky, use binoculars or a telescope to look for the **moons of Jupiter.** See the guide on page 18.

With the dark sky, look for the **Andromeda Galaxy** with binoculars or a telescope. It is a good target until around 10:15 p.m. See the guide on page 11.

Rust-red and bright, **Mars** is at its highest, around 75 degrees, at 10:40 p.m.

One of the brightest stars in the sky, **Vega** rises in the northeast around 11:15 p.m.

The **Great Bear (Big Dipper)** is at its highest, around 65 degrees, at 11:55 p.m. See the illustration on page 10.

The brightest star in the sky, **Sirius** is at its highest in the south, around 40 degrees, around dusk and sets in the west around 12:40 a.m.

Orion sets in the west around 12:50 a.m. Look for the bright star **Aldebaran** and the **Pleiades** near Orion. See the illustration on page 17.

Scorpius rises in the southeast around 1:10 a.m. and is in the eastern sky before dawn. See the illustration on page 19.

The bright star **Altair** rises in the east around 2:25 a.m. and is in the eastern sky before dawn. Altair can be identified by its less bright neighboring stars, one on either side, together called the shaft of Aquila. See page 18.

Bright **Venus** rises in the east around 4:20 a.m. and is in the eastern sky before dawn.

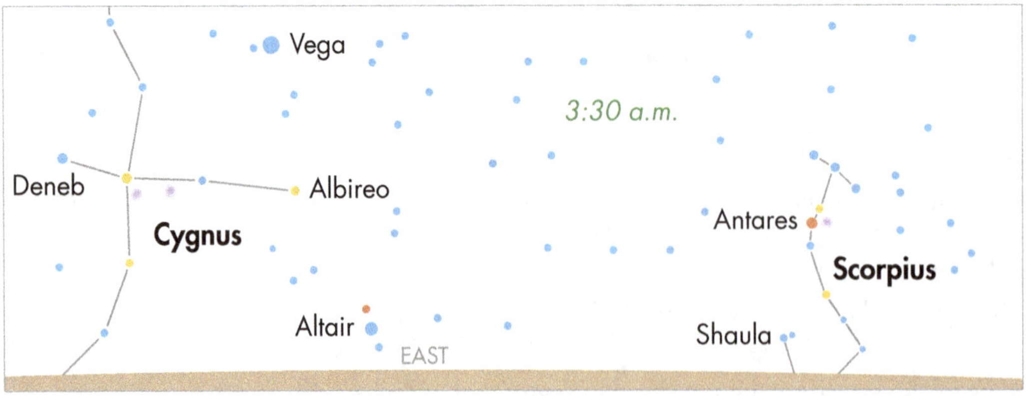

2027
Sunday, March 7

TONIGHT

sunset: **5:46 p.m.**
twilight begins: **6:41 p.m.**

IN THE MORNING

twilight begins: **5:11 a.m.**
sunrise: **6:04 a.m.**

MOON PHASE

moonrise: **6:05 a.m.**
moonset: **5:26 p.m.**

IN THE SKY

The absence of the moon tonight provides excellent **dark-sky** viewing conditions.

Saturn is in the western sky at dusk and sets in the west around 7:40 p.m. Unlike bright Jupiter and Venus and ruddy Mars, Saturn is harder to distinguish from the stars. Stars twinkle and shimmer, while the planets, including Saturn, show a steady flat light.

The bright star **Arcturus** rises in the northeast around 8:10 p.m. and is at its highest, around 75 degrees, at 3 a.m. See page 13 for an easy way to locate the star.

Castor and Pollux, the twin stars in the constellation Gemini, are almost directly overhead at 8:30 p.m. and set in the northwest around 3 a.m. Read how to find Castor and Pollux on page 14.

Look for the **Andromeda Galaxy** with binoculars or a telescope. It is a good target until around 10:10 p.m. See the guide on page 11.

Jupiter is at its highest, around 70 degrees, at 10:10 p.m.

Use binoculars or a telescope to look for the **moons of Jupiter.** See the guide on page 18.

Cassiopeia, shaped like a *Σ*, sets in the northwest around 10:20 p.m. See the illustration on page 11.

Rust-red and bright, **Mars** is at its highest, around 75 degrees, at 10:35 p.m.

One of the brightest stars in the sky, **Vega** rises in the northeast around 11:15 p.m.

The **Great Bear (Big Dipper)** is at its highest, around 65 degrees, at 11:50 p.m. See the illustration on page 10.

The brightest star in the sky, **Sirius** is at its highest in the south, around 40 degrees, around dusk and sets in the west around 12:35 a.m.

Orion sets in the west around 12:45 a.m. Look for the bright star **Aldebaran** and the **Pleiades** near Orion. See the illustration on page 17.

Scorpius rises in the southeast around 1:05 a.m. and is in the eastern sky before dawn. See the illustration on page 19.

The bright star **Altair** rises in the east around 2:20 a.m. and is in the eastern sky before dawn. Altair can be identified by its less bright neighboring stars, one on either side, together called the shaft of Aquila. See page 18.

Bright star **Capella** sets in the northwest around 3 a.m. Read about Capella on page 14.

Just east of **Scorpius,** Sagittarius rises in the southeast around 3:55 a.m. and is in the eastern sky before dawn.

Bright **Venus** rises in the east around 4:25 a.m. and is in the eastern sky before dawn.

2027

Monday, March 8

TONIGHT

sunset: **5:47 p.m.**
twilight begins: **6:42 p.m.**

MOON PHASE

IN THE MORNING

twilight begins: **5:10 a.m.**
sunrise: **6:03 a.m.**

moonrise: **6:05 a.m.**
moonset: **6:25 p.m.**

IN THE SKY

The absence of the moon tonight provides excellent **dark-sky** viewing conditions.

Saturn is in the western sky at dusk and sets in the west around 7:35 p.m. Unlike bright Jupiter and Venus and ruddy Mars, Saturn is harder to distinguish from the stars. Stars twinkle and shimmer, while the planets, including Saturn, show a steady flat light.

The bright star **Arcturus** rises in the northeast around 8:10 p.m. and is at its highest, around 75 degrees, at 3 a.m. See page 13 for an easy way to locate the star.

Castor and Pollux, the twin stars in the constellation Gemini, are almost directly overhead at 8:25 p.m. and set in the northwest around 2:55 a.m. Read how to find Castor and Pollux on page 14.

Look for the **Andromeda Galaxy** with binoculars or a telescope. It is a good target until around 10:05 p.m. See the guide on page 11.

Jupiter is at its highest, around 70 degrees, at 10:05 p.m.

Use binoculars or a telescope to look for the **moons of Jupiter.** See the guide on page 18.

Cassiopeia, shaped like a *Σ,* sets in the northwest around 10:15 p.m. See the illustration on page 11.

Rust-red and bright, Mars is at its highest, around 75 degrees, at 10:30 p.m.

One of the brightest stars in the sky, **Vega** rises in the northeast around 11:10 p.m.

The **Great Bear (Big Dipper)** is at its highest, around 65 degrees, at 11:50 p.m. See the illustration on page 10.

The brightest star in the sky, **Sirius** is at its highest in the south, around 40 degrees, around dusk and sets in the west around 12:35 a.m.

Orion sets in the west around 12:40 a.m. Look for the bright star **Aldebaran** and the **Pleiades** near Orion. See the illustration on page 17.

Scorpius rises in the southeast around 1:05 a.m. and is in the eastern sky before dawn. See the illustration on page 19.

The bright star **Altair** rises in the east around 2:15 a.m. and is in the eastern sky before dawn. Altair can be identified by its less bright neighboring stars, one on either side, together called the shaft of Aquila. See page 18.

Bright star **Capella** sets in the northwest around 2:55 a.m. Read about Capella on page 14.

Just east of **Scorpius,** Sagittarius rises in the southeast around 3:50 a.m. and is in the eastern sky before dawn.

Bright Venus rises in the east around 4:25 a.m. and is in the eastern sky before dawn.

2027
Tuesday, March 9

MOON PHASE

TONIGHT

sunset: **5:48 p.m.**
twilight begins: **6:43 p.m.**

IN THE MORNING

twilight begins: **5:08 a.m.**
sunrise: **6:02 a.m.**

moonrise: **6:31 a.m.**
moonset: **7:25 p.m.**

IN THE SKY

The absence of the moon tonight provides excellent **dark-sky** viewing conditions.

Saturn is in the western sky at dusk and sets in the west around 7:35 p.m. Unlike bright Jupiter and Venus and ruddy Mars, Saturn is harder to distinguish from the stars. Stars twinkle and shimmer, while the planets, including Saturn, show a steady flat light.

After the moon sets, from around 7:45 p.m., look for the **Andromeda Galaxy** with binoculars or a telescope. It is a good target until around 10:05 p.m. See the guide on page 11.

The bright star **Arcturus** rises in the northeast around 8:05 p.m. and is at its highest, around 75 degrees, at 2:55 a.m. See page 13 for an easy way to locate the star.

Castor and Pollux, the twin stars in the constellation Gemini, are almost directly overhead at 8:20 p.m. and set in the northwest around 2:50 a.m. Read how to find Castor and Pollux on page 14.

Jupiter is at its highest, around 70 degrees, at 10 p.m.

After the moon sets, from around 7:45 p.m., use binoculars or a telescope to look for the **moons of Jupiter.** See the guide on page 18.

Cassiopeia, shaped like a Σ, sets in the northwest around 10:10 p.m. See the illustration on page 11.

Rust-red and bright, Mars is at its highest, around 75 degrees, at 10:25 p.m.

One of the brightest stars in the sky, **Vega** rises in the northeast around 11:05 p.m.

The **Great Bear (Big Dipper)** is at its highest, around 65 degrees, at 11:45 p.m. See the illustration on page 10.

The brightest star in the sky, **Sirius** is at its highest in the south, around 40 degrees, around dusk and sets in the west around 12:30 a.m.

Orion sets in the west around 12:35 a.m. Look for the bright star **Aldebaran** and the **Pleiades** near Orion. See the illustration on page 17.

Scorpius rises in the southeast around 1 a.m. and is in the eastern sky before dawn. See the illustration on page 19.

The bright star **Altair** rises in the east around 2:10 a.m. and is in the eastern sky before dawn. Altair can be identified by its less bright neighboring stars, one on either side, together called the shaft of Aquila. See page 18.

Bright star **Capella** sets in the northwest around 2:50 a.m. Read about Capella on page 14.

Just east of **Scorpius**, Sagittarius rises in the southeast around 3:45 a.m. and is in the eastern sky before dawn.

Bright Venus rises in the east around 4:25 a.m. and is in the eastern sky before dawn.

2027

Wednesday, March 10

TONIGHT

sunset: **5:49 p.m.**
twilight begins: **6:43 p.m.**

MOON PHASE

IN THE MORNING

twilight begins: **5:07 a.m.**
sunrise: **6 a.m.**

moonrise: **6:58 a.m.**
moonset: **8:28 p.m.**

IN THE SKY

The bright star **Arcturus** rises in the northeast around 8 p.m. and is at its highest, around 75 degrees, at 2:50 a.m. See page 13 for an easy way to locate the star.

Castor and Pollux, the twin stars in the constellation Gemini, are almost directly overhead at 8:15 p.m. and set in the northwest around 2:50 a.m. Read how to find Castor and Pollux on page 14.

The absence of the moon tonight provides excellent **dark-sky** viewing conditions. Explore the **dark sky** after the moon sets, from around 8:50 p.m.

After the moon sets, from around 8:50 p.m., look for the **Andromeda Galaxy** with binoculars or a telescope. It is a good target until around 10 p.m. See the guide on page 11.

Jupiter is at its highest, around 70 degrees, at 9:55 p.m.

After the moon sets, from around 8:50 p.m., use binoculars or a telescope to look for the **moons of Jupiter.** See the guide on page 18.

Rust-red and bright, **Mars** is at its highest, around 75 degrees, at 10:20 p.m.

Saturn is in the western sky at dusk and sets in the west around 7:30 p.m. Unlike bright Jupiter and Venus and ruddy Mars, Saturn is harder to distinguish from the stars. Stars twinkle and shimmer, while the planets, including Saturn, show a steady flat light.

Cassiopeia, shaped like a Σ, sets in the northwest around 10:10 p.m. See the illustration on page 11.

One of the brightest stars in the sky, **Vega** rises in the northeast around 11 p.m.

The **Great Bear (Big Dipper)** is at its highest, around 65 degrees, at 11:40 p.m. See the illustration on page 10.

The brightest star in the sky, **Sirius** is at its highest in the south, around 40 degrees, around dusk and sets in the west around 12:25 a.m.

Orion sets in the west around 12:35 a.m. Look for the bright star **Aldebaran** and the **Pleiades** near Orion. See the illustration on page 17.

Scorpius rises in the southeast around 12:55 a.m. and is in the eastern sky before dawn. See the illustration on page 19.

The bright star **Altair** rises in the east around 2:10 a.m. and is in the eastern sky before dawn. Altair can be identified by its less bright neighboring stars, one on either side, together called the shaft of Aquila. See page 18.

Bright star **Capella** sets in the northwest around 2:45 a.m. Read about Capella on page 14.

Just east of **Scorpius**, Sagittarius rises in the southeast around 3:45 a.m. and is in the eastern sky before dawn.

Bright **Venus** rises in the east around 4:25 a.m. and is in the eastern sky before dawn.

2027

Thursday, March 11

TONIGHT

sunset: **5:49 p.m.**
twilight begins: **6:44 p.m.**

IN THE MORNING

twilight begins: **5:06 a.m.**
sunrise: **5:59 a.m.**

MOON PHASE

moonrise: **7:28 a.m.**
moonset: **9:33 p.m.**

IN THE SKY

Saturn is in the western sky at dusk and sets in the west around 7:25 p.m. Unlike bright Jupiter and Venus and ruddy Mars, Saturn is harder to distinguish from the stars. Stars twinkle and shimmer, while the planets, including Saturn, show a steady flat light.

The bright star **Arcturus** rises in the northeast around 7:55 p.m. and is at its highest, around 75 degrees, at 2:45 a.m. See page 13 for an easy way to locate the star.

Jupiter is at its highest, around 70 degrees, at 9:50 p.m. and sets in the west around 4:35 a.m.

After the moon sets, from around 9:55 p.m., use binoculars or a telescope to look for the **moons of Jupiter.** See the guide on page 18.

The absence of the moon tonight provides excellent **dark-sky** viewing conditions. Explore the **dark sky** after the moon sets, from around 9:55 p.m.

Cassiopeia, shaped like a *Σ,* sets in the northwest around 10:05 p.m. See the illustration on page 11.

Rust-red and bright, **Mars** is at its highest, around 75 degrees, at 10:15 p.m.

One of the brightest stars in the sky, **Vega** rises in the northeast around 10:55 p.m.

The **Great Bear (Big Dipper)** is at its highest, around 65 degrees, at 11:35 p.m. See the illustration on page 10.

The brightest star in the sky, **Sirius** is at its highest in the south, around 40 degrees, around dusk and sets in the west around 12:20 a.m.

Orion sets in the west around 12:30 a.m. Look for the bright star **Aldebaran** and the **Pleiades** near Orion. See the illustration on page 17.

Scorpius rises in the southeast around 12:50 a.m. and is in the eastern sky before dawn. See the illustration on page 19.

The bright star **Altair** rises in the east around 2:05 a.m. and is in the eastern sky before dawn. Altair can be identified by its less bright neighboring stars, one on either side, together called the shaft of Aquila. See page 18.

Bright star **Capella** sets in the northwest around 2:40 a.m. Read about Capella on page 14.

Castor and Pollux, the twin stars in the constellation Gemini, are almost directly overhead at 8:15 p.m. and set in the northwest around 2:45 a.m. Read how to find Castor and Pollux on page 14.

Just east of **Scorpius,** Sagittarius rises in the southeast around 3:40 a.m. and is in the eastern sky before dawn.

Bright **Venus** rises in the east around 4:25 a.m. and is in the eastern sky before dawn.

2027

Friday, March 12

TONIGHT

sunset: **5:50 p.m.**
twilight begins: **6:45 p.m.**

IN THE MORNING

twilight begins: **5:04 a.m.**
sunrise: **5:58 a.m.**

MOON PHASE

moonrise: **8:03 a.m.**
moonset: **10:40 p.m.**

IN THE SKY

Saturn is in the western sky at dusk and sets in the west around 7:25 p.m. Unlike bright Jupiter and Venus and ruddy Mars, Saturn is harder to distinguish from the stars. Stars twinkle and shimmer, while the planets, including Saturn, show a steady flat light.

The bright star **Arcturus** rises in the northeast around 7:50 p.m. and is at its highest, around 75 degrees, at 2:40 a.m. See page 13 for an easy way to locate the star.

Castor and Pollux, the twin stars in the constellation Gemini, are almost directly overhead at 8:10 p.m. and set in the northwest around 2:40 a.m. Read how to find Castor and Pollux on page 14.

Jupiter is at its highest, around 70 degrees, at 9:50 p.m. and sets in the west around 4:30 a.m.

After the moon sets, from around 11 p.m., use binoculars or a telescope to look for the **moons of Jupiter.** See the guide on page 18.

The **moon** is prominent in the sky early in the evening. Using a telescope or binoculars, take a look. You should be able to see rough mountains and craters along the boundary between light and dark, in astronomy called the terminator.

Cassiopeia, shaped like a *Σ,* sets in the northwest around 10 p.m. See the illustration on page 11.

Rust-red and bright, Mars is at its highest, around 75 degrees, at 10:10 p.m.

One of the brightest stars in the sky, **Vega** rises in the northeast around 10:55 p.m.

Explore the **dark sky** after the moon sets, from around 11 p.m.

The **Great Bear (Big Dipper)** is at its highest, around 65 degrees, at 11:30 p.m. See the illustration on page 10.

The brightest star in the sky, **Sirius** is at its highest in the south, around 40 degrees, around dusk and sets in the west around 12:15 a.m.

Orion sets in the west around 12:25 a.m. Look for the bright star **Aldebaran** and the **Pleiades** near Orion. See the illustration on page 17.

Scorpius rises in the southeast around 12:45 a.m. and is in the eastern sky before dawn. See the illustration on page 19.

The bright star **Altair** rises in the east around 2 a.m. and is in the eastern sky before dawn. Altair can be identified by its less bright neighboring stars, one on either side, together called the shaft of Aquila. See page 18.

Bright star **Capella** sets in the northwest around 2:40 a.m. Read about Capella on page 14.

Just east of **Scorpius,** Sagittarius rises in the southeast around 3:35 a.m. and is in the eastern sky before dawn.

Bright Venus rises in the east around 4:25 a.m. and is in the eastern sky before dawn.

2027

Saturday, March 13

TONIGHT

sunset: **5:51 p.m.**
twilight begins: **6:46 p.m.**

IN THE MORNING

twilight begins: **6:03 a.m.**
sunrise: **6:56 a.m.**

MOON PHASE

moonrise: **8:44 a.m.**
moonset: **11:48 p.m.**

IN THE SKY

The **moon** is prominent in the sky early in the evening. Using a telescope or binoculars, take a look. You should be able to see rough mountains and craters along the boundary between light and dark, in astronomy called the terminator.

The waxing **crescent moon** and the **Pleiades** appear close together in the evening sky. Look west at about 55 degrees elevation around 7 p.m.

Saturn is in the western sky at dusk and sets in the west around 7:20 p.m. Unlike bright Jupiter and Venus and ruddy Mars, Saturn is harder to distinguish from the stars. Stars twinkle and shimmer, while the planets, including Saturn, show a steady flat light.

Jupiter is at its highest, around 70 degrees, at 9:45 p.m. and sets in the west around 4:25 a.m.

Shaped like a *Σ*, **Cassiopeia** sets in the northwest around 9:55 p.m. See the illustration on page 11.

After the moon sets, from around 12:10 a.m., use binoculars or a telescope to look for the **moons of Jupiter.** See the guide on page 18.

Rust-red and bright, Mars is at its highest, around 75 degrees, at 10:05 p.m. and sets in the west around 4:50 a.m.

One of the brightest stars in the sky, **Vega** rises in the northeast around 10:50 p.m.

Explore the **dark sky** after the moon sets, from around 12:10 a.m.

The brightest star in the sky, **Sirius** sets in the west around 12:15 a.m.

Orion sets in the west around 12:20 a.m. Look for the bright star **Aldebaran** and the **Pleiades** near Orion. See the illustration on page 17.

Scorpius rises in the southeast around 12:45 a.m. and is at its highest in the south, around 30 degrees, at 4:50 a.m. See the illustration on page 19.

Bright star **Capella** sets in the northwest around 2:35 a.m. Read about Capella on page 14.

Castor and Pollux, the twin stars in the constellation Gemini, are almost directly overhead at 8:05 p.m. and set in the northwest around 2:35 a.m. Read how to find Castor and Pollux on page 14.

Bright Venus rises in the east around 4:25 a.m. and is in the eastern sky before dawn.

Cassiopeia, shaped like a *W,* rises in the northeast around 5 a.m. It is in the northeastern sky before dawn.

2027

Sunday, March 14

TONIGHT

sunset: **6:52 p.m.**
twilight begins: **7:47 p.m.**

IN THE MORNING

twilight begins: **6:02 a.m.**
sunrise: **6:55 a.m.**

MOON PHASE

moonrise: **10:34 a.m.**
moonset: **1:54 a.m.**

IN THE SKY

The **moon** is prominent in the sky early in the evening. Using a telescope or binoculars, take a look. You should be able to see rough mountains and craters along the boundary between light and dark, in astronomy called the terminator.

Saturn sets in the west around 8:15 p.m. Unlike bright Jupiter and Venus and ruddy Mars, Saturn is harder to distinguish from the stars. Stars twinkle and shimmer, while the planets, including Saturn, show a steady flat light.

The bright star **Arcturus** rises in the northeast around 8:45 p.m. and is at its highest, around 75 degrees, at 3:35 a.m. See page 13 for an easy way to locate the star.

Castor and Pollux, the twin stars in the constellation Gemini, are almost directly overhead at 9 p.m. and set in the northwest around 3:30 a.m. Read how to find Castor and Pollux on page 14.

Jupiter is at its highest, around 70 degrees, at 10:40 p.m. and sets in the west around 5:20 a.m.

Cassiopeia, shaped like a Σ, sets in the northwest around 10:55 p.m. See the illustration on page 11.

Rust-red and bright, **Mars** is at its highest, around 75 degrees, at 11 p.m.

One of the brightest stars in the sky, **Vega** rises in the northeast around 11:45 p.m.

Orion sets in the west around 12:15 a.m. Look for the bright star **Aldebaran** and the **Pleiades** near Orion. See the illustration on page 17.

The **Great Bear (Big Dipper)** is at its highest, around 65 degrees, at 12:25 a.m. See the illustration on page 10.

The brightest star in the sky, **Sirius** sets in the west around 1:10 a.m.

Scorpius rises in the southeast around 1:40 a.m. and is in the eastern sky before dawn. See the illustration on page 19.

Explore the **dark sky** after the moon sets, from around 2:15 a.m.

After the moon sets, from around 2:15 a.m., use binoculars or a telescope to look for the **moons of Jupiter.** See the guide on page 18.

The bright star **Altair** rises in the east around 2:50 a.m. and is in the eastern sky before dawn. Altair can be identified by its less bright neighboring stars, one on either side, together called the shaft of Aquila. See page 18.

Bright star **Capella** sets in the northwest around 3:30 a.m. Read about Capella on page 14.

Just east of **Scorpius**, Sagittarius rises in the southeast around 4:25 a.m. and is in the eastern sky before dawn.

Bright **Venus** rises in the east around 5:25 a.m. and is in the eastern sky before dawn.

2027

Monday, March 15

TONIGHT

sunset: **6:53 p.m.**
twilight begins: **7:47 p.m.**

IN THE MORNING

twilight begins: **6 a.m.**
sunrise: **6:54 a.m.**

MOON PHASE

moonrise: **11:33 a.m.**
moonset: **2:54 a.m.**

IN THE SKY

The **moon** is prominent in the sky early in the evening. Using a telescope or binoculars, take a look. You should be able to see rough mountains and craters along the boundary between light and dark, in astronomy called the terminator.

Saturn sets in the west around 8:15 p.m. Unlike bright Jupiter and Venus and ruddy Mars, Saturn is harder to distinguish from the stars. Stars twinkle and shimmer, while the planets, including Saturn, show a steady flat light.

The bright star **Arcturus** rises in the northeast around 8:40 p.m. and is at its highest, around 75 degrees, at 3:30 a.m. See page 13 for an easy way to locate the star.

Castor and Pollux, the twin stars in the constellation Gemini, are almost directly overhead at 8:55 p.m. and set in the northwest around 3:30 a.m. Read how to find Castor and Pollux on page 14.

Jupiter is at its highest, around 70 degrees, at 10:35 p.m. and sets in the west around 5:15 a.m.

After the moon sets, from around 3:15 a.m., use binoculars or a telescope to look for the **moons of Jupiter.** See the guide on page 18.

Cassiopeia, shaped like a Σ, sets in the northwest around 10:50 p.m. See the illustration on page 11.

Rust-red and bright, **Mars** is at its highest, around 75 degrees, at 10:55 p.m.

One of the brightest stars in the sky, **Vega** rises in the northeast around 11:40 p.m.

Orion sets in the west around 12:15 a.m. Look for the bright star **Aldebaran** and the **Pleiades** near Orion. See the illustration on page 17.

The **Great Bear (Big Dipper)** is at its highest, around 65 degrees, at 12:20 a.m. See the illustration on page 10.

The brightest star in the sky, **Sirius** sets in the west around 1:05 a.m.

Scorpius rises in the southeast around 1:35 a.m. and is in the eastern sky before dawn. See the illustration on page 19.

The bright star **Altair** rises in the east around 2:50 a.m. and is in the eastern sky before dawn. Altair can be identified by its less bright neighboring stars, one on either side, together called the shaft of Aquila. See page 18.

Explore the **dark sky** after the moon sets, from around 3:15 a.m.

Bright star **Capella** sets in the northwest around 3:25 a.m. Read about Capella on page 14.

Just east of **Scorpius**, Sagittarius rises in the southeast around 4:25 a.m. and is in the eastern sky before dawn.

Bright **Venus** rises in the east around 5:25 a.m. and is in the eastern sky before dawn.

2027

Tuesday, March 16

TONIGHT

sunset: **6:53 p.m.**
twilight begins: **7:48 p.m.**

IN THE MORNING

twilight begins: **5:59 a.m.**
sunrise: **6:52 a.m.**

MOON PHASE

moonrise: **12:40 p.m.**
moonset: **3:46 a.m.**

IN THE SKY

The **moon** is prominent in the sky early in the evening. Using a telescope or binoculars, take a look. You should be able to see rough mountains and craters along the boundary between light and dark, in astronomy called the terminator.

Saturn sets in the west around 8:10 p.m. Unlike bright Jupiter and Venus and ruddy Mars, Saturn is harder to distinguish from the stars. Stars twinkle and shimmer, while the planets, including Saturn, show a steady flat light.

The bright star **Arcturus** rises in the northeast around 8:35 p.m. and is at its highest, around 75 degrees, at 3:25 a.m. See page 13 for an easy way to locate the star.

Castor and Pollux, the twin stars in the constellation Gemini, are almost directly overhead at 8:55 p.m. and set in the northwest around 3:25 a.m. Read how to find Castor and Pollux on page 14.

Jupiter is at its highest, around 70 degrees, at 10:30 p.m. and sets in the west around 5:15 a.m.

Cassiopeia, shaped like a Σ, sets in the northwest around 10:45 p.m. See the illustration on page 11.

Rust-red and bright, **Mars** is at its highest, around 75 degrees, at 10:50 p.m.

One of the brightest stars in the sky, **Vega** rises in the northeast around 11:35 p.m.

Orion sets in the west around 12:10 a.m. Look for the bright star **Aldebaran** and the **Pleiades** near Orion. See the illustration on page 17.

The **Great Bear (Big Dipper)** is at its highest, around 65 degrees, at 12:15 a.m. See the illustration on page 10.

The brightest star in the sky, **Sirius** sets in the west around 1 a.m.

Scorpius rises in the southeast around 1:30 a.m. and is in the eastern sky before dawn. See the illustration on page 19.

The bright star **Altair** rises in the east around 2:45 a.m. and is in the eastern sky before dawn. Altair can be identified by its less bright neighboring stars, one on either side, together called the shaft of Aquila. See page 18.

Bright star **Capella** sets in the northwest around 3:25 a.m. Read about Capella on page 14.

Just east of **Scorpius**, Sagittarius rises in the southeast around 4:20 a.m. and is in the eastern sky before dawn.

Bright **Venus** rises in the east around 5:25 a.m. and is in the eastern sky before dawn.

2027
Wednesday, March 17

TONIGHT

sunset: **6:54 p.m.**
twilight begins: **7:49 p.m.**

MOON PHASE

IN THE MORNING

twilight begins: **5:58 a.m.**
sunrise: **6:51 a.m.**

moonrise: **1:52 p.m.**
moonset: **4:30 a.m.**

IN THE SKY

Saturn sets in the west around 8:05 p.m. Unlike bright Jupiter and Venus and ruddy Mars, Saturn is harder to distinguish from the stars. Stars twinkle and shimmer, while the planets, including Saturn, show a steady flat light.

The bright star **Arcturus** rises in the northeast around 8:30 p.m. and is at its highest, around 75 degrees, at 3:20 a.m. See page 13 for an easy way to locate the star.

Castor and Pollux, the twin stars in the constellation Gemini, are almost directly overhead at 8:50 p.m. and set in the northwest around 3:20 a.m. Read how to find Castor and Pollux on page 14.

Jupiter is at its highest, around 70 degrees, at 10:25 p.m. and sets in the west around 5:10 a.m.

Cassiopeia, shaped like a *Σ,* sets in the northwest around 10:40 p.m. See the illustration on page 11.

One of the brightest stars in the sky, **Vega** rises in the northeast around 11:35 p.m.

Rust-red and bright, **Mars** is at its highest, around 75 degrees, at 10:45 p.m.

Orion sets in the west around 12:05 a.m. Look for the bright star **Aldebaran** and the **Pleiades** near Orion. See the illustration on page 17.

The **Great Bear (Big Dipper)** is at its highest, around 65 degrees, at 12:10 a.m. See the illustration on page 10.

The brightest star in the sky, **Sirius** sets in the west around 1 a.m.

Scorpius rises in the southeast around 1:25 a.m. and is at its highest in the south, around 30 degrees, around dawn. See the illustration on page 19.

The bright star **Altair** rises in the east around 2:40 a.m. and is in the eastern sky before dawn. Altair can be identified by its less bright neighboring stars, one on either side, together called the shaft of Aquila. See page 18.

Bright star **Capella** sets in the northwest around 3:20 a.m. Read about Capella on page 14.

Just east of **Scorpius,** Sagittarius rises in the southeast around 4:15 a.m. and is in the eastern sky before dawn.

Bright **Venus** rises in the east around 5:25 a.m. and is in the eastern sky before dawn.

Look for **Mercury** in the morning sky, just before sunrise in the east. It's difficult but possible to see the tiny dot in the glare of the sun. It's easier to find with the naked eye, not using binoculars or a telescope. Never look directly at the sun or point binoculars or a telescope at it.

2027

Thursday, March 18

TONIGHT

sunset: **6:55 p.m.**
twilight begins: **7:50 p.m.**

IN THE MORNING

twilight begins: **5:56 a.m.**
sunrise: **6:49 a.m.**

MOON PHASE

moonrise: **3:04 p.m.**
moonset: **5:07 a.m.**

IN THE SKY

Castor and Pollux, the twin stars in the constellation Gemini, are almost directly overhead around dusk and set in the northwest around 3:15 a.m. Read how to find Castor and Pollux on page 14.

Saturn sets in the west around 8:05 p.m. Unlike bright Jupiter and Venus and ruddy Mars, Saturn is harder to distinguish from the stars. Stars twinkle and shimmer, while the planets, including Saturn, show a steady flat light.

The bright star **Arcturus** rises in the northeast around 8:30 p.m. and is at its highest, around 75 degrees, at 3:20 a.m. See page 13 for an easy way to locate the star.

Jupiter is at its highest, around 70 degrees, at 10:20 p.m. and sets in the west around 5:05 a.m.

Cassiopeia, shaped like a Σ, sets in the northwest around 10:35 p.m. See the illustration on page 11.

Rust-red and bright, **Mars** is at its highest, around 75 degrees, at 10:40 p.m.

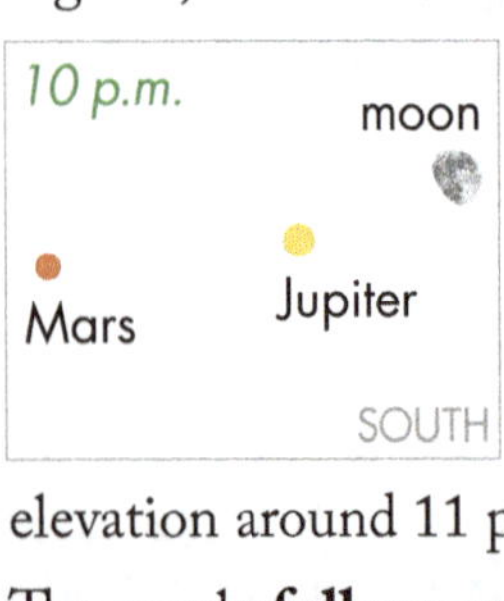

The nearly **full moon, Jupiter,** and **Mars** appear close together in the evening sky. Look southwest at about 70 degrees elevation around 11 p.m.

The nearly **full moon** tonight acts like a spotlight. Look around at the desert landscape at how much illumination the moon provides.

One of the brightest stars in the sky, **Vega** rises in the northeast around 11:30 p.m.

Orion sets in the west around midnight. Look for the bright star **Aldebaran** and the **Pleiades** near Orion. See the illustration on page 17.

The **Great Bear (Big Dipper)** is at its highest, around 65 degrees, at 12:10 a.m. See the illustration on page 10.

The brightest star in the sky, **Sirius** sets in the west around 12:55 a.m.

The bright star **Altair** rises in the east around 2:35 a.m. and is in the eastern sky before dawn. Altair can be identified by its less bright neighboring stars, one on either side, together called the shaft of Aquila. See page 18.

Bright star **Capella** sets in the northwest around 3:15 a.m. Read about Capella on page 14.

Just east of **Scorpius,** Sagittarius rises in the southeast around 4:10 a.m. and is in the eastern sky before dawn.

Bright **Venus** rises in the east around 5:25 a.m. and is in the eastern sky before dawn.

Look for **Mercury** in the morning sky, just before sunrise in the east. It's difficult but possible to see the tiny dot in the glare of the sun. It's easier to find with the naked eye, not using binoculars or a telescope. Never look directly at the sun or point binoculars or a telescope at it.

2027

Friday, March 19

TONIGHT

sunset: **6:56 p.m.**
twilight begins: **7:51 p.m.**

IN THE MORNING

twilight begins: **5:55 a.m.**
sunrise: **6:48 a.m.**

MOON PHASE

moonrise: **4:16 p.m.**
moonset: **5:40 a.m.**

IN THE SKY

Castor and Pollux, the twin stars in the constellation Gemini, are almost directly overhead around dusk and set in the northwest around 3:10 a.m. Read how to find Castor and Pollux on page 14.

The bright star **Arcturus** rises in the northeast around 8:25 p.m. and is at its highest, around 75 degrees, at 3:15 a.m. See page 13 for an easy way to locate the star.

The **full moon** tonight acts like a spotlight. Look around at the desert landscape at how much illumination the moon provides.

Jupiter is at its highest, around 70 degrees, at 10:20 p.m. and sets in the west around 5 a.m.

Rust-red and bright, **Mars** is at its highest, around 75 degrees, at 10:35 p.m. and sets in the west around 5:25 a.m.

Cassiopeia, shaped like a Σ, sets in the northwest around 10:35 p.m. See the illustration on page 11.

One of the brightest stars in the sky, **Vega** rises in the northeast around 11:25 p.m.

The **Great Bear (Big Dipper)** is at its highest, around 65 degrees, at 12:05 a.m. See the illustration on page 10.

The brightest star in the sky, **Sirius** sets in the west around 12:45 a.m.

Orion sets in the west around 1 a.m. Look for the bright star **Aldebaran** and the **Pleiades** near Orion. See the illustration on page 17.

Scorpius rises in the southeast around 1:15 a.m. and is at its highest in the south, around 30 degrees, around dawn. See the illustration on page 19.

The bright star **Altair** rises in the east around 2:30 a.m. and is in the eastern sky before dawn. Altair can be identified by its less bright neighboring stars, one on either side, together called the shaft of Aquila. See page 18.

Bright star **Capella** sets in the northwest around 3:10 a.m. Read about Capella on page 14.

Just east of **Scorpius**, Sagittarius rises in the southeast around 4:05 a.m. and is in the eastern sky before dawn.

Bright **Venus** rises in the east around 5:25 a.m. and is in the eastern sky before dawn.

Look for **Mercury** in the morning sky, just before sunrise in the east. It's difficult but possible to see the tiny dot in the glare of the sun. It's easier to find with the naked eye, not using binoculars or a telescope. Never look directly at the sun or point binoculars or a telescope at it.

2027

Saturday, March 20

TONIGHT

sunset: **6:57 p.m.**
twilight begins: **7:52 p.m.**

IN THE MORNING

twilight begins: **5:53 a.m.**
sunrise: **6:47 a.m.**

MOON PHASE

moonrise: **5:26 p.m.**
moonset: **6:10 a.m.**

IN THE SKY

Castor and Pollux, the twin stars in the constellation Gemini, are almost directly overhead around dusk and set in the northwest around 3:10 a.m. Read how to find Castor and Pollux on page 14.

The **full moon** tonight acts like a spotlight. Look around at the desert landscape at how much illumination the moon provides.

Jupiter is at its highest, around 70 degrees, at 10:15 p.m. and sets in the west around 4:55 a.m.

Cassiopeia, shaped like a Σ, sets in the northwest around 10:30 p.m. See the illustration on page 11.

Rust-red and bright, **Mars** is at its highest, around 75 degrees, at 10:35 p.m. and sets in the west around 5:20 a.m.

One of the brightest stars in the sky, **Vega** rises in the northeast around 11:20 p.m.

The **Great Bear (Big Dipper)** is at its highest, around 65 degrees, at midnight. See the illustration on page 10.

The brightest star in the sky, **Sirius** sets in the west around 12:40 a.m.

Orion sets in the west around 12:55 a.m. Look for the bright star **Aldebaran** and the **Pleiades** near Orion. See the illustration on page 17.

Scorpius rises in the southeast around 1:10 a.m. and is at its highest in the south, around 30 degrees, around dawn. See the illustration on page 19.

The bright star **Altair** rises in the east around 2:30 a.m. and is in the eastern sky before dawn. Altair can be identified by its less bright neighboring stars, one on either side, together called the shaft of Aquila. See page 18.

Bright star **Capella** sets in the northwest around 3:05 a.m. Read about Capella on page 14.

Just east of **Scorpius,** Sagittarius rises in the southeast around 4:05 a.m. and is in the eastern sky before dawn.

Bright **Venus** rises in the east around 5:25 a.m.

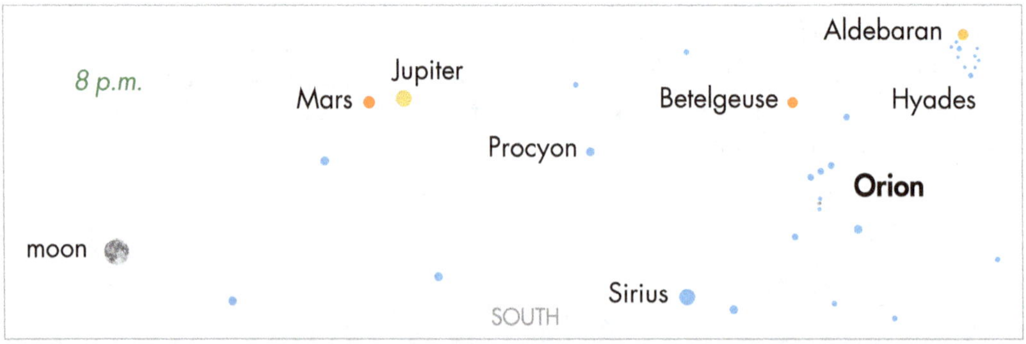

2027

Sunday, March 21

MOON PHASE

TONIGHT

sunset: **6:57 p.m.**
twilight begins: **7:52 p.m.**

IN THE MORNING

twilight begins: **5:52 a.m.**
sunrise: **6:45 a.m.**

moonrise: **6:34 p.m.**
moonset: **6:39 a.m.**

IN THE SKY

Castor and Pollux, the twin stars in the constellation Gemini, are almost directly overhead around dusk and set in the northwest around 3:05 a.m. Read how to find Castor and Pollux on page 14.

The **full moon** tonight acts like a spotlight. Look around at the desert landscape at how much illumination the moon provides.

Jupiter is at its highest, around 70 degrees, at 10:10 p.m. and sets in the west around 4:50 a.m.

Shaped like a *Σ,* **Cassiopeia** sets in the northwest around 10:25 p.m. See the illustration on page 11.

Rust-red and bright, **Mars** is at its highest, around 75 degrees, at 10:30 p.m. and sets in the west around 5:15 a.m.

One of the brightest stars in the sky, **Vega** rises in the northeast around 11:15 p.m.

The **Great Bear (Big Dipper)** is at its highest, around 65 degrees, at 11:55 p.m. See the illustration on page 10.

The brightest star in the sky, **Sirius** sets in the west around 12:40 a.m.

Orion sets in the west around 12:50 a.m. Look for the bright star **Aldebaran** and the **Pleiades** near Orion. See the illustration on page 17.

Scorpius rises in the southeast around 1:10 a.m. and is at its highest in the south, around 30 degrees, around dawn. See the illustration on page 19.

The bright star **Altair** rises in the east around 2:25 a.m. and is in the eastern sky before dawn. Altair can be identified by its less bright neighboring stars, one on either side, together called the shaft of Aquila. See page 18.

Bright star **Capella** sets in the northwest around 3:05 a.m. Read about Capella on page 14.

The bright star **Arcturus** is in the eastern sky at dusk and is at its highest, around 75 degrees, at 3:05 a.m. See page 13 for an easy way to locate the star.

Just east of **Scorpius,** Sagittarius rises in the southeast around 4 a.m. and is in the eastern sky before dawn.

Bright **Venus** rises in the east around 5:25 a.m.

Cassiopeia, shaped like a *W,* rises in the northeast around 5:30 a.m. It is in the northeastern sky before dawn.

2027

Monday, March 22

TONIGHT

sunset: **6:58 p.m.**
twilight begins: **7:53 p.m.**

IN THE MORNING

twilight begins: **5:51 a.m.**
sunrise: **6:44 a.m.**

MOON PHASE

moonrise: **7:40 p.m.**
moonset: **7:09 a.m.**

IN THE SKY

Castor and Pollux, the twin stars in the constellation Gemini, are almost directly overhead around dusk and set in the northwest around 3 a.m. Read how to find Castor and Pollux on page 14.

The **full moon** tonight acts like a spotlight. Look around at the desert landscape at how much illumination the moon provides.

Jupiter is at its highest, around 70 degrees, at 10:05 p.m. and sets in the west around 4:50 a.m.

Shaped like a *Σ*, **Cassiopeia** sets in the northwest around 10:20 p.m. See the illustration on page 11.

Rust-red and bright, **Mars** is at its highest, around 75 degrees, at 10:25 p.m. and sets in the west around 5:10 a.m.

One of the brightest stars in the sky, **Vega** rises in the northeast around 11:15 p.m.

The **Great Bear (Big Dipper)** is at its highest, around 65 degrees, at 11:50 p.m. See the illustration on page 10.

The brightest star in the sky, **Sirius** sets in the west around 12:35 a.m.

Orion sets in the west around 12:45 a.m. Look for the bright star **Aldebaran** and the **Pleiades** near Orion. See the illustration on page 17.

Scorpius rises in the southeast around 1:05 a.m. and is at its highest in the south, around 30 degrees, around dawn. See the illustration on page 19.

The bright star **Arcturus** is in the eastern sky at dusk and is at its highest, around 75 degrees, at 3:05 a.m. See page 13 for an easy way to locate the star.

The bright star **Altair** rises in the east around 2:20 a.m. and is in the eastern sky before dawn. Altair can be identified by its less bright neighboring stars, one on either side, together called the shaft of Aquila. See page 18.

Bright star **Capella** sets in the northwest around 3 a.m. Read about Capella on page 14.

Just east of **Scorpius,** Sagittarius rises in the southeast around 3:55 a.m. and is in the eastern sky before dawn.

Bright **Venus** rises in the east around 5:20 a.m.

Cassiopeia, shaped like a *W*, rises in the northeast around 5:25 a.m. It is in the northeastern sky before dawn.

2027

Tuesday, March 23

TONIGHT

sunset: **6:59 p.m.**
twilight begins: **7:54 p.m.**

MOON PHASE 

IN THE MORNING

twilight begins: **5:49 a.m.**
sunrise: **6:43 a.m.**

moonrise: **8:46 p.m.**
moonset: **7:09 a.m.**

IN THE SKY

Castor and Pollux, the twin stars in the constellation Gemini, are almost directly overhead around dusk and set in the northwest around 2:55 a.m. Read how to find Castor and Pollux on page 14.

The **full moon** tonight acts like a spotlight. Look around at the desert landscape at how much illumination the moon provides.

Jupiter is at its highest, around 70 degrees, at 10 p.m. and sets in the west around 4:45 a.m.

Shaped like a *Σ,* **Cassiopeia** sets in the northwest around 10:15 p.m. See the illustration on page 11.

Rust-red and bright, **Mars** is at its highest, around 75 degrees, at 10:20 p.m. and sets in the west around 5:05 a.m.

One of the brightest stars in the sky, **Vega** rises in the northeast around 11:10 p.m.

The **Great Bear (Big Dipper)** is at its highest, around 65 degrees, at 11:50 p.m. See the illustration on page 10.

The brightest star in the sky, **Sirius** sets in the west around 12:30 a.m.

Orion sets in the west around 12:40 a.m. Look for the bright star **Aldebaran** and the **Pleiades** near Orion. See the illustration on page 17.

Scorpius rises in the southeast around 1 a.m. and is at its highest in the south, around 30 degrees, around dawn. See the illustration on page 19.

The bright star **Altair** rises in the east around 2:15 a.m. and is in the eastern sky before dawn. Altair can be identified by its less bright neighboring stars, one on either side, together called the shaft of Aquila. See page 18.

Bright star **Capella** sets in the northwest around 2:55 a.m. Read about Capella on page 14.

The bright star **Arcturus** is in the eastern sky at dusk and is at its highest, around 75 degrees, at 3 a.m. See page 13 for an easy way to locate the star.

Just east of **Scorpius**, Sagittarius rises in the southeast around 3:50 a.m. and is in the eastern sky before dawn.

Bright **Venus** rises in the east around 5:20 a.m.

Cassiopeia, shaped like a *W,* rises in the northeast around 5:20 a.m. It is in the northeastern sky before dawn.

2027

Wednesday, March 24

TONIGHT

sunset: **7 p.m.**
twilight begins: **7:55 p.m.**

IN THE MORNING

twilight begins: **5:48 a.m.**
sunrise: **6:41 a.m.**

MOON PHASE

moonrise: **9:52 p.m.**
moonset: **7:41 a.m.**

IN THE SKY

Castor and Pollux, the twin stars in the constellation Gemini, are almost directly overhead around dusk and set in the northwest around 2:55 a.m. Read how to find Castor and Pollux on page 14.

Jupiter is at its highest, around 70 degrees, at 9:55 p.m. and sets in the west around 4:40 a.m.

Rust-red and bright, **Mars** is at its highest, around 75 degrees, at 10:15 p.m. and sets in the west around 5 a.m.

The nearly **full moon** tonight acts like a spotlight. Look around at the desert landscape at how much illumination the moon provides.

Shaped like a *Σ*, **Cassiopeia** sets in the northwest around 10:15 p.m. See the illustration on page 11.

One of the brightest stars in the sky, **Vega** rises in the northeast around 11:05 p.m.

The **Great Bear (Big Dipper)** is at its highest, around 65 degrees, at 11:45 p.m. See the illustration on page 10.

Orion sets in the west around 11:35 p.m. Look for the bright star **Aldebaran** and the **Pleiades** near Orion. See the illustration on page 17.

The brightest star in the sky, **Sirius** sets in the west around 12:25 a.m.

Scorpius rises in the southeast around 12:55 a.m. and is at its highest in the south, around 30 degrees, around dawn. See the illustration on page 19.

The bright star **Altair** rises in the east around 2:15 a.m. and is in the eastern sky before dawn. Altair can be identified by its less bright neighboring stars, one on either side, together called the shaft of Aquila. See page 18.

Bright star **Capella** sets in the northwest around 2:50 a.m. Read about Capella on page 14.

The bright star **Arcturus** is in the eastern sky at dusk and is at its highest, around 75 degrees, at 2:55 a.m. See page 13 for an easy way to locate the star.

Just east of **Scorpius**, Sagittarius rises in the southeast around 3:50 a.m. and is in the eastern sky before dawn.

Bright **Venus** rises in the east around 5:20 a.m.

Cassiopeia, shaped like a *W*, rises in the northeast around 5:20 a.m. It is in the northeastern sky before dawn.

2027
Thursday, March 25

TONIGHT

sunset: **7 p.m.**
twilight begins: **7:56 p.m.**

IN THE MORNING

twilight begins: **5:46 a.m.**
sunrise: **6:40 a.m.**

MOON PHASE

moonrise: **10:56 p.m.**
moonset: **8:15 a.m.**

IN THE SKY

Castor and Pollux, the twin stars in the constellation Gemini, are almost directly overhead around dusk and set in the northwest around 2:50 a.m. Read how to find Castor and Pollux on page 14.

Jupiter is at its highest, around 70 degrees, at 9:55 p.m. and sets in the west around 4:35 a.m.

Rust-red and bright, **Mars** is at its highest, around 75 degrees, at 10:10 p.m. and sets in the west around 4:55 a.m.

Shaped like a *Σ,* **Cassiopeia** sets in the northwest around 10:10 p.m. See the illustration on page 11.

One of the brightest stars in the sky, **Vega** rises in the northeast around 11 p.m.

Orion sets in the west around 11:30 p.m. Look for the bright star **Aldebaran** and the **Pleiades** near Orion. See the illustration on page 17.

The **Great Bear (Big Dipper)** is at its highest, around 65 degrees, at 11:40 p.m. See the illustration on page 10.

The brightest star in the sky, **Sirius** sets in the west around 12:20 a.m.

Scorpius rises in the southeast around 12:50 a.m. and is at its highest in the south, around 30 degrees, around dawn. See the illustration on page 19.

The bright star **Altair** rises in the east around 2:10 a.m. and is in the eastern sky before dawn. Altair can be identified by its less bright neighboring stars, one on either side, together called the shaft of Aquila. See page 18.

Bright star **Capella** sets in the northwest around 2:45 a.m. Read about Capella on page 14.

The bright star **Arcturus** is in the eastern sky at dusk and is at its highest, around 75 degrees, at 2:50 a.m. See page 13 for an easy way to locate the star.

Just east of **Scorpius,** Sagittarius rises in the southeast around 3:45 a.m. and is in the eastern sky before dawn.

Cassiopeia, shaped like a *W,* rises in the northeast around 5:15 a.m. It is in the northeastern sky before dawn.

Bright **Venus** rises in the east around 5:20 a.m.

2027

Friday, March 26

TONIGHT

sunset: **7:01 p.m.**
twilight begins: **7:56 p.m.**

IN THE MORNING

twilight begins: **5:45 a.m.**
sunrise: **6:38 a.m.**

MOON PHASE

moonrise: **11:58 p.m.**
moonset: **8:55 a.m.**

IN THE SKY

Explore the **dark sky** after twilight and before the moon rises, around 11:40 p.m.

Jupiter is at its highest, around 70 degrees, at 9:50 p.m. and sets in the west around 4:30 a.m.

Before the moon rises, until around 11:40 p.m., use binoculars or a telescope to look for the **moons of Jupiter.** See the guide on page 18.

Rust-red and bright, **Mars** is at its highest, around 75 degrees, at 10:05 p.m. and sets in the west around 4:55 a.m.

Look for the **Andromeda Galaxy** with binoculars or a telescope. It is a good target until around 9:55 p.m. See the guide on page 11.

Shaped like a *Σ,* **Cassiopeia** sets in the northwest around 10:05 p.m. See the illustration on page 11.

One of the brightest stars in the sky, **Vega** rises in the northeast around 11 p.m.

Orion sets in the west around 11:25 p.m. Look for the bright star **Aldebaran** and the **Pleiades** near Orion. See the illustration on page 17.

The **Great Bear (Big Dipper)** is at its highest, around 65 degrees, at 11:35 p.m. See the illustration on page 10.

The brightest star in the sky, **Sirius** sets in the west around 12:20 a.m.

Scorpius rises in the southeast around 12:50 a.m. and is at its highest in the south, around 30 degrees, around dawn. See the illustration on page 19.

The waning gibbous moon and the bright yellowish star **Antares** appear close together in the morning sky. Look southeast near the horizon around 2 a.m.

The bright star **Altair** rises in the east around 2:05 a.m. and is in the eastern sky before dawn. Altair can be identified by its less bright neighboring stars, one on either side, together called the shaft of Aquila. See page 18.

Bright star **Capella** sets in the northwest around 2:45 a.m. Read about Capella on page 14.

Castor and Pollux, the twin stars in the constellation Gemini, set in the northwest around 2:45 a.m. Read how to find Castor and Pollux on page 14.

The bright star **Arcturus** is in the eastern sky at dusk and is at its highest, around 75 degrees, at 2:45 a.m. See page 13 for an easy way to locate the star.

Just east of **Scorpius,** Sagittarius rises in the southeast around 3:40 a.m. and is in the eastern sky before dawn.

Cassiopeia, shaped like a *W,* rises in the northeast around 5:10 a.m. It is in the northeastern sky before dawn.

Bright **Venus** rises in the east around 5:20 a.m.

2027
Saturday, March 27

TONIGHT

sunset: **7:02 p.m.**
twilight begins: **7:57 p.m.**

IN THE MORNING

twilight begins: **5:43 a.m.**
sunrise: **6:37 a.m.**

MOON PHASE

moonrise: **12:55 a.m.**
moonset: **9:39 a.m.**

IN THE SKY

Explore the **dark sky** after twilight and before the moon rises, around 12:35 a.m.

Jupiter is at its highest, around 70 degrees, at 9:45 p.m. and sets in the west around 4:25 a.m.

Before the moon rises, until around 12:35 a.m., use binoculars or a telescope to look for the **moons of Jupiter.** See the guide on page 18.

Look for the **Andromeda Galaxy** with binoculars or a telescope in the northwestern sky. It is a good target until around 9:50 p.m. See page 11.

Rust-red and bright, **Mars** is at its highest, around 75 degrees, at 10 p.m. and sets in the west around 4:50 a.m.

Shaped like a *Σ*, **Cassiopeia** sets in the northwest around 10 p.m. See the illustration on page 11.

One of the brightest stars in the sky, **Vega** rises in the northeast around 10:55 p.m.

Orion sets in the west around 11:20 p.m. Look for the bright star **Aldebaran** and the **Pleiades** near Orion. See the illustration on page 17.

The **Great Bear (Big Dipper)** is at its highest, around 65 degrees, at 11:35 p.m. See the illustration on page 10.

The brightest star in the sky, **Sirius** sets in the west around 12:15 a.m.

Scorpius rises in the southeast around 12:45 a.m. and is at its highest in the south, around 30 degrees, around dawn. See the illustration on page 19.

Bright star **Capella** sets in the northwest around 2:40 a.m. Read about Capella on page 14.

Castor and Pollux, the twin stars in the constellation Gemini, set in the northwest around 2:40 a.m. Read how to find Castor and Pollux on page 14.

The bright star **Arcturus** is in the eastern sky at dusk and is at its highest, around 75 degrees, at 2:45 a.m. See page 13 for an easy way to locate the star.

Cassiopeia, shaped like a *W*, rises in the northeast around 5:05 a.m. It is in the northeastern sky before dawn.

Bright **Venus** rises in the east around 5:20 a.m.

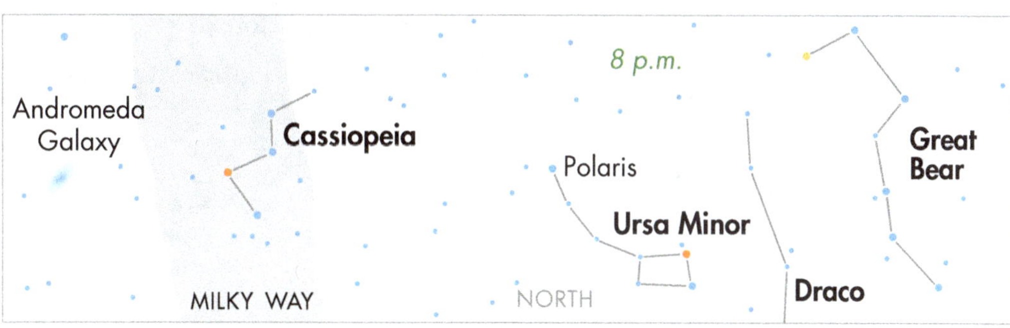

2027

Sunday, March 28

TONIGHT

sunset: **7:03 p.m.**
twilight begins: **7:58 p.m.**

MOON PHASE

IN THE MORNING

twilight begins: **5:42 a.m.**
sunrise: **6:36 a.m.**

moonrise: **1:47 a.m.**
moonset: **10:29 a.m.**

IN THE SKY

Explore the **dark sky** after twilight and before the moon rises, around 1:25 a.m.

Jupiter is at its highest, around 70 degrees, at 9:40 p.m. and sets in the west around 4:25 a.m.

Before the moon rises, until around 1:25 a.m., use binoculars or a telescope to look for the **moons of Jupiter.** See the guide on page 18.

Look for the **Andromeda Galaxy** with binoculars or a telescope. It is in the northwestern sky in the evening. It is a good target until around 9:50 p.m. See the guide on page 11.

Rust-red and bright, Mars is at its highest, around 75 degrees, at 10 p.m. and sets in the west around 4:45 a.m.

Shaped like a *Σ*, **Cassiopeia** sets in the northwest around 10 p.m. See the illustration on page 11.

One of the brightest stars in the sky, **Vega** rises in the northeast around 10:50 p.m.

Orion sets in the west around 11:20 p.m. Look for the bright star **Aldebaran** and the **Pleiades** near Orion. See the illustration on page 17.

The **Great Bear (Big Dipper)** is at its highest, around 65 degrees, at 11:30 p.m. See the illustration on page 10.

The brightest star in the sky, **Sirius** sets in the west around 12:10 a.m.

Scorpius rises in the southeast around 12:40 a.m. and is at its highest in the south, around 30 degrees, around dawn. See the illustration on page 19.

The bright star **Altair** rises in the east around 1:55 a.m. and is in the eastern sky before dawn. Altair can be identified by its less bright neighboring stars, one on either side, together called the shaft of Aquila. See page 18.

Bright star **Capella** sets in the northwest around 2:35 a.m. Read about Capella on page 14.

Castor and Pollux, the twin stars in the constellation Gemini, set in the northwest around 2:35 a.m. Read how to find Castor and Pollux on page 14.

The bright star **Arcturus** is in the eastern sky at dusk and is at its highest, around 75 degrees, at 2:40 a.m. See page 13 for an easy way to locate the star.

Just east of **Scorpius**, Sagittarius rises in the southeast around 3:30 a.m. and is in the eastern sky before dawn.

Cassiopeia, shaped like a *W,* rises in the northeast around 5:05 a.m. It is in the northeastern sky before dawn.

Bright Venus rises in the east around 5:20 a.m.

2027

Monday, March 29

TONIGHT

sunset: **7:04 p.m.**
twilight begins: **7:59 p.m.**

IN THE MORNING

twilight begins: **5:41 a.m.**
sunrise: **6:34 a.m.**

MOON PHASE

moonrise: **2:32 a.m.**
moonset: **11:23 a.m.**

IN THE SKY

Jupiter is at its highest, around 70 degrees, at 9:35 p.m. and sets in the west around 4:20 a.m.

Look for the **Andromeda Galaxy** with binoculars or a telescope. It is in the northwestern sky in the evening. It is a good target until around 9:45 p.m. See the guide on page 11.

Rust-red and bright, Mars is at its highest, around 75 degrees, at 9:55 p.m. and sets in the west around 4:40 a.m.

Before the moon rises, around 2:20 a.m., use binoculars or a telescope to look for the **moons of Jupiter.** See the guide on page 18.

The **Great Bear (Big Dipper)** is at its highest, around 65 degrees, at 11:25 p.m. See the illustration on page 10.

Shaped like a *Σ*, **Cassiopeia** sets in the northwest around 9:55 p.m. See the illustration on page 11.

One of the brightest stars in the sky, **Vega** rises in the northeast around 10:45 p.m.

Orion sets in the west around 11:15 p.m. Look for the bright star **Aldebaran** and the **Pleiades** near Orion. See the illustration on page 17.

The brightest star in the sky, **Sirius** sets in the west around 12:05 a.m.

Scorpius rises in the southeast around 12:35 a.m. and is at its highest in the south, around 30 degrees, around dawn. See the illustration on page 19.

The bright star **Altair** rises in the east around 1:55 a.m. and is in the eastern sky before dawn. Altair can be identified by its less bright neighboring stars, one on either side, together called the shaft of Aquila. See page 18.

Bright star **Capella** sets in the northwest around 2:30 a.m. Read about Capella on page 14.

Castor and Pollux, the twin stars in the constellation Gemini, set in the northwest around 2:35 a.m. Read how to find Castor and Pollux on page 14.

The bright star **Arcturus** is in the eastern sky at dusk and is at its highest, around 75 degrees, at 2:35 a.m. See page 13 for an easy way to locate the star.

Just east of **Scorpius**, Sagittarius rises in the southeast around 3:30 a.m. and is in the eastern sky before dawn.

Cassiopeia, shaped like a *W*, rises in the northeast around 5 a.m. It is in the northwestern sky at dusk and is in the northeastern sky before dawn.

Bright Venus rises in the east around 5:20 a.m.

2027

Tuesday, March 30

TONIGHT

sunset: 7:04 p.m.
twilight begins: 8 p.m.

IN THE MORNING

twilight begins: 5:39 a.m.
sunrise: 6:33 a.m.

MOON PHASE

moonrise: 3:11 a.m.
moonset: 12:20 p.m.

IN THE SKY

Jupiter is at its highest, around 70 degrees, at 9:30 p.m. and sets in the west around 4:15 a.m.

Until the moon rises, around 3 a.m., use binoculars or a telescope to look for the **moons of Jupiter.** See the guide on page 18.

Look for the **Andromeda Galaxy** with binoculars or a telescope. It is in the northwestern sky in the evening. It is a good target until around 9:40 p.m. See the guide on page 11.

Rust-red and bright, **Mars** is at its highest, around 75 degrees, at 9:50 p.m. and sets in the west around 4:35 a.m.

Shaped like a *Σ*, **Cassiopeia** sets in the northwest around 9:50 p.m. See the illustration on page 11.

One of the brightest stars in the sky, **Vega** rises in the northeast around 10:40 p.m.

Jupiter and **Mars** appear close together in the evening sky. Look southwest at about 65 degrees elevation around 11 p.m.

Orion sets in the west around 11:10 p.m. Look for the bright star **Aldebaran** and the **Pleiades** near Orion. See the illustration on page 17.

The **Great Bear (Big Dipper)** is at its highest, around 65 degrees, at 11:20 p.m. See the illustration on page 10.

The brightest star in the sky, **Sirius** sets in the west around 12:05 a.m.

Scorpius rises in the southeast around 12:35 a.m. and is at its highest in the south, around 30 degrees, around dawn. See the illustration on page 19.

The bright star **Altair** rises in the east around 1:50 a.m. and is in the eastern sky before dawn. Altair can be identified by its less bright neighboring stars, one on either side, together called the shaft of Aquila. See page 18.

Bright star **Capella** sets in the northwest around 2:30 a.m. Read about Capella on page 14.

Castor and Pollux, the twin stars in the constellation Gemini, set in the northwest around 2:30 a.m. Read how to find Castor and Pollux on page 14.

The bright star **Arcturus** is in the eastern sky at dusk and is at its highest, around 75 degrees, at 2:30 a.m. See page 13 for an easy way to locate the star.

Just east of **Scorpius,** Sagittarius rises in the southeast around 3:25 a.m. and is in the eastern sky before dawn.

Cassiopeia, shaped like a *W,* rises in the northeast around 4:55 a.m. It is in the northwestern sky at dusk and is in the northeastern sky before dawn.

Bright **Venus** rises in the east around 5:20 a.m.

2027

Wednesday, March 31

TONIGHT

sunset: **7:05 p.m.**
twilight begins: **8:01 p.m.**

MOON PHASE

IN THE MORNING

twilight begins: **5:38 a.m.**
sunrise: **6:32 a.m.**

moonrise: **3:44 a.m.**
moonset: **1:19 p.m.**

IN THE SKY

Jupiter is at its highest, around 70 degrees, at 9:30 p.m. and sets in the west around 4:10 a.m.

Until the moon rises, around 3:40 a.m., use binoculars or a telescope to look for the **moons of Jupiter.** See the guide on page 18.

Look for the **Andromeda Galaxy** with binoculars or a telescope. It is in the northwestern sky in the evening. It is a good target until around 9:35 p.m. See the guide on page 11.

Rust-red and bright, **Mars** is at its highest, around 75 degrees, at 9:45 p.m. and sets in the west around 4:30 a.m.

Shaped like a *Σ,* **Cassiopeia** sets in the northwest around 9:45 p.m. See the illustration on page 11.

One of the brightest stars in the sky, **Vega** rises in the northeast around 10:40 p.m.

The **Pleiades** sets in the northwest around 10:55 p.m. Read about the cluster on page 18.

Orion sets in the west around 11:05 p.m. See the illustration on page 17.

The **Great Bear (Big Dipper)** is at its highest, around 65 degrees, at 11:15 p.m. See the illustration on page 10.

The brightest star in the sky, **Sirius** sets in the west around midnight.

Scorpius rises in the southeast around 12:30 a.m. and is at its highest in the south, around 30 degrees, around dawn. See the illustration on page 19.

The bright star **Altair** rises in the east around 1:45 a.m. and is in the eastern sky before dawn. Altair can be identified by its less bright neighboring stars, one on either side, together called the shaft of Aquila. See page 18.

Bright star **Capella** sets in the northwest around 2:25 a.m. Read about Capella on page 14.

Castor and Pollux, the twin stars in the constellation Gemini, set in the northwest around 2:25 a.m. Read how to find Castor and Pollux on page 14.

The bright star **Arcturus** is in the eastern sky at dusk and is at its highest, around 75 degrees, at 2:25 a.m. See page 13 for an easy way to locate the star.

Just east of **Scorpius,** Sagittarius rises in the southeast around 3:20 a.m. and is in the eastern sky before dawn.

Cassiopeia, shaped like a *W,* rises in the northeast around 4:50 a.m. It is in the northwestern sky at dusk and is in the northeastern sky before dawn.

Bright **Venus** rises in the east around 5:20 a.m.

2027

Thursday, April 1

TONIGHT

sunset: **7:06 p.m.**
twilight begins: **8:01 p.m.**

IN THE MORNING

twilight begins: **5:36 a.m.**
sunrise: **6:30 a.m.**

MOON PHASE

moonrise: **4:14 a.m.**
moonset: **2:17 p.m.**

IN THE SKY

The absence of the moon tonight provides excellent **dark-sky** viewing conditions.

Jupiter is at its highest, around 70 degrees, at 9:25 p.m. and sets in the west around 4:05 a.m.

With the dark sky, use binoculars or a telescope to look for the **moons of Jupiter.** See the guide on page 18.

Look for the **Andromeda Galaxy** with binoculars or a telescope. It is in the northwestern sky in the evening. It is a good target until around 9:30 p.m. See the guide on page 11.

Rust-red and bright, Mars is at its highest, around 75 degrees, at 9:40 p.m. and sets in the west around 4:25 a.m.

Shaped like a *Σ*, **Cassiopeia** sets in the northwest around 9:40 p.m. See the illustration on page 11.

One of the brightest stars in the sky, **Vega** rises in the northeast around 10:35 p.m.

The **Pleiades** sets in the northwest around 10:55 p.m. Read about the cluster on page 18.

Orion sets in the west around 11 p.m. See the illustration on page 17.

The **Great Bear (Big Dipper)** is at its highest, around 65 degrees, at 11:15 p.m. See the illustration on page 10.

The brightest star in the sky, **Sirius** sets in the west around 11:55 p.m.

Scorpius rises in the southeast around 12:25 a.m. and is at its highest in the south, around 30 degrees, around dawn. See the illustration on page 19.

The bright star **Altair** rises in the east around 1:40 a.m. and is in the eastern sky before dawn. Altair can be identified by its less bright neighboring stars, one on either side, together called the shaft of Aquila. See page 18.

Bright star **Capella** sets in the northwest around 2:20 a.m. Read about Capella on page 14.

Castor and Pollux, the twin stars in the constellation Gemini, set in the northwest around 2:20 a.m. Read how to find Castor and Pollux on page 14.

The bright star **Arcturus** is in the eastern sky at dusk and is at its highest, around 75 degrees, at 2:25 a.m. See page 13 for an easy way to locate the star.

Just east of **Scorpius,** Sagittarius rises in the southeast around 3:15 a.m. and is in the eastern sky before dawn.

Cassiopeia, shaped like a *W*, rises in the northeast around 4:45 a.m. It is in the northwestern sky at dusk and is in the northeastern sky before dawn.

Bright Venus rises in the east around 5:20 a.m.

2027
Friday, April 2

TONIGHT

sunset: **7:07 p.m.**
twilight begins: **8:02 p.m.**

IN THE MORNING

twilight begins: **5:35 a.m.**
sunrise: **6:29 a.m.**

MOON PHASE

moonrise: **4:41 a.m.**
moonset: **3:16 p.m.**

IN THE SKY

Jupiter is at its highest, around 70 degrees, at 9:20 p.m. and sets in the west around 4:05 a.m.

With the dark sky, use binoculars or a telescope to look for the **moons of Jupiter.** See the guide on page 18.

Look for the **Andromeda Galaxy** with binoculars or a telescope in the northwestern sky. It is a good target until around 9:30 p.m. See page 11.

Rust-red and bright, **Mars** is at its highest, around 75 degrees, at 9:40 p.m. and sets in the west around 4:25 a.m.

Shaped like a *Σ,* **Cassiopeia** sets in the northwest around 9:40 p.m. See the illustration on page 11.

The absence of the moon tonight provides excellent **dark-sky** viewing conditions.

One of the brightest stars in the sky, **Vega** rises in the northeast around 10:30 p.m.

The **Pleiades** sets in the northwest around 10:50 p.m. Read about the cluster on page 18.

Orion sets in the west around 11 p.m. See the illustration on page 17.

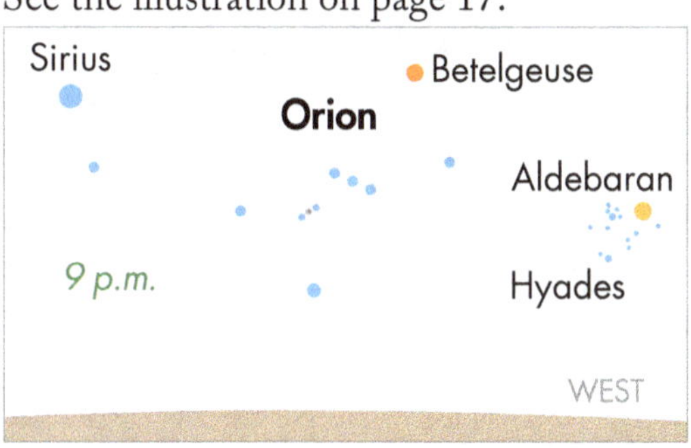

The **Great Bear (Big Dipper)** is at its highest, around 65 degrees, at 11:10 p.m. See the illustration on page 10.

The brightest star in the sky, **Sirius** sets in the west around 11:50 p.m.

Scorpius rises in the southeast around 12:20 a.m. and is at its highest in the south, around 30 degrees, at 4:35 a.m. See the illustration on page 19.

The bright star **Altair** rises in the east around 1:35 a.m. and is in the eastern sky before dawn. Altair can be identified by its less bright neighboring stars, one on either side, together called the shaft of Aquila. See page 18.

Bright star **Capella** sets in the northwest around 2:15 a.m. Read about Capella on page 14.

Castor and Pollux, the twin stars in the constellation Gemini, set in the northwest around 2:15 a.m. Read how to find Castor and Pollux on page 14.

The bright star **Arcturus** is at its highest, around 75 degrees, at 2:20 a.m. See page 13 for an easy way to locate the star.

Just east of **Scorpius,** Sagittarius rises in the southeast around 3:10 a.m. and is in the eastern sky before dawn.

Cassiopeia, shaped like a *W,* rises in the northeast around 4:45 a.m. It is in the northwestern sky at dusk and is in the northeastern sky before dawn.

Bright **Venus** rises in the east around 5:20 a.m.

2027

Saturday, April 3

TONIGHT

sunset: **7:07 p.m.**
twilight begins: **8:03 p.m.**

MOON PHASE

IN THE MORNING

twilight begins: **5:33 a.m.**
sunrise: **6:28 a.m.**

moonrise: **5:07 a.m.**
moonset: **4:14 p.m.**

IN THE SKY

Jupiter is at its highest, around 75 degrees, at 9:15 p.m. and sets in the west around 4 a.m.

With the dark sky, use binoculars or a telescope to look for the **moons of Jupiter.** See the guide on page 18.

With the dark sky, look for the **Andromeda Galaxy** with binoculars or a telescope. It is in the northwestern sky in the evening. It is a good target until around 9:25 p.m. See the guide on page 11.

Rust-red and bright, **Mars** is at its highest, around 75 degrees, at 9:35 p.m. and sets in the west around 4:20 a.m.

The absence of the moon tonight provides excellent **dark-sky** viewing conditions.

Shaped like a *Σ,* **Cassiopeia** sets in the northwest around 9:35 p.m. See the illustration on page 11.

One of the brightest stars in the sky, **Vega** rises in the northeast around 10:25 p.m.

The **Pleiades** sets in the northwest around 10:45 p.m. Read about the cluster on page 18.

Orion sets in the west around 10:55 p.m. See the illustration on page 17.

The **Great Bear (Big Dipper)** is at its highest, around 65 degrees, at 11:05 p.m. See the illustration on page 10.

The brightest star in the sky, **Sirius** sets in the west around 11:45 p.m.

Scorpius rises in the southeast around 12:15 a.m. and is at its highest in the south, around 30 degrees, at 4:30 a.m. See the illustration on page 19.

The bright star **Altair** rises in the east around 1:35 a.m. and is in the eastern sky before dawn. Altair can be identified by its less bright neighboring stars, one on either side, together called the shaft of Aquila. See page 18.

Bright star **Capella** sets in the northwest around 2:10 a.m. Read about Capella on page 14.

The bright star **Arcturus** is at its highest, around 75 degrees, at 2:15 a.m. See page 13 for an easy way to locate the star.

Castor and Pollux, the twin stars in the constellation Gemini, set in the northwest around 2:15 a.m. Read how to find Castor and Pollux on page 14.

Just east of **Scorpius,** Sagittarius rises in the southeast around 3:10 a.m. and is in the eastern sky before dawn.

Cassiopeia, shaped like a *W,* rises in the northeast around 4:40 a.m. It is in the northwestern sky at dusk and is in the northeastern sky before dawn.

Bright **Venus** rises in the east around 5:20 a.m.

The new moon and **Venus** appear close together in the morning sky. Look east near the horizon around 5:40 a.m.

2027

Sunday, April 4

TONIGHT

sunset: **7:08 p.m.**
twilight begins: **8:04 p.m.**

IN THE MORNING

twilight begins: **5:32 a.m.**
sunrise: **6:26 a.m.**

MOON PHASE

moonrise: **5:33 a.m.**
moonset: **5:13 p.m.**

IN THE SKY

Jupiter is at its highest, around 75 degrees, at 9:10 p.m. and sets in the west around 3:55 a.m.

With the dark sky, use binoculars or a telescope to look for the **moons of Jupiter.** See the guide on page 18.

With the dark sky, look for the **Andromeda Galaxy** with binoculars or a telescope. It is in the northwestern sky in the evening. It is a good target until around 9:20 p.m. See the guide on page 11.

Rust-red and bright, **Mars** is at its highest, around 75 degrees, at 9:30 p.m. and sets in the west around 4:15 a.m.

The absence of the moon tonight provides excellent **dark-sky** viewing conditions.

Shaped like a *Σ*, **Cassiopeia** sets in the northwest around 9:30 p.m. See the illustration on page 11.

One of the brightest stars in the sky, **Vega** rises in the northeast around 10:20 p.m.

The **Pleiades** sets in the northwest around 10:40 p.m. Read about the cluster on page 18.

Orion sets in the west around 10:50 p.m. See the illustration on page 17.

The **Great Bear (Big Dipper)** is at its highest, around 65 degrees, at 11 p.m. See the illustration on page 10.

The brightest star in the sky, **Sirius** sets in the west around 11:45 p.m.

Scorpius rises in the southeast around 12:15 a.m. and is at its highest in the south, around 30 degrees, at 4:25 a.m. See the illustration on page 19.

The bright star **Altair** rises in the east around 1:30 a.m. and is in the eastern sky before dawn. Altair can be identified by its less bright neighboring stars, one on either side, together called the shaft of Aquila. See page 18.

Bright star **Capella** sets in the northwest around 2:10 a.m. Read about Capella on page 14.

Castor and Pollux, the twin stars in the constellation Gemini, set in the northwest around 2:10 a.m. Read how to find Castor and Pollux on page 14.

The bright star **Arcturus** is at its highest, around 75 degrees, at 2:10 a.m. See page 13 for an easy way to locate the star.

Just east of **Scorpius**, Sagittarius rises in the southeast around 3:05 a.m. and is in the eastern sky before dawn.

Cassiopeia, shaped like a *W*, rises in the northeast around 4:35 a.m. It is in the northwestern sky at dusk and is in the northeastern sky before dawn.

Bright **Venus** rises in the east around 5:15 a.m.

2027

Monday, April 5

TONIGHT

sunset: **7:09 p.m.**
twilight begins: **8:05 p.m.**

MOON PHASE

IN THE MORNING

twilight begins: **5:30 a.m.**
sunrise: **6:25 a.m.**

moonrise: **6 a.m.**
moonset: **6:13 p.m.**

IN THE SKY

Jupiter is at its highest, around 75 degrees, at 9:10 p.m. and sets in the west around 3:50 a.m.

With the dark sky, use binoculars or a telescope to look for the **moons of Jupiter.** See the guide on page 18.

With the dark sky, look for the **Andromeda Galaxy** with binoculars or a telescope. It is in the northwestern sky in the evening. It is a good target until around 9:15 p.m. See the guide on page 11.

Rust-red and bright, **Mars** is at its highest, around 75 degrees, at 9:25 p.m. and sets in the west around 4:10 a.m.

The absence of the moon tonight provides excellent **dark-sky** viewing conditions.

Shaped like a *Σ*, **Cassiopeia** sets in the northwest around 9:25 p.m. See the illustration on page 11.

One of the brightest stars in the sky, **Vega** rises in the northeast around 10:20 p.m.

The **Pleiades** sets in the northwest around 10:35 p.m. Read about the cluster on page 18.

Orion sets in the west around 10:45 p.m. See the illustration on page 17.

The **Great Bear (Big Dipper)** is at its highest, around 65 degrees, at 10:55 p.m. See the illustration on page 10.

The brightest star in the sky, **Sirius** sets in the west around 11:40 p.m.

Scorpius rises in the southeast around 12:10 a.m. and is at its highest in the south, around 30 degrees, at 4:20 a.m. See the illustration on page 19.

The bright star **Altair** rises in the east around 1:25 a.m. and is in the eastern sky before dawn. Altair can be identified by its less bright neighboring stars, one on either side, together called the shaft of Aquila. See page 18.

Bright star **Capella** sets in the northwest around 2:05 a.m. Read about Capella on page 14.

Castor and Pollux, the twin stars in the constellation Gemini, set in the northwest around 2:05 a.m. Read how to find Castor and Pollux on page 14.

The bright star **Arcturus** is at its highest, around 75 degrees, at 2:10 a.m. See page 13 for an easy way to locate the star.

Just east of **Scorpius**, Sagittarius rises in the southeast around 3 a.m. and is in the eastern sky before dawn.

Cassiopeia, shaped like a *W*, rises in the northeast around 4:30 a.m. It is in the northwestern sky at dusk and is in the northeastern sky before dawn.

Bright **Venus** rises in the east around 5:15 a.m.

2027
Tuesday, April 6

TONIGHT

sunset: **7:10 p.m.**
twilight begins: **8:06 p.m.**

IN THE MORNING

twilight begins: **5:29 a.m.**
sunrise: **6:24 a.m.**

MOON PHASE

moonrise: **6:29 a.m.**
moonset: **7:16 p.m.**

IN THE SKY

Jupiter is at its highest, around 75 degrees, around dusk and sets in the west around 3:45 a.m.

After the moon sets, from around 7:35 p.m., use binoculars or a telescope to look for the **moons of Jupiter.** See the guide on page 18.

After the moon sets, from around 7:35 p.m., look for the **Andromeda Galaxy** with binoculars or a telescope. It is in the northwestern sky in the evening. It is a good target until around 9:15 p.m. See the guide on page 11.

The absence of the moon tonight provides excellent **dark-sky** viewing conditions.

Shaped like a *Σ,* **Cassiopeia** sets in the northwest around 9:20 p.m. See the illustration on page 11.

Rust-red and bright, Mars is at its highest, around 75 degrees, at 9:25 p.m. and sets in the west around 4:05 a.m.

One of the brightest stars in the sky, **Vega** rises in the northeast around 10:15 p.m.

The **Pleiades** sets in the northwest around 10:35 p.m. Read about the cluster on page 18.

Orion sets in the west around 10:45 p.m. See the illustration on page 17.

The **Great Bear (Big Dipper)** is at its highest, around 65 degrees, at 10:55 p.m. See the illustration on page 10.

The brightest star in the sky, **Sirius** sets in the west around 11:35 p.m.

Scorpius rises in the southeast around 12:05 a.m. and is at its highest in the south, around 30 degrees, at 4:15 a.m. See the illustration on page 19.

The bright star **Arcturus** is at its highest, around 75 degrees, at 2:05 a.m. See page 13 for an easy way to locate the star.

The bright star **Altair** rises in the east around 1:20 a.m. and is in the eastern sky before dawn. Altair can be identified by its less bright neighboring stars, one on either side, together called the shaft of Aquila. See page 18.

Bright star **Capella** sets in the northwest around 2 a.m. Read about Capella on page 14.

Castor and Pollux, the twin stars in the constellation Gemini, set in the northwest around 2 a.m. Read how to find Castor and Pollux on page 14.

Just east of **Scorpius,** Sagittarius rises in the southeast around 2:55 a.m. and is in the eastern sky before dawn.

Cassiopeia, shaped like a *W,* rises in the northeast around 4:25 a.m. It is in the northwestern sky at dusk and is in the northeastern sky before dawn.

Bright Venus rises in the east around 5:15 a.m.

2027

Wednesday, April 7

TONIGHT

sunset: **7:10 p.m.**
twilight begins: **8:07 p.m.**

IN THE MORNING

twilight begins: **5:28 a.m.**
sunrise: **6:22 a.m.**

MOON PHASE

moonrise: **7:03 a.m.**
moonset: **8:21 p.m.**

IN THE SKY

Jupiter is at its highest, around 75 degrees, around dusk and sets in the west around 3:45 a.m.

After the moon sets, from around 8:40 p.m., use binoculars or a telescope to look for the **moons of Jupiter.** See the guide on page 18.

After the moon sets, from around 8:40 p.m., look for the **Andromeda Galaxy** with binoculars or a telescope. It is in the northwestern sky in the evening. It is a good target until around 9:10 p.m. See the guide on page 11.

Rust-red and bright, **Mars** is at its highest, around 75 degrees, at 9:20 p.m. and sets in the west around 4:05 a.m.

The absence of the moon tonight provides excellent **dark-sky** viewing conditions.

Shaped like a *Σ*, **Cassiopeia** sets in the northwest around 9:20 p.m. See the illustration on page 11.

One of the brightest stars in the sky, **Vega** rises in the northeast around 10:10 p.m.

The **Pleiades** sets in the northwest around 10:30 p.m. Read about the cluster on page 18.

Orion sets in the west around 10:40 p.m. See the illustration on page 17.

The **Great Bear (Big Dipper)** is at its highest, around 65 degrees, at 10:50 p.m. See the illustration on page 10.

The brightest star in the sky, **Sirius** sets in the west around 11:30 p.m.

The bright star **Altair** rises in the east around 1:20 a.m. and is in the eastern sky before dawn. Altair can be identified by its less bright neighboring stars, one on either side, together called the shaft of Aquila. See page 18.

Bright star **Capella** sets in the northwest around 1:55 a.m. Read about Capella on page 14.

Castor and Pollux, the twin stars in the constellation Gemini, set in the northwest around 2 a.m. Read how to find Castor and Pollux on page 14.

The bright star **Arcturus** is at its highest, around 75 degrees, at 2 a.m. See page 13 for an easy way to locate the star.

Just east of **Scorpius,** Sagittarius rises in the southeast around 2:55 a.m. and is in the eastern sky before dawn.

Scorpius rises in the southeast around midnight and is at its highest in the south, around 30 degrees, at 4:15 a.m. See the illustration on page 19.

Cassiopeia, shaped like a *W,* rises in the northeast around 4:25 a.m. It is in the northwestern sky at dusk and is in the northeastern sky before dawn.

Bright **Venus** rises in the east around 5:15 a.m.

2027
Thursday, April 8

TONIGHT

sunset: **7:11 p.m.**
twilight begins: **8:07 p.m.**

IN THE MORNING

twilight begins: **5:26 a.m.**
sunrise: **6:21 a.m.**

MOON PHASE

moonrise: **7:03 a.m.**
moonset: **9:30 p.m.**

IN THE SKY

Jupiter is at its highest, around 75 degrees, around dusk and sets in the west around 3:40 a.m.

After the moon sets, from around 9:50 p.m., use binoculars or a telescope to look for the **moons of Jupiter.** See the guide on page 18.

Rust-red and bright, Mars is at its highest, around 75 degrees, at 9:15 p.m. and sets in the west around 4 a.m.

Shaped like a *Σ,* **Cassiopeia** sets in the northwest around 9:15 p.m. See the illustration on page 11.

The absence of the moon tonight provides excellent **dark-sky** viewing conditions.

One of the brightest stars in the sky, **Vega** rises in the northeast around 10:05 p.m.

The **Pleiades** sets in the northwest around 10:25 p.m. Read about the cluster on page 18.

Orion sets in the west around 10:35 p.m. See the illustration on page 17.

The **Great Bear (Big Dipper)** is at its highest, around 65 degrees, at 10:45 p.m. See the illustration on page 10.

The brightest star in the sky, **Sirius** sets in the west around 11:25 p.m.

Scorpius rises in the southeast around 11:55 p.m. and is at its highest in the south, around 30 degrees, at 4:10 a.m. See the illustration on page 19.

The bright star **Altair** rises in the east around 1:15 a.m. and is in the eastern sky before dawn. Altair can be identified by its less bright neighboring stars, one on either side, together called the shaft of Aquila. See page 18.

Bright star **Capella** sets in the northwest around 1:50 a.m. Read about Capella on page 14.

Castor and Pollux, the twin stars in the constellation Gemini, set in the northwest around 1:55 a.m. Read how to find Castor and Pollux on page 14.

The bright star **Arcturus** is at its highest, around 75 degrees, at 1:55 a.m. See page 13 for an easy way to locate the star.

Just east of **Scorpius,** Sagittarius rises in the southeast around 2:50 a.m. and is in the eastern sky before dawn.

Cassiopeia, shaped like a *W,* rises in the northeast around 4:20 a.m. It is in the northwestern sky at dusk and is in the northeastern sky before dawn.

Bright Venus rises in the east around 5:15 a.m.

2027

Friday, April 9

TONIGHT

sunset: **7:12 p.m.**
twilight begins: **8:08 p.m.**

MOON PHASE

IN THE MORNING

twilight begins: **5:25 a.m.**
sunrise: **6:20 a.m.**

moonrise: **7:43 a.m.**
moonset: **10:39 p.m.**

IN THE SKY

Jupiter is at its highest, around 75 degrees, around dusk and sets in the west around 3:35 a.m.

After the moon sets, from around 11 p.m., use binoculars or a telescope to look for the **moons of Jupiter.** See the guide on page 18.

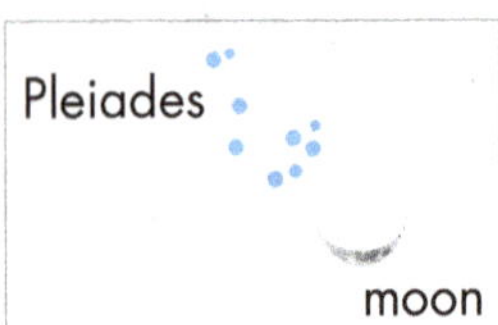

The waxing **sliver-crescent moon** and the **Pleiades** appear close together in the evening sky. Look west near the horizon around 8:30 p.m.

Rust-red and bright, Mars is at its highest, around 75 degrees, at 9:10 p.m. and sets in the west around 3:55 a.m.

Shaped like a *Σ*, **Cassiopeia** sets in the northwest around 9:10 p.m. See the illustration on page 11.

One of the brightest stars in the sky, **Vega** rises in the northeast around 10:05 p.m.

The **Pleiades** sets in the northwest around 10:20 p.m. Read about the cluster on page 18.

Orion sets in the west around 10:30 p.m. See the illustration on page 17.

The **Great Bear (Big Dipper)** is at its highest, around 65 degrees, at 10:40 p.m. See the illustration on page 10.

The brightest star in the sky, **Sirius** sets in the west around 11:25 p.m.

Scorpius rises in the southeast around 11:55 p.m. and is at its highest in the south, around 30 degrees, at 4:05 a.m. See the illustration on page 19.

The bright star **Altair** rises in the east around 1:05 a.m. Altair can be identified by its less bright neighboring stars, one on either side, together called the shaft of Aquila. See page 18.

Bright star **Capella** sets in the northwest around 1:50 a.m. Read about Capella on page 14.

Castor and Pollux, the twin stars in the constellation Gemini, set in the northwest around 1:50 a.m. Read how to find Castor and Pollux on page 14.

The bright star **Arcturus** is at its highest, around 75 degrees, at 1:50 a.m. See page 13 for an easy way to locate the star.

Just east of **Scorpius,** Sagittarius rises in the southeast around 2:45 a.m. and is in the eastern sky before dawn.

Cassiopeia, shaped like a *W,* rises in the northeast around 4:15 a.m. It is in the northwestern sky at dusk and is in the northeastern sky before dawn.

Bright Venus rises in the east around 5:15 a.m.

2027

Saturday, April 10

TONIGHT

sunset: **7:13 p.m.**
twilight begins: **8:09 p.m.**

MOON PHASE

IN THE MORNING

twilight begins: **5:23 a.m.**
sunrise: **6:18 a.m.**

moonrise: **8:31 a.m.**
moonset: **11:46 p.m.**

IN THE SKY

The **moon** is prominent in the sky early in the evening. Using a telescope or binoculars, take a look. You should be able to see rough mountains and craters along the boundary between light and dark, in astronomy called the terminator.

Jupiter is at its highest, around 75 degrees, around dusk and sets in the west around 3:30 a.m.

After the moon sets, from around 12:05 a.m., use binoculars or a telescope to look for the **moons of Jupiter.** See the guide on page 18.

Rust-red and bright, **Mars** is at its highest, around 75 degrees, around dusk and sets in the west around 3:50 a.m.

Shaped like a *Σ*, **Cassiopeia** sets in the northwest around 9:05 p.m. See the illustration on page 11.

One of the brightest stars in the sky, **Vega** rises in the northeast around 10 p.m.

The **Pleiades** sets in the northwest around 10:15 p.m. Read about the cluster on page 18.

Orion sets in the west around 10:25 p.m. See the illustration on page 17.

The **Great Bear (Big Dipper)** is at its highest, around 65 degrees, at 10:40 p.m. See the illustration on page 10.

The brightest star in the sky, **Sirius** sets in the west around 11:20 p.m.

Scorpius rises in the southeast around 11:50 p.m. and is at its highest in the south, around 30 degrees, at 4 a.m. See the illustration on page 19.

Explore the **dark sky** after the moon sets, from around 12:05 a.m.

The bright star **Altair** rises in the east around 1 a.m. Altair can be identified by its less bright neighboring stars, one on either side, together called the shaft of Aquila. See page 18.

Bright star **Capella** sets in the northwest around 1:45 a.m. Read about Capella on page 14.

Castor and Pollux, the twin stars in the constellation Gemini, set in the northwest around 1:45 a.m. Read how to find Castor and Pollux on page 14.

The bright star **Arcturus** is at its highest, around 75 degrees, at 1:50 a.m. See page 13 for an easy way to locate the star.

Just east of **Scorpius,** Sagittarius rises in the southeast around 2:40 a.m. and is in the eastern sky before dawn.

Cassiopeia, shaped like a *W*, rises in the northeast around 4:10 a.m. and is in the northeastern sky before dawn.

Bright **Venus** rises in the east around 5:15 a.m.

2027

Sunday, April 11

TONIGHT

sunset: **7:13 p.m.**
twilight begins: **8:10 p.m.**

IN THE MORNING

twilight begins: **5:22 a.m.**
sunrise: **6:17 a.m.**

MOON PHASE

moonrise: **9:28 a.m.**
moonset: **12:49 a.m.**

IN THE SKY

Rust-red and bright, Mars is at its highest, around 75 degrees, around dusk and sets in the west around 3:50 a.m.

Jupiter is at its highest, around 75 degrees, around dusk and sets in the west around 3:25 a.m.

After the moon sets, from around 1:10 a.m., use binoculars or a telescope to look for the **moons of Jupiter.** See the guide on page 18.

The **moon** is prominent in the sky early in the evening. Using a telescope or binoculars, take a look. You should be able to see rough mountains and craters along the boundary between light and dark, in astronomy called the terminator.

Shaped like a *Σ,* **Cassiopeia** sets in the northwest around 9 p.m. See the illustration on page 11.

One of the brightest stars in the sky, **Vega** rises in the northeast around 9:55 p.m.

The **Pleiades** sets in the northwest around 10:15 p.m. Read about the cluster on page 18.

Orion sets in the west around 10:25 p.m. See the illustration on page 17.

The **Great Bear (Big Dipper)** is at its highest, around 65 degrees, at 10:35 p.m. See the illustration on page 10.

The brightest star in the sky, **Sirius** sets in the west around 11:15 p.m.

Scorpius rises in the southeast around 11:45 p.m. and is at its highest in the south, around 30 degrees, at 4 a.m. See the illustration on page 19.

The bright star **Altair** rises in the east around 1 a.m. Altair can be identified by its less bright neighboring stars, one on either side, together called the shaft of Aquila. See page 18.

Explore the **dark sky** after the moon sets, from around 1:10 a.m.

Bright star **Capella** sets in the northwest around 1:40 a.m. Read about Capella on page 14.

Castor and Pollux, the twin stars in the constellation Gemini, set in the northwest around 1:40 a.m. Read how to find Castor and Pollux on page 14.

The bright star **Arcturus** is at its highest, around 75 degrees, at 1:45 a.m. See page 13 for an easy way to locate the star.

Just east of **Scorpius,** Sagittarius rises in the southeast around 2:35 a.m. and is in the eastern sky before dawn.

Cassiopeia, shaped like a *W,* rises in the northeast around 4:10 a.m. It is in the northeastern sky before dawn.

Bright Venus rises in the east around 5:15 a.m.

2027

Monday, April 12

TONIGHT

sunset: **7:14 p.m.**
twilight begins: **8:11 p.m.**

IN THE MORNING

twilight begins: **5:21 a.m.**
sunrise: **6:16 a.m.**

MOON PHASE

moonrise: **10:33 a.m.**
moonset: **1:43 a.m.**

IN THE SKY

Rust-red and bright, Mars is at its highest, around 75 degrees, around dusk and sets in the west around 3:45 a.m.

Jupiter is at its highest, around 75 degrees, around dusk and sets in the west around 3:25 a.m.

After the moon sets, from around 2:05 a.m., use binoculars or a telescope to look for the **moons of Jupiter.** See the guide on page 18.

The **moon** is prominent in the sky early in the evening. Using a telescope or binoculars, take a look. You should be able to see rough mountains and craters along the boundary between light and dark, in astronomy called the terminator.

Shaped like a *Σ,* **Cassiopeia** sets in the northwest around 9 p.m. See the illustration on page 11.

One of the brightest stars in the sky, **Vega** rises in the northeast around 9:50 p.m.

The **Pleiades** sets in the northwest around 10:10 p.m. Read about the cluster on page 18.

Orion sets in the west around 10:20 p.m. See the illustration on page 17.

The **Great Bear (Big Dipper)** is at its highest, around 65 degrees, at 10:30 p.m. See the illustration on page 10.

The brightest star in the sky, **Sirius** sets in the west around 11:10 p.m.

Scorpius rises in the southeast around 11:40 p.m. and is at its highest in the south, around 30 degrees, at 3:55 a.m. See the illustration on page 19.

The bright star **Altair** rises in the east around 12:55 a.m. Altair can be identified by its less bright neighboring stars, one on either side, together called the shaft of Aquila. See page 18.

Bright star **Capella** sets in the northwest around 1:35 a.m. Read about Capella on page 14.

Castor and Pollux, the twin stars in the constellation Gemini, set in the northwest around 1:40 a.m. Read how to find Castor and Pollux on page 14.

The bright star **Arcturus** is at its highest, around 75 degrees, at 1:40 a.m. See page 13 for an easy way to locate the star.

Explore the **dark sky** after the moon sets, from around 2:05 a.m.

Just east of **Scorpius**, Sagittarius rises in the southeast around 2:35 a.m. and is in the eastern sky before dawn.

Cassiopeia, shaped like a *W,* rises in the northeast around 4:05 a.m. and is in the northeastern sky before dawn.

Bright Venus rises in the east around 5:10 a.m.

2027

Tuesday, April 13

TONIGHT

sunset: **7:15 p.m.**
twilight begins: **8:12 p.m.**

IN THE MORNING

twilight begins: **5:19 a.m.**
sunrise: **6:15 a.m.**

MOON PHASE

moonrise: **11:43 a.m.**
moonset: **2:28 a.m.**

IN THE SKY

Rust-red and bright, Mars is at its highest, around 75 degrees, around dusk and sets in the west around 3:40 a.m.

Jupiter is at its highest, around 75 degrees, around dusk and sets in the west around 3:20 a.m.

The **moon** is prominent in the sky early in the evening. Using a telescope or binoculars, take a look. You should be able to see rough mountains and craters along the boundary between light and dark, in astronomy called the terminator.

Shaped like a *Σ*, **Cassiopeia** sets in the northwest around 8:55 p.m. See the illustration on page 11.

One of the brightest stars in the sky, **Vega** rises in the northeast around 9:45 p.m.

The **Pleiades** sets in the northwest around 10:05 p.m. Read about the cluster on page 18.

Orion sets in the west around 10:15 p.m. See the illustration on page 17.

The **Great Bear (Big Dipper)** is at its highest, around 65 degrees, at 10:25 p.m. See the illustration on page 10.

The brightest star in the sky, **Sirius** sets in the west around 11:10 p.m.

Scorpius rises in the southeast around 11:40 p.m. and is at its highest in the south, around 30 degrees, at 3:50 a.m. See the illustration on page 19.

The bright star **Altair** rises in the east around 12:50 a.m. Altair can be identified by its less bright neighboring stars, one on either side, together called the shaft of Aquila. See page 18.

Bright star **Capella** sets in the northwest around 1:35 a.m. Read about Capella on page 14.

Castor and Pollux, the twin stars in the constellation Gemini, set in the northwest around 1:35 a.m. Read how to find Castor and Pollux on page 14.

The bright star **Arcturus** is at its highest, around 75 degrees, at 1:35 a.m. See page 13 for an easy way to locate the star.

Just east of **Scorpius,** Sagittarius rises in the southeast around 2:30 a.m. and is in the eastern sky before dawn.

Explore the **dark sky** after the moon sets, from around 2:50 a.m.

Cassiopeia, shaped like a *W*, rises in the northeast around 4 a.m. and is in the northeastern sky before dawn.

Bright Venus rises in the east around 5:10 a.m.

2027

Wednesday, April 14

TONIGHT

sunset: **7:16 p.m.**
twilight begins: **8:13 p.m.**

IN THE MORNING

twilight begins: **5:18 a.m.**
sunrise: **6:13 a.m.**

MOON PHASE

moonrise: **12:54 p.m.**
moonset: **3:07 a.m.**

IN THE SKY

Rust-red and bright, Mars is at its highest, around 75 degrees, around dusk and sets in the west around 3:35 a.m.

The **moon** is prominent in the sky early in the evening. Using a telescope or binoculars, take a look. You should be able to see rough mountains and craters along the boundary between light and dark, in astronomy called the terminator.

Shaped like a *Σ*, **Cassiopeia** sets in the northwest around 8:50 p.m. See the illustration on page 11.

One of the brightest stars in the sky, **Vega** rises in the northeast around 9:45 p.m.

The **Pleiades** sets in the northwest around 10 p.m. Read about the cluster on page 18.

Orion is in the western sky at dusk and sets in the west around 10:10 p.m. See the illustration on page 17.

The **Great Bear (Big Dipper)** is at its highest, around 65 degrees, at 10:20 p.m. See the illustration on page 10.

The waxing gibbous **moon,** Jupiter, and Mars appear close together in the evening sky. Look southwest at about 60 degrees elevation around 10:30 p.m.

The brightest star in the sky, **Sirius** sets in the west around 11:05 p.m.

Scorpius rises in the southeast around 11:35 p.m. and is at its highest in the south, around 30 degrees, at 3:45 a.m. See the illustration on page 19.

The bright star **Altair** rises in the east around 12:45 a.m. Altair can be identified by its less bright neighboring stars, one on either side, together called the shaft of Aquila. See page 18.

Bright star **Capella** sets in the northwest around 1:30 a.m. Read about Capella on page 14.

Castor and Pollux, the twin stars in the constellation Gemini, set in the northwest around 1:30 a.m. Read how to find Castor and Pollux on page 14.

The bright star **Arcturus** is at its highest, around 75 degrees, at 1:30 a.m. See page 13 for an easy way to locate the star.

Just east of **Scorpius**, Sagittarius rises in the southeast around 2:25 a.m. and is in the eastern sky before dawn.

Jupiter sets in the west around 3:15 a.m.

Cassiopeia, shaped like a *W*, rises in the northeast around 3:55 a.m. It is in the northeastern sky before dawn.

Bright Venus rises in the east around 5:10 a.m.

2027

Thursday, April 15

TONIGHT

sunset: **7:17 p.m.**
twilight begins: **8:14 p.m.**

MOON PHASE

IN THE MORNING

twilight begins: **5:16 a.m.**
sunrise: **6:12 a.m.**

moonrise: **2:04 p.m.**
moonset: **3:40 a.m.**

IN THE SKY

Rust-red and bright, **Mars** is at its highest, around 70 degrees, around dusk and sets in the west around 3:35 a.m.

Shaped like a *Σ*, **Cassiopeia** sets in the northwest around 8:45 p.m. See the illustration on page 11.

One of the brightest stars in the sky, **Vega** rises in the northeast around 9:40 p.m.

The waxing gibbous moon, **Jupiter,** and **Mars** appear close together in the evening sky. Look southwest at about 65 degrees elevation around 9:45 p.m.

The **Pleiades** sets in the northwest around 10 p.m. Read about the cluster on page 18.

Orion is in the western sky at dusk and sets in the west around 10:05 p.m. See the illustration on page 17.

The **Great Bear (Big Dipper)** is at its highest, around 65 degrees, at 10:20 p.m. See the illustration on page 10.

The brightest star in the sky, **Sirius** sets in the west around 11 p.m.

Scorpius rises in the southeast around 11:30 p.m. and is at its highest in the south, around 30 degrees, at 3:40 a.m. See the illustration on page 19.

The bright star **Altair** rises in the east around 12:40 a.m. Altair can be identified by its less bright neighboring stars, one on either side, together called the shaft of Aquila. See page 18.

Bright star **Capella** sets in the northwest around 1:25 a.m. Read about Capella on page 14.

Castor and Pollux, the twin stars in the constellation Gemini, set in the northwest around 1:25 a.m. Read how to find Castor and Pollux on page 14.

The bright star **Arcturus** is at its highest, around 75 degrees, at 1:30 a.m. See page 13 for an easy way to locate the star.

Just east of **Scorpius,** Sagittarius rises in the southeast around 2:20 a.m. and is in the eastern sky before dawn.

Jupiter sets in the west around 3:10 a.m.

Cassiopeia, shaped like a *W*, rises in the northeast around 3:50 a.m. It is in the northeastern sky before dawn.

Bright **Venus** rises in the east around 5:10 a.m.

2027

Friday, April 16

TONIGHT

sunset: **7:17 p.m.**
twilight begins: **8:15 p.m.**

IN THE MORNING

twilight begins: **5:15 a.m.**
sunrise: **6:11 a.m.**

MOON PHASE

moonrise: **3:12 p.m.**
moonset: **4:11 a.m.**

IN THE SKY

Rust-red and bright, **Mars** is at its highest, around 70 degrees, around dusk and sets in the west around 3:30 a.m.

Shaped like a *Σ,* **Cassiopeia** sets in the northwest around 8:45 p.m. See the illustration on page 11.

One of the brightest stars in the sky, **Vega** rises in the northeast around 9:35 p.m.

The **Pleiades** is in the western sky at dusk and sets in the northwest around 9:55 p.m. Read about the cluster on page 18.

Orion is in the western sky at dusk and sets in the west around 10:05 p.m. See the illustration on page 17.

The **Great Bear (Big Dipper)** is at its highest, around 65 degrees, at 10:15 p.m. See the illustration on page 10.

The brightest star in the sky, **Sirius** sets in the west around 10:55 p.m.

Scorpius rises in the southeast around 11:25 p.m. and is at its highest in the south, around 30 degrees, at 3:40 a.m. See the illustration on page 19.

The bright star **Altair** rises in the east around 12:40 a.m. Altair can be identified by its less bright neighboring stars, one on either side, together called the shaft of Aquila. See page 18.

Bright star **Capella** sets in the northwest around 1:20 a.m. Read about Capella on page 14.

Castor and Pollux, the twin stars in the constellation Gemini, set in the northwest around 1:20 a.m. Read how to find Castor and Pollux on page 14.

The bright star **Arcturus** is at its highest, around 75 degrees, at 1:25 a.m. See page 13 for an easy way to locate the star.

Just east of **Scorpius,** Sagittarius rises in the southeast around 2:15 a.m. and is in the eastern sky before dawn.

Jupiter sets in the west around 3:05 a.m.

Cassiopeia, shaped like a *W,* rises in the northeast around 3:50 a.m. It is in the northeastern sky before dawn.

Bright **Venus** rises in the east around 5:10 a.m.

2027

Saturday, April 17

TONIGHT

sunset: **7:18** p.m.
twilight begins: **8:16** p.m.

IN THE MORNING

twilight begins: **5:14** a.m.
sunrise: **6:10** a.m.

MOON PHASE

moonrise: **4:18** p.m.
moonset: **4:39** a.m.

IN THE SKY

Rust-red and bright, Mars is at its highest, around 70 degrees, around dusk and sets in the west around 3:25 a.m.

Mars and Jupiter are close to each other in the constellation in **Leo.** Look high in the sky, above the moon, around 9 p.m.

The **Pleiades** is in the western sky at dusk and sets in the northwest around 9:50 p.m. Read about the cluster on page 18.

The **full moon** tonight acts like a spotlight. Look around at the desert landscape at how much illumination the moon provides.

Shaped like a Σ, **Cassiopeia** sets in the northwest around 8:40 p.m. See the illustration on page 11.

One of the brightest stars in the sky, **Vega** rises in the northeast around 9:30 p.m.

Orion is in the western sky at dusk and sets in the west around 10 p.m. See the illustration on page 17.

The **Great Bear (Big Dipper)** is at its highest, around 65 degrees, at 10:10 p.m. See the illustration on page 10.

The brightest star in the sky, **Sirius** sets in the west around 10:50 p.m.

The bright star **Altair** rises in the east around 12:35 a.m. Altair can be identified by its less bright neighboring stars, one on either side, together called the shaft of Aquila. See page 18.

Bright star **Capella** sets in the northwest around 1:15 a.m. Read about Capella on page 14.

Castor and Pollux, the twin stars in the constellation Gemini, set in the northwest around 1:20 a.m. Read how to find Castor and Pollux on page 14.

The bright star **Arcturus** is at its highest, around 75 degrees, at 1:20 a.m. See page 13 for an easy way to locate the star.

Jupiter sets in the west around 3:05 a.m.

Bright Venus rises in the east around 5:10 a.m.

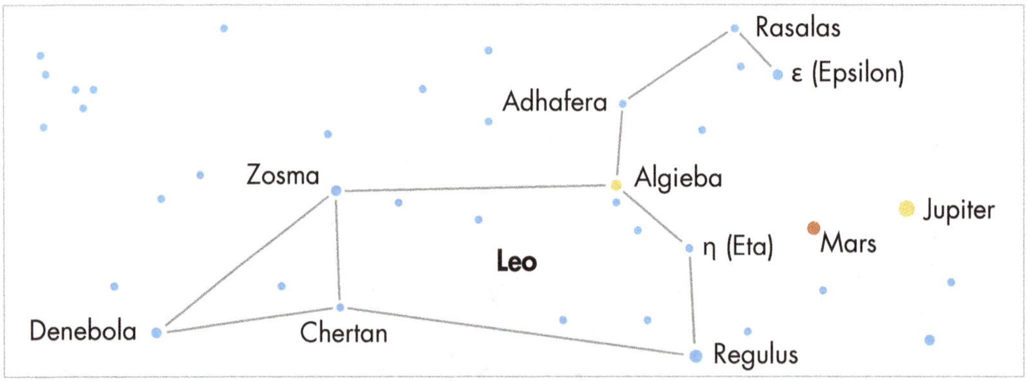

2027

Sunday, April 18

MOON PHASE

TONIGHT

sunset: 7:19 p.m.
twilight begins: 8:16 p.m.

IN THE MORNING

twilight begins: 5:12 a.m.
sunrise: 6:08 a.m.

moonrise: 5:24 p.m.
moonset: 5:08 a.m.

IN THE SKY

Rust-red and bright, Mars is at its highest, around 70 degrees, around dusk and sets in the west around 3:20 a.m.

The **full moon** tonight acts like a spotlight. Look around at the desert landscape at how much illumination the moon provides.

One of the brightest stars in the sky, **Vega** rises in the northeast around 9:25 p.m.

The **Pleiades** is in the western sky at dusk and sets in the northwest around 9:45 p.m. Read about the cluster on page 18.

Orion is in the western sky at dusk and sets in the west around 9:55 p.m. See the illustration on page 17.

The **Great Bear (Big Dipper)** is at its highest, around 65 degrees, at 10:05 p.m. See the illustration on page 10.

The brightest star in the sky, **Sirius** sets in the west around 10:50 p.m.

Scorpius rises in the southeast around 11:20 p.m. and is at its highest in the south, around 30 degrees, at 3:30 a.m. See the illustration on page 19.

The bright star **Altair** rises in the east around 12:30 a.m. Altair can be identified by its less bright neighboring stars, one on either side, together called the shaft of Aquila. See page 18.

Bright star **Capella** sets in the northwest around 1:15 a.m. Read about Capella on page 14.

Castor and Pollux, the twin stars in the constellation Gemini, set in the northwest around 1:15 a.m. Read how to find Castor and Pollux on page 14.

The bright star **Arcturus** is at its highest, around 75 degrees, at 1:15 a.m. See page 13 for an easy way to locate the star.

Just east of **Scorpius**, Sagittarius rises in the southeast around 2:05 a.m. and is in the eastern sky before dawn.

Jupiter sets in the west around 3 a.m.

Cassiopeia, shaped like a *W*, rises in the northeast around 3:40 a.m. It is in the northeastern sky before dawn. See the illustration on page 11.

Bright Venus rises in the east around 5:10 a.m.

2027

Monday, April 19

TONIGHT

sunset: **7:20 p.m.**
twilight begins: **8:17 p.m.**

MOON PHASE

IN THE MORNING

twilight begins: **5:11 a.m.**
sunrise: **6:07 a.m.**

moonrise: **6:29 p.m.**
moonset: **5:38 a.m.**

IN THE SKY

Rust-red and bright, **Mars** is at its highest, around 70 degrees, around dusk and sets in the west around 3:20 a.m.

The **full moon** tonight acts like a spotlight. Look around at the desert landscape at how much illumination the moon provides.

One of the brightest stars in the sky, **Vega** rises in the northeast around 9:25 p.m.

The **Pleiades** is in the western sky at dusk and sets in the northwest around 9:40 p.m. Read about the cluster on page 18.

Orion is in the western sky at dusk and sets in the west around 9:50 p.m. See the illustration on page 17.

The **Great Bear (Big Dipper)** is at its highest, around 65 degrees, at 10 p.m. See the illustration on page 10.

The brightest star in the sky, **Sirius** sets in the west around 10:45 p.m.

Scorpius rises in the southeast around 11:15 p.m. and is at its highest in the south, around 30 degrees, at 3:25 a.m. See the illustration on page 19.

The bright star **Altair** rises in the east around 12:25 a.m. Altair can be identified by its less bright neighboring stars, one on either side, together called the shaft of Aquila. See page 18.

Bright star **Capella** sets in the northwest around 1:10 a.m. Read about Capella on page 14.

Castor and Pollux, the twin stars in the constellation Gemini, set in the northwest around 1:10 a.m. Read how to find Castor and Pollux on page 14.

The bright star **Arcturus** is at its highest, around 75 degrees, at 1:15 a.m. See page 13 for an easy way to locate the star.

Just east of **Scorpius,** Sagittarius rises in the southeast around 2 a.m. and is in the eastern sky before dawn.

Jupiter sets in the west around 2:55 a.m.

Cassiopeia, shaped like a *W,* rises in the northeast around 3:35 a.m. It is in the northeastern sky before dawn. See the illustration on page 11.

Bright **Venus** rises in the east around 5:05 a.m.

2027

Tuesday, April 20

TONIGHT

sunset: **7:20 p.m.**
twilight begins: **8:18 p.m.**

IN THE MORNING

twilight begins: **5:10 a.m.**
sunrise: **6:06 a.m.**

MOON PHASE

moonrise: **7:34 p.m.**
moonset: **6:12 a.m.**

IN THE SKY

Jupiter sets in the west around 2:50 a.m.

The **full moon** tonight acts like a spotlight. Look around at the desert landscape at how much illumination the moon provides.

One of the brightest stars in the sky, **Vega** rises in the northeast around 9:20 p.m.

The **Pleiades** is in the western sky at dusk and sets in the northwest around 9:40 p.m. Read about the cluster on page 18.

Orion is in the western sky at dusk and sets in the west around 9:50 p.m. See the illustration on page 17.

The **Great Bear (Big Dipper)** is at its highest, around 65 degrees, at 10 p.m. See the illustration on page 10.

The brightest star in the sky, **Sirius** sets in the west around 10:40 p.m.

Scorpius rises in the southeast around 11:10 p.m. and is at its highest in the south, around 30 degrees, at 3:20 a.m. See the illustration on page 19.

The bright star **Altair** rises in the east around 12:25 a.m. Altair can be identified by its less bright neighboring stars, one on either side, together called the shaft of Aquila. See page 18.

Bright star **Capella** sets in the northwest around 1:05 a.m. Read about Capella on page 14.

Castor and Pollux, the twin stars in the constellation Gemini, set in the northwest around 1:05 a.m. Read how to find Castor and Pollux on page 14.

The bright star **Arcturus** is at its highest, around 75 degrees, at 1:10 a.m. See page 13 for an easy way to locate the star.

Just east of **Scorpius**, Sagittarius rises in the southeast around 2 a.m. and is in the eastern sky before dawn.

Rust-red and bright, **Mars** is in the western sky before dawn and sets in the west around 3:15 a.m.

Cassiopeia, shaped like a *W,* rises in the northeast around 3:30 a.m. It is in the northeastern sky before dawn. See the illustration on page 11.

Bright **Venus** rises in the east around 5:05 a.m.

2027

Wednesday, April 21

TONIGHT

sunset: **7:21 p.m.**
twilight begins: **8:19 p.m.**

IN THE MORNING

twilight begins: **5:08 a.m.**
sunrise: **6:05 a.m.**

MOON PHASE

moonrise: **8:39 p.m.**
moonset: **6:49 a.m.**

IN THE SKY

The **Lyrids meteor shower** happens tonight. A good shower with up to 20 meteors per hour, the best time to watch is after midnight. Tonight unfortunately moonlight washes out the shower.

The **full moon** tonight acts like a spotlight. Look around at the desert landscape at how much illumination the moon provides.

One of the brightest stars in the sky, **Vega** rises in the northeast around 9:15 p.m.

The **Pleiades** is in the western sky at dusk and sets in the northwest around 9:35 p.m. Read about the cluster on page 18.

Orion is in the western sky at dusk and sets in the west around 9:45 p.m. See the illustration on page 17.

The **Great Bear (Big Dipper)** is at its highest, around 65 degrees, at 9:55 p.m. See the illustration on page 10.

The brightest star in the sky, **Sirius** sets in the west around 10:35 p.m.

Scorpius rises in the southeast around 11:05 p.m. and is at its highest in the south, around 30 degrees, at 3:20 a.m. See the illustration on page 19.

The bright star **Altair** rises in the east around 12:20 a.m. Altair can be identified by its less bright neighboring stars, one on either side, together called the shaft of Aquila. See page 18.

Bright star **Capella** sets in the northwest around 1 a.m. Read about Capella on page 14.

Castor and Pollux, the twin stars in the constellation Gemini, set in the northwest around 1 a.m. Read how to find Castor and Pollux on page 14.

The bright star **Arcturus** is at its highest, around 75 degrees, at 1:05 a.m. See page 13 for an easy way to locate the star.

Just east of **Scorpius**, Sagittarius rises in the southeast around 1:55 a.m. and is in the eastern sky before dawn.

Jupiter sets in the west around 2:50 a.m.

Rust-red and bright, **Mars** is in the western sky before dawn and sets in the west around 3:10 a.m.

Cassiopeia, shaped like a *W,* rises in the northeast around 3:30 a.m. It is in the northeastern sky before dawn. See the illustration on page 11.

Bright **Venus** rises in the east around 5:05 a.m.

2027

Thursday, April 22

TONIGHT

sunset: **7:22 p.m.**
twilight begins: **8:20 p.m.**

IN THE MORNING

twilight begins: **5:07 a.m.**
sunrise: **6:04 a.m.**

MOON PHASE

moonrise: **9:43 p.m.**
moonset: **7:32 a.m.**

IN THE SKY

The **Lyrids meteor shower** happens tonight. A good shower with up to 20 meteors per hour, the best time to watch is after midnight. Tonight unfortunately moonlight washes out the shower.

The **full moon** tonight acts like a spotlight. Look around at the desert landscape at how much illumination the moon provides.

One of the brightest stars in the sky, **Vega** rises in the northeast around 9:10 p.m.

The **Pleiades** is in the western sky at dusk and sets in the northwest around 9:30 p.m. Read about the cluster on page 18.

Orion is in the western sky at dusk and sets in the west around 9:40 p.m. See the illustration on page 17.

The **Great Bear (Big Dipper)** is at its highest, around 65 degrees, at 9:50 p.m. See the illustration on page 10.

The brightest star in the sky, **Sirius** sets in the west around 10:30 p.m.

Scorpius rises in the southeast around 11 p.m. and is at its highest in the south, around 30 degrees, at 3:15 a.m. See the illustration on page 19.

The bright star **Altair** rises in the east around 12:15 a.m. Altair can be identified by its less bright neighboring stars, one on either side, together called the shaft of Aquila. See page 18.

Bright star **Capella** sets in the northwest around 12:55 a.m. Read about Capella on page 14.

Castor and Pollux, the twin stars in the constellation Gemini, set in the northwest around 1 a.m. Read how to find Castor and Pollux on page 14.

The bright star **Arcturus** is at its highest, around 75 degrees, at 1 a.m. See page 13 for an easy way to locate the star.

Just east of **Scorpius**, Sagittarius rises in the southeast around 1:50 a.m. and is in the eastern sky before dawn.

Jupiter sets in the west around 2:45 a.m.

Rust-red and bright, **Mars** is in the western sky before dawn and sets in the west around 3:10 a.m.

Cassiopeia, shaped like a *W*, rises in the northeast around 3:25 a.m. It is in the northeastern sky before dawn. See the illustration on page 11.

Bright **Venus** rises in the east around 5:05 a.m.

2027

Friday, April 23

TONIGHT

sunset: **7:23 p.m.**
twilight begins: **8:21 p.m.**

IN THE MORNING

twilight begins: **5:06 a.m.**
sunrise: **6:03 a.m.**

MOON PHASE

moonrise: **10:43 p.m.**
moonset: **7:32 a.m.**

IN THE SKY

The **Lyrids meteor shower** happens tonight. A good shower with up to 20 meteors per hour, the best time to watch is after midnight. Tonight unfortunately moonlight washes out the shower.

The nearly **full moon** tonight acts like a spotlight. Look around at the desert landscape at how much illumination the moon provides.

One of the brightest stars in the sky, **Vega** rises in the northeast around 9:10 p.m. and is in the eastern sky at dusk.

The **Pleiades** is in the western sky at dusk and sets in the northwest around 9:25 p.m. Read about the cluster on page 18.

Orion is in the western sky at dusk and sets in the west around 9:35 p.m. See the illustration on page 17.

The **Great Bear (Big Dipper)** is at its highest, around 65 degrees, at 9:45 p.m. See the illustration on page 10.

The brightest star in the sky, **Sirius** sets in the west around 10:30 p.m.

Scorpius rises in the southeast around 11 p.m. and is at its highest in the south, around 30 degrees, at 3:10 a.m. See the illustration on page 19.

The nearly **full moon** and the bright yellowish star **Antares** appear close together in the evening sky. Look southeast near the horizon around 11:50 p.m.

The bright star **Altair** rises in the east around 12:10 a.m. Altair can be identified by its less bright neighboring stars, one on either side, together called the shaft of Aquila. See page 18.

Bright star **Capella** sets in the northwest around 12:55 a.m. Read about Capella on page 14.

Castor and Pollux, the twin stars in the constellation Gemini, set in the northwest around 12:55 a.m. Read how to find Castor and Pollux on page 14.

The bright star **Arcturus** is at its highest, around 75 degrees, at 12:55 a.m. See page 13 for an easy way to locate the star.

Just east of **Scorpius,** Sagittarius rises in the southeast around 1:45 a.m. and is in the eastern sky before dawn.

Jupiter sets in the west around 2:40 a.m.

Rust-red and bright, **Mars** is in the western sky before dawn and sets in the west around 3:05 a.m.

Cassiopeia, shaped like a *W,* rises in the northeast around 3:20 a.m. It is in the northeastern sky before dawn. See the illustration on page 11.

Bright **Venus** rises in the east around 5:05 a.m.

2027

Saturday, April 24

MOON PHASE

TONIGHT

sunset: **7:24 p.m.**
twilight begins: **8:22 p.m.**

IN THE MORNING

twilight begins: **5:04 a.m.**
sunrise: **6:01 a.m.**

moonrise: **11:37 p.m.**
moonset: **8:20 a.m.**

IN THE SKY

The **Pleiades** is in the western sky at dusk and sets in the northwest around 9:20 p.m. Read about the cluster on page 18.

Orion is in the western sky at dusk and sets in the west around 9:30 p.m. See the illustration on page 17.

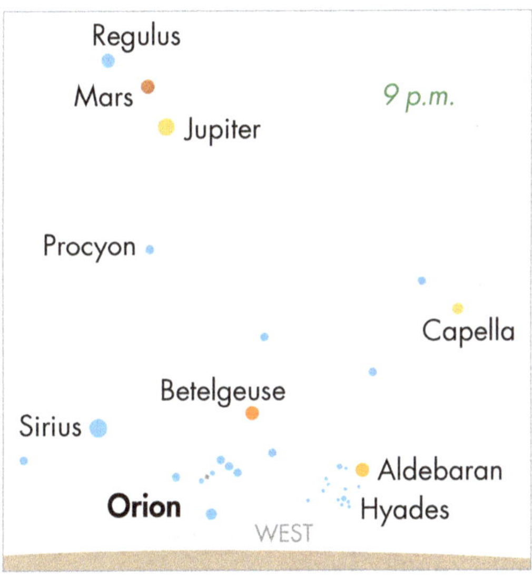

One of the brightest stars in the sky, **Vega** rises in the northeast around 9:05 p.m. and is in the eastern sky at dusk.

The **Great Bear (Big Dipper)** is at its highest, around 65 degrees, at 9:45 p.m. See the illustration on page 10.

The brightest star in the sky, **Sirius** sets in the west around 10:25 p.m.

Scorpius rises in the southeast around 10:55 p.m. and is at its highest in the south, around 30 degrees, at 3:05 a.m. See the illustration on page 19.

The bright star **Altair** rises in the east around 12:05 a.m. Altair can be identified by its less bright neighboring stars, one on either side, together called the shaft of Aquila. See page 18.

Bright star **Capella** sets in the northwest around 12:45 a.m. Read about Capella on page 14.

Castor and Pollux, the twin stars in the constellation Gemini, set in the northwest around 12:50 a.m. Read how to find Castor and Pollux on page 14.

The bright star **Arcturus** is at its highest, around 75 degrees, at 12:50 a.m. See page 13 for an easy way to locate the star.

Just east of **Scorpius,** Sagittarius rises in the southeast around 1:40 a.m. and is in the eastern sky before dawn.

Jupiter sets in the west around 2:35 a.m.

Rust-red and bright, **Mars** is in the western sky before dawn and sets in the west around 3 a.m.

Cassiopeia, shaped like a *W,* rises in the northeast around 3:15 a.m. It is in the northeastern sky before dawn. See the illustration on page 11.

Bright **Venus** rises in the east around 5:05 a.m.

2027

Sunday, April 25

TONIGHT

sunset: **7:24 p.m.**
twilight begins: **8:23 p.m.**

IN THE MORNING

twilight begins: **5:03 a.m.**
sunrise: **6 a.m.**

MOON PHASE

moonrise: **12:25 a.m.**
moonset: **9:13 a.m.**

IN THE SKY

One of the brightest stars in the sky, **Vega** rises in the northeast around 9 p.m. and is in the eastern sky at dusk.

The **Pleiades** is in the western sky at dusk and sets in the northwest around 9:20 p.m. Read about the cluster on page 18.

Orion is in the western sky at dusk and sets in the west around 9:30 p.m. See the illustration on page 17.

The **Great Bear (Big Dipper)** is at its highest, around 65 degrees, at 9:40 p.m. See the illustration on page 10.

The brightest star in the sky, **Sirius** sets in the west around 10:20 p.m.

Scorpius rises in the southeast around 10:50 p.m. and is at its highest in the south, around 30 degrees, at 3:05 a.m. See the illustration on page 19.

Explore the **dark sky** after twilight and before the moon rises, around 12:05 a.m.

Before the moon rises, until around 12:05 a.m., use binoculars or a telescope to look for the **moons of Jupiter.** See the guide on page 18.

The bright star **Altair** rises in the east around 12:05 a.m. Altair can be identified by its less bright neighboring stars, one on either side, together called the shaft of Aquila. See page 18.

Bright star **Capella** sets in the northwest around 12:40 a.m. Read about Capella on page 14.

The bright star **Arcturus** is at its highest, around 75 degrees, at 12:45 a.m. See page 13 for an easy way to locate the star.

Castor and Pollux, the twin stars in the constellation Gemini, set in the northwest around 12:45 a.m. Read how to find Castor and Pollux on page 14.

Just east of **Scorpius**, Sagittarius rises in the southeast around 1:40 a.m. and is in the eastern sky before dawn.

Jupiter sets in the west around 2:35 a.m.

Rust-red and bright, **Mars** is in the western sky before dawn and sets in the west around 3 a.m.

Cassiopeia, shaped like a *W*, rises in the northeast around 3:15 a.m. It is in the northeastern sky before dawn. See the illustration on page 11.

Bright **Venus** rises in the east around 5:05 a.m.

2027

Monday, April 26

TONIGHT

sunset: **7:25 p.m.**
twilight begins: **8:24 p.m.**

IN THE MORNING

twilight begins: **5:02 a.m.**
sunrise: **5:59 a.m.**

MOON PHASE

moonrise: **1:07 a.m.**
moonset: **10:09 a.m.**

IN THE SKY

One of the brightest stars in the sky, **Vega** rises in the northeast around 8:55 p.m. and is in the eastern sky at dusk.

The **Pleiades** is in the western sky at dusk and sets in the northwest around 9:15 p.m. Read about the cluster on page 18.

Orion is in the western sky at dusk and sets in the west around 9:25 p.m. See the illustration on page 17.

The **Great Bear (Big Dipper)** is at its highest, around 65 degrees, at 9:35 p.m. See the illustration on page 10.

The brightest star in the sky, **Sirius** sets in the west around 10:15 p.m.

Scorpius is visible most of the night. It rises in the southeast around 10:45 p.m. and is at its highest in the south, around 30 degrees, at 3 a.m. See the illustration on page 19.

The bright star **Altair** rises in the east around midnight. Altair can be identified by its less bright neighboring stars, one on either side, together called the shaft of Aquila. See page 18.

The bright star **Arcturus** is at its highest, around 75 degrees, at 12:40 a.m. See page 13 for an easy way to locate the star.

Bright star **Capella** sets in the northwest around 12:40 a.m. Read about Capella on page 14.

Castor and Pollux, the twin stars in the constellation Gemini, set in the northwest around 12:45 a.m. Read how to find Castor and Pollux on page 14.

Before the moon rises, until around 12:45 a.m., use binoculars or a telescope to look for the **moons of Jupiter.** See the guide on page 18.

Explore the **dark sky** after twilight and before the moon rises, around 12:45 a.m.

Just east of **Scorpius**, Sagittarius rises in the southeast around 1:35 a.m. and is in the eastern sky before dawn.

Jupiter sets in the west around 2:30 a.m.

Rust-red and bright, **Mars** is in the western sky before dawn and sets in the west around 2:55 a.m.

Cassiopeia, shaped like a *W,* rises in the northeast around 3:10 a.m. It is in the northeastern sky before dawn. See the illustration on page 11.

Bright **Venus** rises in the east around 5 a.m.

2027

Tuesday, April 27

TONIGHT

sunset: **7:26 p.m.**
twilight begins: **8:25 p.m.**

IN THE MORNING

twilight begins: **5:01 a.m.**
sunrise: **5:58 a.m.**

MOON PHASE

moonrise: **1:42 a.m.**
moonset: **11:07 a.m.**

IN THE SKY

One of the brightest stars in the sky, **Vega** is in the eastern sky at dusk.

The **Pleiades** is in the western sky at dusk and sets in the northwest around 9:10 p.m. Read about the cluster on page 18.

Orion is in the western sky at dusk and sets in the west around 9:20 p.m. See the illustration on page 17.

The **Great Bear (Big Dipper)** is at its highest, around 65 degrees, at 9:30 p.m. See the illustration on page 10.

The brightest star in the sky, **Sirius** is in the western sky at dusk and sets in the west around 10:15 p.m.

Scorpius is visible most of the night. It rises in the southeast around 10:45 p.m. and is at its highest in the south, around 30 degrees, at 2:55 a.m. See the illustration on page 19.

The bright star **Altair** rises in the east around 11:55 p.m. Altair can be identified by its less bright neighboring stars, one on either side, together called the shaft of Aquila. See page 18.

The bright star **Arcturus** is at its highest, around 75 degrees, at 12:35 a.m. See page 13 for an easy way to locate the star.

Bright star **Capella** sets in the northwest around 12:35 a.m. Read about Capella on page 14.

Castor and Pollux, the twin stars in the constellation Gemini, set in the northwest around 12:40 a.m. Read how to find Castor and Pollux on page 14.

Before the moon rises, until around 1:20 a.m., use binoculars or a telescope to look for the **moons of Jupiter.** See the guide on page 18.

Explore the **dark sky** after twilight and before the moon rises, around 1:20 a.m.

Just east of **Scorpius**, Sagittarius rises in the southeast around 1:30 a.m. and is in the eastern sky before dawn.

Jupiter sets in the west around 2:25 a.m.

Rust-red and bright, **Mars** sets in the west around 2:50 a.m.

Cassiopeia, shaped like a *W,* rises in the northeast around 3:05 a.m. It is in the northeastern sky before dawn. See the illustration on page 11.

Bright **Venus** rises in the east around 5 a.m.

2027

Wednesday, April 28

TONIGHT

sunset: **7:27 p.m.**
twilight begins: **8:26 p.m.**

MOON PHASE

IN THE MORNING

twilight begins: **4:59 a.m.**
sunrise: **5:57 a.m.**

moonrise: **2:13 a.m.**
moonset: **12:06 p.m.**

IN THE SKY

One of the brightest stars in the sky, **Vega** is in the eastern sky at dusk.

The **Pleiades** is in the western sky at dusk and sets in the northwest around 9:05 p.m. Read about the cluster on page 18.

Orion is in the western sky at dusk and sets in the west around 9:15 p.m. See the illustration on page 17.

The **Great Bear (Big Dipper)** is at its highest, around 65 degrees, at 9:25 p.m. See the illustration on page 10.

The brightest star in the sky, **Sirius** is in the western sky at dusk and sets in the west around 10:10 p.m.

Scorpius is visible most of the night. It rises in the southeast around 10:40 p.m. and is at its highest in the south, around 30 degrees, at 2:50 a.m. See the illustration on page 19.

The bright star **Altair** rises in the east around 11:50 p.m. Altair can be identified by its less bright neighboring stars, one on either side, together called the shaft of Aquila. See page 18.

Bright star **Capella** sets in the northwest around 12:30 a.m. Read about Capella on page 14.

Castor and Pollux, the twin stars in the constellation Gemini, set in the northwest around 12:35 a.m. Read how to find Castor and Pollux on page 14.

The bright star **Arcturus** is at its highest, around 75 degrees, at 12:35 a.m. See page 13 for an easy way to locate the star.

Just east of **Scorpius**, Sagittarius rises in the southeast around 1:25 a.m. and is in the eastern sky before dawn.

Before the moon rises, around 2 a.m., use binoculars or a telescope to look for the **moons of Jupiter.** See the guide on page 18.

Jupiter sets in the west around 2:20 a.m.

Rust-red and bright, **Mars** sets in the west around 2:50 a.m.

Cassiopeia, shaped like a *W,* rises in the northeast around 3 a.m. It is in the northeastern sky before dawn. See the illustration on page 11.

Bright **Venus** rises in the east around 5 a.m.

2027

Thursday, April 29

TONIGHT

sunset: **7:28 p.m.**
twilight begins: **8:27 p.m.**

IN THE MORNING

twilight begins: **4:58 a.m.**
sunrise: **5:56 a.m.**

MOON PHASE

moonrise: **2:41 a.m.**
moonset: **1:04 p.m.**

IN THE SKY

One of the brightest stars in the sky, **Vega** is in the eastern sky at dusk.

The **Great Bear (Big Dipper)** is at its highest, around 65 degrees, around dusk. See the illustration on page 10.

The **Pleiades** is in the western sky at dusk and sets in the northwest around 9:05 p.m. Read about the cluster on page 18.

Orion is in the western sky at dusk and sets in the west around 9:10 p.m. See the illustration on page 17.

The brightest star in the sky, **Sirius** is in the western sky at dusk and sets in the west around 10:05 p.m.

Scorpius is visible most of the night. It rises in the southeast around 10:35 p.m. and is at its highest in the south, around 30 degrees, at 2:45 a.m. See the illustration on page 19.

The bright star **Altair** rises in the east around 11:45 p.m. Altair can be identified by its less bright neighboring stars, one on either side, together called the shaft of Aquila. See page 18.

Bright star **Capella** sets in the northwest around 12:25 a.m. Read about Capella on page 14.

Castor and Pollux, the twin stars in the constellation Gemini, set in the northwest around 12:30 a.m. Read how to find Castor and Pollux on page 14.

The bright star **Arcturus** is at its highest, around 75 degrees, at 12:30 a.m. See page 13 for an easy way to locate the star.

Just east of **Scorpius**, Sagittarius rises in the southeast around 1:20 a.m. and is at its highest in the south, around 25 degrees, around dawn.

Jupiter sets in the west around 2:15 a.m.

Before the moon rises, use binoculars or a telescope to look for the **moons of Jupiter.** See the guide on page 18.

Rust-red and bright, **Mars** sets in the west around 2:45 a.m.

Cassiopeia, shaped like a *W,* rises in the northeast around 2:55 a.m. It is in the northeastern sky before dawn. See the illustration on page 11.

Bright **Venus** rises in the east around 5 a.m.

2027
Friday, April 30

TONIGHT

sunset: **7:28 p.m.**
twilight begins: **8:28 p.m.**

IN THE MORNING

twilight begins: **4:57 a.m.**
sunrise: **5:55 a.m.**

MOON PHASE

moonrise: **3:07 a.m.**
moonset: **2:01 p.m.**

IN THE SKY

One of the brightest stars in the sky, **Vega** is in the eastern sky at dusk.

The **Great Bear (Big Dipper)** is at its highest, around 65 degrees, around dusk. See the illustration on page 10.

The **Pleiades** is in the western sky at dusk and sets in the northwest around 9 p.m. Read about the cluster on page 18.

Orion sets in the west around 9:10 p.m. See the illustration on page 17.

The brightest star in the sky, **Sirius** is in the western sky at dusk and sets in the west around 10 p.m.

Scorpius is visible most of the night. It rises in the southeast around 10:30 p.m. and is at its highest in the south, around 30 degrees, at 2:45 a.m. See the illustration on page 19.

The bright star **Altair** rises in the east around 11:45 p.m. Altair can be identified by its less bright neighboring stars, one on either side, together called the shaft of Aquila. See page 18.

Bright star **Capella** sets in the northwest around 12:20 a.m. Read about Capella on page 14.

The bright star **Arcturus** is at its highest, around 75 degrees, at 12:25 a.m. See page 13 for an easy way to locate the star.

Castor and Pollux, the twin stars in the constellation Gemini, set in the northwest around 12:25 a.m. Read how to find Castor and Pollux on page 14.

Just east of **Scorpius**, Sagittarius rises in the southeast around 1:20 a.m. and is at its highest in the south, around 25 degrees, around dawn.

Jupiter sets in the west around 2:15 a.m.

With the dark sky, use binoculars or a telescope to look for the **moons of Jupiter.** See the guide on page 18.

Rust-red and bright, Mars sets in the west around 2:40 a.m.

Cassiopeia, shaped like a *W,* rises in the northeast around 2:55 a.m. It is in the northeastern sky before dawn. See the illustration on page 11.

Bright Venus rises in the east around 5 a.m.

2027

Saturday, May 1

TONIGHT

sunset: **7:29 p.m.**
twilight begins: **8:29 p.m.**

MOON PHASE

IN THE MORNING

twilight begins: **4:56 a.m.**
sunrise: **5:54 a.m.**

moonrise: **3:33 a.m.**
moonset: **2:59 p.m.**

IN THE SKY

One of the brightest stars in the sky, **Vega** is in the eastern sky at dusk.

The **Great Bear (Big Dipper)** is at its highest, around 65 degrees, around dusk. See the illustration on page 10.

The absence of the moon tonight provides excellent **dark-sky** viewing conditions.

The **Pleiades** is in the western sky at dusk and sets in the northwest around 8:55 p.m. Read about the cluster on page 18.

Orion sets in the west around 9:05 p.m. See the illustration on page 17.

The brightest star in the sky, **Sirius** is in the western sky at dusk and sets in the west around 9:55 p.m.

Scorpius is visible most of the night. It rises in the southeast around 10:25 p.m. and is at its highest in the south, around 30 degrees, at 2:40 a.m. See the illustration on page 19.

The bright star **Altair** rises in the east around 11:40 p.m. Altair can be identified by its less bright neighboring stars, one on either side, together called the shaft of Aquila. See page 18.

Bright star **Capella** sets in the northwest around 12:20 a.m. Read about Capella on page 14.

The bright star **Arcturus** is at its highest, around 75 degrees, at 12:20 a.m. See page 13 for an easy way to locate the star.

Castor and Pollux, the twin stars in the constellation Gemini, set in the northwest around 12:25 a.m. Read how to find Castor and Pollux on page 14.

Just east of **Scorpius**, Sagittarius rises in the southeast around 1:15 a.m. and is at its highest in the south, around 25 degrees, around dawn.

Jupiter sets in the west around 2:10 a.m.

With the dark sky, use binoculars or a telescope to look for the **moons of Jupiter.** See the guide on page 18.

Rust-red and bright, **Mars** sets in the west around 2:40 a.m.

Cassiopeia, shaped like a *W*, rises in the northeast around 2:50 a.m. It is in the northeastern sky before dawn. See the illustration on page 11.

Bright **Venus** rises in the east around 5 a.m.

2027

Sunday, May 2

TONIGHT

sunset: **7:30 p.m.**
twilight begins: **8:30 p.m.**

IN THE MORNING

twilight begins: **4:55 a.m.**
sunrise: **5:53 a.m.**

MOON PHASE

moonrise: **3:59 a.m.**
moonset: **3:58 p.m.**

IN THE SKY

One of the brightest stars in the sky, **Vega** is in the eastern sky at dusk.

The **Great Bear (Big Dipper)** is at its highest, around 65 degrees, around dusk. See the illustration on page 10.

The absence of the moon tonight provides excellent **dark-sky** viewing conditions.

The **Pleiades** is in the western sky at dusk and sets in the northwest around 8:50 p.m. Read about the cluster on page 18.

Orion sets in the west around 9 p.m. See the illustration on page 17.

The brightest star in the sky, **Sirius** is in the western sky at dusk and sets in the west around 9:55 p.m.

Scorpius is visible most of the night. It rises in the southeast around 10:25 p.m. and is at its highest in the south, around 30 degrees, at 2:35 a.m. See the illustration on page 19.

The bright star **Altair** rises in the east around 11:35 p.m. Altair can be identified by its less bright neighboring stars, one on either side, together called the shaft of Aquila. See page 18.

The bright star **Arcturus** is at its highest, around 75 degrees, at 12:15 a.m. See page 13 for an easy way to locate the star.

Bright star **Capella** sets in the northwest around 12:15 a.m. Read about Capella on page 14.

Castor and Pollux, the twin stars in the constellation Gemini, set in the northwest around 12:20 a.m. Read how to find Castor and Pollux on page 14.

Just east of **Scorpius**, Sagittarius rises in the southeast around 1:10 a.m. and is at its highest in the south, around 25 degrees, around dawn.

Jupiter sets in the west around 2:05 a.m.

With the dark sky, use binoculars or a telescope to look for the **moons of Jupiter.** See the guide on page 18.

Rust-red and bright, **Mars** sets in the west around 2:35 a.m.

Cassiopeia, shaped like a *W,* rises in the northeast around 2:45 a.m. It is in the northeastern sky before dawn. See the illustration on page 11.

Bright **Venus** rises in the east around 5 a.m.

2027

Monday, May 3

TONIGHT

sunset: **7:31 p.m.**
twilight begins: **8:31 p.m.**

MOON PHASE

IN THE MORNING

twilight begins: **4:53 a.m.**
sunrise: **5:52 a.m.**

moonrise: **4:28 a.m.**
moonset: **5 p.m.**

IN THE SKY

The brightest star in the sky, **Sirius** is in the western sky at dusk and sets in the west around 9:50 p.m.

The **Great Bear (Big Dipper)** is at its highest, around 65 degrees, around dusk. See the illustration on page 10.

One of the brightest stars in the sky, **Vega** is in the eastern sky at dusk.

The absence of the moon tonight provides excellent **dark-sky** viewing conditions.

With the dark sky, use binoculars or a telescope to look for the **moons of Jupiter.** See the guide on page 18.

The **Pleiades** sets in the northwest around 8:45 p.m. Read about the cluster on page 18.

Orion sets in the west around 8:55 p.m. See the illustration on page 17.

Scorpius is visible most of the night. It rises in the southeast around 10:20 p.m. and is at its highest in the south, around 30 degrees, at 2:30 a.m. See the illustration on page 19.

The bright star **Altair** rises in the east around 11:30 p.m. Altair can be identified by its less bright neighboring stars, one on either side, together called the shaft of Aquila. See page 18.

Bright star **Capella** sets in the northwest around 12:10 a.m. Read about Capella on page 14.

Castor and Pollux, the twin stars in the constellation Gemini, set in the northwest around 12:15 a.m. Read how to find Castor and Pollux on page 14.

The bright star **Arcturus** is at its highest, around 75 degrees, at 12:15 a.m. See page 13 for an easy way to locate the star.

Just east of **Scorpius**, Sagittarius rises in the southeast around 1:05 a.m. and is at its highest in the south, around 25 degrees, around dawn.

Jupiter sets in the west around 2 a.m.

Rust-red and bright, **Mars** sets in the west around 2:30 a.m.

Cassiopeia, shaped like a *W*, rises in the northeast around 2:40 a.m. It is in the northeastern sky before dawn. See the illustration on page 11.

Bright **Venus** rises in the east around 4:55 a.m.

2027

Tuesday, May 4

TONIGHT

sunset: **7:31 p.m.**
twilight begins: **8:32 p.m.**

IN THE MORNING

twilight begins: **4:52 a.m.**
sunrise: **5:51 a.m.**

MOON PHASE

moonrise: **5 a.m.**
moonset: **6:04 p.m.**

IN THE SKY

The **Great Bear (Big Dipper)** is at its highest, around 65 degrees, around dusk. See the illustration on page 10.

One of the brightest stars in the sky, **Vega** is in the eastern sky at dusk.

The absence of the moon tonight provides excellent **dark-sky** viewing conditions.

With the dark sky, use binoculars or a telescope to look for the **moons of Jupiter.** See the guide on page 18.

The **Pleiades** sets in the northwest around 8:45 p.m. Read about the cluster on page 18.

Orion sets in the west around 8:55 p.m. See the illustration on page 17.

The brightest star in the sky, **Sirius** is in the western sky at dusk and sets in the west around 9:45 p.m.

Scorpius is visible most of the night. It rises in the southeast around 10:15 p.m. and is at its highest in the south, around 30 degrees, at 2:25 a.m. See the illustration on page 19.

The bright star **Altair** rises in the east around 11:30 p.m. Altair can be identified by its less bright neighboring stars, one on either side, together called the shaft of Aquila. See page 18.

Bright star **Capella** sets in the northwest around 12:05 a.m. Read about Capella on page 14.

The bright star **Arcturus** is at its highest, around 75 degrees, at 12:10 a.m. See page 13 for an easy way to locate the star.

Just east of **Scorpius,** Sagittarius rises in the southeast around 1:05 a.m. and is at its highest in the south, around 25 degrees, around dawn.

Castor and Pollux, the twin stars in the constellation Gemini, set in the northwest around 12:10 a.m. Read how to find Castor and Pollux on page 14.

Jupiter sets in the west around 2 a.m.

Rust-red and bright, Mars sets in the west around 2:30 a.m.

Cassiopeia, shaped like a *W,* rises in the northeast around 2:35 a.m. It is in the northeastern sky before dawn. See the illustration on page 11.

Bright Venus rises in the east around 4:55 a.m.

2027

Wednesday, May 5

TONIGHT

sunset: 7:32 p.m.
twilight begins: 8:33 p.m.

MOON PHASE

IN THE MORNING

twilight begins: 4:51 a.m.
sunrise: 5:50 a.m.

moonrise: 5:38 a.m.
moonset: 7:12 p.m.

IN THE SKY

The **Eta Aquarids meteor shower** happens tonight. A good shower with up to 30 meteors per hour, the best time to watch is after midnight. Tonight has moonless dark skies all night.

The **Great Bear (Big Dipper)** is at its highest, around 65 degrees, around dusk. See the illustration on page 10.

After the moon sets, from around 7:30 p.m., use binoculars or a telescope to look for the **moons of Jupiter.** See the guide on page 18.

The absence of the moon tonight provides excellent **dark-sky** viewing conditions.

The **Pleiades** sets in the northwest around 8:40 p.m. Read about the cluster on page 18.

The brightest star in the sky, **Sirius** is in the western sky at dusk and sets in the west around 9:40 p.m.

Scorpius is visible most of the night. It rises in the southeast around 10:10 p.m. and is at its highest in the south, around 30 degrees, at 2:25 a.m. See the illustration on page 19.

The **summer triangle**'s three bright stars are **Vega, Altair,** and **Deneb. Vega** is in the eastern sky at dusk and is almost directly overhead around dawn. **Altair** rises in the east around 11:25 p.m. Altair can be identified by its less bright neighboring stars, one on either side, together called the shaft of Aquila. **Deneb** is in the band of the Milky Way. See the illustration of the summer triangle on page 19.

Bright star **Capella** sets in the northwest around midnight. Read about Capella on page 14.

The bright star **Arcturus** is at its highest, around 75 degrees, at 12:05 a.m. See page 13 for an easy way to locate the star.

Castor and Pollux, the twin stars in the constellation Gemini, set in the northwest around 12:05 a.m. Read how to find Castor and Pollux on page 14.

Just east of **Scorpius,** Sagittarius rises in the southeast around 1 a.m. and is at its highest in the south, around 25 degrees, around dawn.

Jupiter sets in the west around 1:55 a.m.

Rust-red and bright, **Mars** sets in the west around 2:25 a.m.

Cassiopeia, shaped like a *W,* rises in the northeast around 2:35 a.m. It is in the northeastern sky before dawn. See the illustration on page 11.

Bright **Venus** rises in the east around 4:55 a.m.

2027

Thursday, May 6

TONIGHT

sunset: **7:33 p.m.**
twilight begins: **8:34 p.m.**

IN THE MORNING

twilight begins: **4:50 a.m.**
sunrise: **5:49 a.m.**

MOON PHASE

moonrise: **6:24 a.m.**
moonset: **8:23 p.m.**

IN THE SKY

The **Eta Aquarids meteor shower** happens tonight. A good shower with up to 30 meteors per hour, the best time to watch is after midnight. Tonight has moonless dark skies all night.

The **Great Bear (Big Dipper)** is at its highest, around 65 degrees, around dusk. See the illustration on page 10.

The **Pleiades** sets in the northwest around 8:35 p.m. Read about the cluster on page 18.

The absence of the moon tonight provides excellent **dark-sky** viewing conditions.

After the moon sets, from around 8:45 p.m., use binoculars or a telescope to look for the **moons of Jupiter.** See the guide on page 18.

The brightest star in the sky, **Sirius** is in the western sky at dusk and sets in the west around 9:35 p.m.

Scorpius is visible most of the night. It rises in the southeast around 10:05 p.m. and is at its highest in the south, around 30 degrees, at 2:20 a.m. See the illustration on page 19.

The **summer triangle**'s three bright stars are **Vega, Altair,** and **Deneb. Vega** is in the eastern sky at dusk and is almost directly overhead around dawn. **Altair** rises in the east around 11:20 p.m. Altair can be identified by its less bright neighboring stars, one on either side, together called the shaft of Aquila. **Deneb** is in the band of the Milky Way. See the illustration of the summer triangle on page 19.

Bright star **Capella** sets in the northwest around midnight. Read about Capella on page 14.

The bright star **Arcturus** is at its highest, around 75 degrees, at midnight. See page 13 for an easy way to locate the star.

Castor and Pollux, the twin stars in the constellation Gemini, set in the northwest around 12:05 a.m. Read how to find Castor and Pollux on page 14.

Just east of **Scorpius,** Sagittarius rises in the southeast around 12:55 a.m. and is at its highest in the south, around 25 degrees, around dawn.

Jupiter sets in the west around 1:50 a.m.

Rust-red and bright, **Mars** sets in the west around 2:20 a.m.

Cassiopeia, shaped like a *W,* rises in the northeast around 2:30 a.m. It is in the northeastern sky before dawn. See the illustration on page 11.

Bright **Venus** rises in the east around 4:55 a.m.

2027

Friday, May 7

TONIGHT

sunset: **7:34 p.m.**
twilight begins: **8:35 p.m.**

IN THE MORNING

twilight begins: **4:49 a.m.**
sunrise: **5:48 a.m.**

MOON PHASE

moonrise: **7:19 a.m.**
moonset: **9:33 p.m.**

IN THE SKY

The **Eta Aquarids meteor shower** happens tonight. A good shower with up to 30 meteors per hour, the best time to watch is after midnight. Tonight has moonless dark skies all night.

The absence of the moon tonight provides excellent **dark-sky** viewing conditions.

The brightest star in the sky, **Sirius** is in the western sky at dusk and sets in the west around 9:35 p.m.

After the moon sets, from around 9:55 p.m., use binoculars or a telescope to look for the **moons of Jupiter.** See the guide on page 18.

Scorpius is visible most of the night. It rises in the southeast around 10:05 p.m. and is at its highest in the south, around 30 degrees, at 2:15 a.m. See the illustration on page 19.

The **summer triangle**'s three bright stars are **Vega, Altair,** and **Deneb. Vega** is in the eastern sky at dusk and is almost directly overhead around dawn. **Altair** rises in the east around 11:15 p.m. Altair can be identified by its less bright neighboring stars, one on either side, together called the shaft of Aquila. **Deneb** is in the band of the Milky Way. See the illustration of the summer triangle on page 19.

Bright star **Capella** sets in the northwest around 11:55 p.m. Read about Capella on page 14.

Castor and Pollux, the twin stars in the constellation Gemini, set in the northwest around 11:55 p.m. Read how to find Castor and Pollux on page 14.

The bright star **Arcturus** is at its highest, around 75 degrees, at midnight.

Just east of **Scorpius,** Sagittarius rises in the southeast around 12:50 a.m. and is at its highest in the south, around 25 degrees, around dawn.

Jupiter sets in the west around 1:45 a.m.

Rust-red and bright, **Mars** sets in the west around 2:20 a.m.

Cassiopeia, shaped like a *W,* rises in the northeast around 2:25 a.m. It is in the northeastern sky before dawn. See the illustration on page 11.

Bright **Venus** rises in the east around 4:55 a.m.

2027

Saturday, May 8

TONIGHT

sunset: **7:35 p.m.**
twilight begins: **8:36 p.m.**

IN THE MORNING

twilight begins: **4:48 a.m.**
sunrise: **5:48 a.m.**

MOON PHASE

moonrise: **7:19 a.m.**
moonset: **10:39 p.m.**

IN THE SKY

The brightest star in the sky, **Sirius** is in the western sky at dusk and sets in the west around 9:30 p.m.

The absence of the moon tonight provides excellent **dark-sky** viewing conditions. Explore the **dark sky** after the moon sets, from around 11 p.m.

Scorpius is visible most of the night. It rises in the southeast around 10 p.m. and is at its highest in the south, around 30 degrees, at 2:10 a.m. See the illustration on page 19.

After the moon sets, from around 11 p.m., use binoculars or a telescope to look for the **moons of Jupiter.** See the guide on page 18.

The **summer triangle**'s three bright stars are **Vega, Altair,** and **Deneb. Vega** is in the eastern sky at dusk and is almost directly overhead around dawn. **Altair** rises in the east around 11:10 p.m. Altair can be identified by its less bright neighboring stars, one on either side, together called the shaft of Aquila. **Deneb** is in the band of the Milky Way. See the illustration of the summer triangle on page 19.

Bright star **Capella** sets in the northwest around 11:50 p.m. Read about Capella on page 14.

Castor and Pollux, the twin stars in the constellation Gemini, set in the northwest around 11:50 p.m. Read how to find Castor and Pollux on page 14.

The bright star **Arcturus** is at its highest, around 75 degrees, at 11:55 p.m.

Just east of **Scorpius**, Sagittarius rises in the southeast around 12:45 a.m. and is at its highest in the south, around 25 degrees, around dawn.

Jupiter sets in the west around 1:45 a.m.

Rust-red and bright, **Mars** sets in the west around 2:15 a.m.

Cassiopeia, shaped like a *W,* rises in the northeast around 2:20 a.m. It is in the northeastern sky before dawn. See the illustration on page 11.

Bright **Venus** rises in the east around 4:55 a.m.

2027

Sunday, May 9

TONIGHT

sunset: **7:35 p.m.**
twilight begins: **8:36 p.m.**

IN THE MORNING

twilight begins: **4:47 a.m.**
sunrise: **5:47 a.m.**

MOON PHASE

moonrise: **8:23 a.m.**
moonset: **11:37 p.m.**

IN THE SKY

After the moon sets, from around 11:55 p.m., use binoculars or a telescope to look for the **moons of Jupiter.** See the guide on page 18.

Explore the **dark sky** after the moon sets, from around 11:55 p.m.

Cassiopeia, shaped like a *W,* rises in the northeast around 2:20 a.m. It is in the northeastern sky before dawn. See the illustration on page 11.

Scorpius is visible most of the night. It rises in the southeast around 9:55 p.m. and is at its highest in the south, around 30 degrees, at 2:10 a.m. See the illustration on page 19.

The brightest star in the sky, **Sirius** is in the western sky at dusk and sets in the west around 9:25 p.m.

The **summer triangle**'s three bright stars are **Vega, Altair,** and **Deneb. Vega** is almost directly overhead around dawn. **Altair** rises in the east around 11:10 p.m. Altair can be identified by its less bright neighboring stars, one on either side, together called the shaft of Aquila. **Deneb** is in the band of the Milky Way. See the illustration of the summer triangle on page 19.

Bright star **Capella** sets in the northwest around 11:45 p.m. Read about Capella on page 14.

Castor and Pollux, the twin stars in the constellation Gemini, set in the northwest around 11:50 p.m. Read how to find Castor and Pollux on page 14.

The bright star **Arcturus** is at its highest, around 75 degrees, at 11:50 p.m.

Just east of **Scorpius**, Sagittarius rises in the southeast around 12:45 a.m. and is at its highest in the south, around 25 degrees, around dawn.

Jupiter sets in the west around 1:40 a.m.

Rust-red and bright, **Mars** sets in the west around 2:15 a.m.

Bright **Venus** rises in the east around 4:55 a.m.

2027
Monday, May 10

TONIGHT

sunset: **7:36 p.m.**
twilight begins: **8:37 p.m.**

IN THE MORNING

twilight begins: **4:46 a.m.**
sunrise: **5:46 a.m.**

MOON PHASE

moonrise: **9:33 a.m.**
moonset: **12:27 a.m.**

IN THE SKY

The **moon** is prominent in the sky early in the evening. Using a telescope or binoculars, take a look. You should be able to see rough mountains and craters along the boundary between light and dark, in astronomy called the terminator.

The brightest star in the sky, **Sirius** is in the western sky at dusk and sets in the west around 9:20 p.m.

The waxing crescent moon, Castor, and Pollux appear close together in the evening sky. Look west at about 30 degrees elevation around 9:45 p.m.

Scorpius is visible most of the night. It rises in the southeast around 9:50 p.m. and is at its highest in the south, around 30 degrees, at 2:05 a.m. See the illustration on page 19.

The **summer triangle**'s three bright stars are **Vega, Altair,** and **Deneb. Vega** is almost directly overhead around dawn. **Altair** rises in the east around 11:05 p.m. Altair can be identified by its less bright neighboring stars, one on either side, together called the shaft of Aquila. **Deneb** is in the band of the Milky Way. See the illustration of the summer triangle on page 19.

Castor and Pollux, the twin stars in the constellation Gemini, set in the northwest around 11:45 p.m. Read how to find Castor and Pollux on page 14.

Bright star **Capella** sets in the northwest around 11:45 p.m. Read about Capella on page 14.

The bright star **Arcturus** is at its highest, around 75 degrees, at 11:45 p.m.

Just east of **Scorpius**, Sagittarius rises in the southeast around 12:40 a.m. and is at its highest in the south, around 25 degrees, around dawn.

Explore the **dark sky** after the moon sets, from around 12:45 a.m.

After the moon sets, from around 12:45 a.m., use binoculars or a telescope to look for the **moons of Jupiter.** See the guide on page 18.

Jupiter sets in the west around 1:35 a.m.

Rust-red and bright, **Mars** sets in the west around 2:10 a.m.

Cassiopeia, shaped like a *W,* rises in the northeast around 2:15 a.m. It is in the northeastern sky before dawn. See the illustration on page 11.

Bright **Venus** rises in the east around 4:50 a.m.

2027

Tuesday, May 11

TONIGHT

sunset: **7:37 p.m.**
twilight begins: **8:38 p.m.**

MOON PHASE

IN THE MORNING

twilight begins: **4:45 a.m.**
sunrise: **5:45 a.m.**

moonrise: **10:45 a.m.**
moonset: **1:08 a.m.**

IN THE SKY

The **moon** is prominent in the sky early in the evening. Using a telescope or binoculars, take a look. You should be able to see rough mountains and craters along the boundary between light and dark, in astronomy called the terminator.

The waxing first-quarter **moon, Jupiter,** and **Mars** appear lined up in the evening sky. Look west at about 50 degrees elevation around 9 p.m.

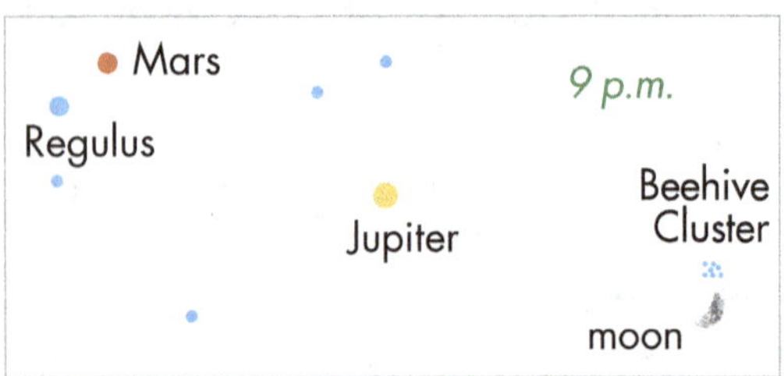

The brightest star in the sky, **Sirius** is in the western sky at dusk and sets in the west around 9:20 p.m.

Scorpius is visible most of the night. It rises in the southeast around 9:50 p.m. and is at its highest in the south, around 30 degrees, at 2 a.m. See the illustration on page 19.

The **summer triangle**'s three bright stars are **Vega, Altair,** and **Deneb. Vega** is almost directly overhead around dawn. **Altair** rises in the east around 11 p.m. Altair can be identified by its less bright neighboring stars, one on either side, together called the shaft of Aquila. **Deneb** is in the band of the Milky Way. See the illustration of the summer triangle on page 19.

The bright star **Arcturus** is at its highest, around 75 degrees, at 11:40 p.m.

Bright star **Capella** sets in the northwest around 11:40 p.m. Read about Capella on page 14.

Castor and Pollux, the twin stars in the constellation Gemini, set in the northwest around 11:40 p.m. Read how to find Castor and Pollux on page 14.

Just east of **Scorpius**, Sagittarius rises in the southeast around 12:35 a.m. and is at its highest in the south, around 25 degrees, around dawn.

Jupiter sets in the west around 1:30 a.m.

Explore the **dark sky** after the moon sets, from around 1:30 a.m.

Rust-red and bright, **Mars** sets in the west around 2:05 a.m.

Cassiopeia, shaped like a *W,* rises in the northeast around 2:10 a.m. It is in the northeastern sky before dawn. See the illustration on page 11.

Bright **Venus** rises in the east around 4:50 a.m.

2027

Wednesday, May 12

TONIGHT

sunset: **7:38 p.m.**
twilight begins: **8:39 p.m.**

IN THE MORNING

twilight begins: **4:44 a.m.**
sunrise: **5:44 a.m.**

MOON PHASE

moonrise: **11:56 a.m.**
moonset: **1:42 a.m.**

IN THE SKY

The **moon** is prominent in the sky early in the evening. Using a telescope or binoculars, take a look. You should be able to see rough mountains and craters along the boundary between light and dark, in astronomy called the terminator.

The brightest star in the sky, **Sirius** is in the western sky at dusk and sets in the west around 9:15 p.m.

The waxing first-quarter **moon, Jupiter,** and **Mars** appear close together in the evening sky. Look west at about 50 degrees elevation around 9:30 p.m.

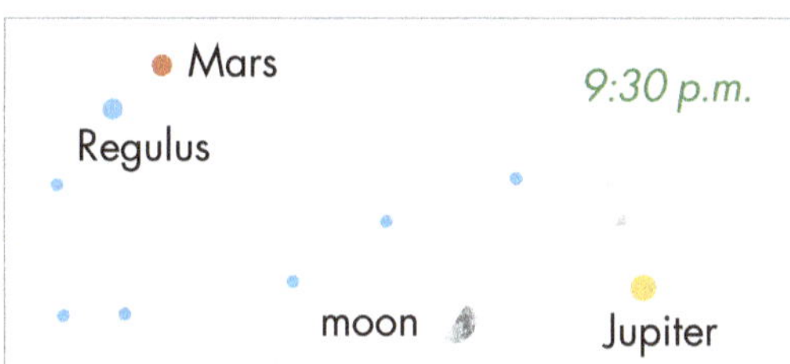

Scorpius is visible most of the night. It rises in the southeast around 9:45 p.m. and is at its highest in the south, around 30 degrees, at 1:55 a.m. See the illustration on page 19.

The **summer triangle**'s three bright stars are **Vega, Altair,** and **Deneb. Vega** is almost directly overhead around dawn. **Altair** rises in the east around 10:55 p.m. Altair can be identified by its less bright neighboring stars, one on either side, together called the shaft of Aquila. **Deneb** is in the band of the Milky Way. See the illustration of the summer triangle on page 19.

Bright star **Capella** sets in the northwest around 11:35 p.m. Read about Capella on page 14.

Castor and Pollux, the twin stars in the constellation Gemini, set in the northwest around 11:35 p.m. Read how to find Castor and Pollux on page 14.

The bright star **Arcturus** is at its highest, around 75 degrees, at 11:40 p.m.

Just east of **Scorpius**, Sagittarius rises in the southeast around 12:30 a.m. and is at its highest in the south, around 25 degrees, around dawn.

Jupiter sets in the west around 1:30 a.m.

Explore the **dark sky** after the moon sets, from around 2 a.m.

Cassiopeia, shaped like a *W*, rises in the northeast around 2:05 a.m. It is in the northeastern sky before dawn. See the illustration on page 11.

Rust-red and bright, **Mars** sets in the west around 2:05 a.m.

Bright **Venus** rises in the east around 4:50 a.m.

2027

Thursday, May 13

TONIGHT

sunset: **7:38 p.m.**
twilight begins: **8:40 p.m.**

IN THE MORNING

twilight begins: **4:43 a.m.**
sunrise: **5:44 a.m.**

MOON PHASE

moonrise: **1:05 p.m.**
moonset: **2:13 a.m.**

IN THE SKY

The **moon** is prominent in the sky early in the evening. Using a telescope or binoculars, take a look. You should be able to see rough mountains and craters along the boundary between light and dark, in astronomy called the terminator.

The brightest star in the sky, **Sirius** sets in the west around 9:10 p.m.

The waxing gibbous moon, **Jupiter,** and **Mars** appear close together in the evening sky. Look southwest at about 55 degrees elevation around 9:30 p.m.

Scorpius is visible most of the night. It rises in the southeast around 9:40 p.m. and is at its highest in the south, around 30 degrees, at 1:50 a.m. See the illustration on page 19.

The **summer triangle**'s three bright stars are **Vega, Altair,** and **Deneb. Vega** is almost directly overhead around dawn. **Altair** rises in the east around 10:50 p.m. Altair can be identified by its less bright neighboring stars, one on either side, together called the shaft of Aquila. **Deneb** is in the band of the Milky Way. See the illustration of the summer triangle on page 19.

Bright star **Capella** sets in the northwest around 11:30 p.m. Read about Capella on page 14.

Castor and Pollux, the twin stars in the constellation Gemini, set in the northwest around 11:30 p.m. Read how to find Castor and Pollux on page 14.

The bright star **Arcturus** is at its highest, around 75 degrees, at 11:35 p.m.

Just east of **Scorpius**, Sagittarius rises in the southeast around 12:25 a.m. and is at its highest in the south, around 25 degrees, at 3:40 a.m.

Jupiter sets in the west around 1:25 a.m.

Rust-red and bright, **Mars** sets in the west around 2 a.m.

Cassiopeia, shaped like a *W,* rises in the northeast around 2 a.m. It is in the northeastern sky before dawn. See the illustration on page 11.

Bright **Venus** rises in the east around 4:50 a.m.

2027

Friday, May 14

TONIGHT

sunset: **7:39 p.m.**
twilight begins: **8:41 p.m.**

IN THE MORNING

twilight begins: **4:42 a.m.**
sunrise: **5:43 a.m.**

MOON PHASE

moonrise: **2:11 p.m.**
moonset: **2:42 a.m.**

IN THE SKY

The brightest star in the sky, **Sirius** sets in the west around 9:05 p.m.

Scorpius is visible most of the night. It rises in the southeast around 9:35 p.m. and is at its highest in the south, around 30 degrees, at 1:50 a.m. See the illustration on page 19.

The **summer triangle**'s three bright stars are **Vega, Altair,** and **Deneb. Vega** is almost directly overhead around dawn. **Altair** rises in the east around 10:50 p.m. Altair can be identified by its less bright neighboring stars, one on either side, together called the shaft of Aquila. **Deneb** is in the band of the Milky Way. See the illustration of the summer triangle on page 19.

Bright star **Capella** sets in the northwest around 11:25 p.m. Read about Capella on page 14.

Castor and Pollux, the twin stars in the constellation Gemini, set in the northwest around 11:30 p.m. Read how to find Castor and Pollux on page 14.

The bright star **Arcturus** is at its highest, around 75 degrees, at 11:30 p.m.

Just east of **Scorpius**, Sagittarius rises in the southeast around 12:25 a.m. and is at its highest in the south, around 25 degrees, at 3:40 a.m.

Jupiter sets in the west around 1:20 a.m.

Cassiopeia, shaped like a *W,* rises in the northeast around 1:55 a.m. It is in the northeastern sky before dawn. See the illustration on page 11.

Rust-red and bright, **Mars** sets in the west around 2 a.m.

Bright **Venus** rises in the east around 4:50 a.m.

2027

Saturday, May 15

TONIGHT

sunset: **7:40 p.m.**
twilight begins: **8:42 p.m.**

MOON PHASE

IN THE MORNING

twilight begins: **4:41 a.m.**
sunrise: **5:42 a.m.**

moonrise: **3:15 p.m.**
moonset: **3:10 a.m.**

IN THE SKY

The brightest star in the sky, **Sirius** sets in the west around 9 p.m.

Scorpius is visible most of the night. It rises in the southeast around 9:30 p.m. and is at its highest in the south, around 30 degrees, at 1:45 a.m. See the illustration on page 19.

The **summer triangle**'s three bright stars are **Vega, Altair,** and **Deneb. Vega** is almost directly overhead around dawn. **Altair** rises in the east around 10:45 p.m. Altair can be identified by its less bright neighboring stars, one on either side, together called the shaft of Aquila. **Deneb** is in the band of the Milky Way. See the illustration of the summer triangle on page 19.

The bright star **Arcturus** is at its highest, around 75 degrees, at 11:25 p.m.

Bright star **Capella** sets in the northwest around 11:25 p.m. Read about Capella on page 14.

Castor and Pollux, the twin stars in the constellation Gemini, set in the northwest around 11:25 p.m. Read how to find Castor and Pollux on page 14.

Just east of **Scorpius,** Sagittarius rises in the southeast around 12:20 a.m. and is at its highest in the south, around 25 degrees, at 3:35 a.m.

Jupiter sets in the west around 1:20 a.m.

Cassiopeia, shaped like a *W,* rises in the northeast around 1:50 a.m. It is in the northeastern sky before dawn. See the illustration on page 11.

Rust-red and bright, **Mars** sets in the west around 1:55 a.m.

Saturn rises in the east around 4:30 a.m. and is in the eastern sky before dawn. Unlike bright Jupiter and Venus and ruddy Mars, Saturn is harder to distinguish from the stars. Stars twinkle and shimmer, while the planets, including Saturn, show a steady flat light.

Bright **Venus** rises in the east around 4:50 a.m.

2027
Sunday, May 16

TONIGHT

sunset: 7:41 p.m.
twilight begins: 8:43 p.m.

IN THE MORNING

twilight begins: 4:40 a.m.
sunrise: 5:41 a.m.

MOON PHASE

moonrise: 4:19 p.m.
moonset: 3:39 a.m.

IN THE SKY

The nearly **full moon** tonight acts like a spotlight. Look around at the desert landscape at how much illumination the moon provides.

The brightest star in the sky, **Sirius** sets in the west around 9 p.m.

Scorpius is visible most of the night. It rises in the southeast around 9:30 p.m. and is at its highest in the south, around 30 degrees, at 1:40 a.m. See the illustration on page 19.

The bright star **Arcturus** is at its highest, around 75 degrees, at 11:20 p.m.

The **summer triangle**'s three bright stars are **Vega, Altair,** and **Deneb. Vega** is almost directly overhead around dawn. **Altair** rises in the east around 10:40 p.m. Altair can be identified by its less bright neighboring stars, one on either side, together called the shaft of Aquila. **Deneb** is in the band of the Milky Way. See the illustration of the summer triangle on page 19.

Bright star **Capella** sets in the northwest around 11:20 p.m. Read about Capella on page 14.

Castor and Pollux, the twin stars in the constellation Gemini, set in the northwest around 11:20 p.m. Read how to find Castor and Pollux on page 14.

Just east of **Scorpius**, Sagittarius rises in the southeast around 12:15 a.m. and is at its highest in the south, around 25 degrees, at 3:30 a.m.

Jupiter sets in the west around 1:15 a.m.

Cassiopeia, shaped like a *W,* rises in the northeast around 1:45 a.m. It is in the northeastern sky before dawn. See the illustration on page 11.

Rust-red and bright, **Mars** sets in the west around 1:50 a.m.

Saturn rises in the east around 4:25 a.m. and is in the eastern sky before dawn. Unlike bright Jupiter and Venus and ruddy Mars, Saturn is harder to distinguish from the stars. Stars twinkle and shimmer, while the planets, including Saturn, show a steady flat light.

Bright **Venus** rises in the east around 4:50 a.m.

2027

Monday, May 17

TONIGHT

sunset: **7:41 p.m.**
twilight begins: **8:44 p.m.**

IN THE MORNING

twilight begins: **4:40 a.m.**
sunrise: **5:41 a.m.**

MOON PHASE

moonrise: **5:22 p.m.**
moonset: **4:11 a.m.**

IN THE SKY

The **full moon** tonight acts like a spotlight. Look around at the desert landscape at how much illumination the moon provides.

Scorpius is visible most of the night. It rises in the southeast around 9:25 p.m. and is at its highest in the south, around 30 degrees, at 1:35 a.m. See the illustration on page 19.

The **summer triangle**'s three bright stars are **Vega, Altair,** and **Deneb. Vega** is almost directly overhead around dawn. **Altair** rises in the east around 10:35 p.m. Altair can be identified by its less bright neighboring stars, one on either side, together called the shaft of Aquila. **Deneb** is in the band of the Milky Way. See the illustration of the summer triangle on page 19.

Bright star **Capella** sets in the northwest around 11:15 p.m. Read about Capella on page 14.

Castor and Pollux, the twin stars in the constellation Gemini, set in the northwest around 11:15 p.m. Read how to find Castor and Pollux on page 14.

The bright star **Arcturus** is at its highest, around 75 degrees, at 11:20 p.m.

Just east of **Scorpius**, Sagittarius rises in the southeast around 12:10 a.m. and is at its highest in the south, around 25 degrees, at 3:25 a.m.

Jupiter sets in the west around 1:10 a.m.

Cassiopeia, shaped like a *W*, rises in the northeast around 1:40 a.m. It is in the northeastern sky before dawn. See the illustration on page 11.

Rust-red **Mars** sets in the west around 1:50 a.m.

Saturn rises in the east around 4:20 a.m. and is in the eastern sky before dawn. Unlike bright Jupiter and Venus and ruddy Mars, Saturn is harder to distinguish from the stars. Stars twinkle and shimmer, while the planets, including Saturn, show a steady flat light.

Bright **Venus** rises in the east around 4:50 a.m.

2027

Tuesday, May 18

TONIGHT

sunset: **7:42 p.m.**
twilight begins: **8:45 p.m.**

IN THE MORNING

twilight begins: **4:39 a.m.**
sunrise: **5:40 a.m.**

MOON PHASE

moonrise: **6:26 p.m.**
moonset: **4:46 a.m.**

IN THE SKY

The **full moon** tonight acts like a spotlight. Look around at the desert landscape at how much illumination the moon provides.

Scorpius is visible most of the night. It rises in the southeast around 9:20 p.m. and is at its highest in the south, around 30 degrees, at 1:30 a.m. See the illustration on page 19.

Bright star **Capella** sets in the northwest around 11:10 p.m. Read about Capella on page 14.

The bright star **Arcturus** is at its highest, around 75 degrees, at 11:15 p.m.

Castor and Pollux, the twin stars in the constellation Gemini, set in the northwest around 11:10 p.m. Read how to find Castor and Pollux on page 14.

Just east of **Scorpius**, Sagittarius rises in the southeast around 12:10 a.m. and is at its highest in the south, around 25 degrees, at 3:25 a.m.

Jupiter sets in the west around 1:05 a.m.

Cassiopeia, shaped like a *W,* rises in the northeast around 1:40 a.m. It is in the northeastern sky before dawn. See the illustration on page 11.

Rust-red **Mars** sets in the west around 1:45 a.m.

The **summer triangle**'s three bright stars are **Vega, Altair,** and **Deneb. Vega** is almost directly overhead at 3:40 a.m. **Altair** rises in the east around 10:35 p.m. Altair can be identified by its less bright neighboring stars, one on either side, together called the shaft of Aquila. **Deneb** is in the band of the Milky Way. See the illustration of the summer triangle on page 19.

Saturn rises in the east around 4:20 a.m. and is in the eastern sky before dawn. Unlike bright Jupiter and Venus and ruddy Mars, Saturn is harder to distinguish from the stars. Stars twinkle and shimmer, while the planets, including Saturn, show a steady flat light.

Bright **Venus** rises in the east around 4:50 a.m.

2027

Wednesday, May 19

TONIGHT

sunset: **7:43 p.m.**
twilight begins: **8:46 p.m.**

IN THE MORNING

twilight begins: **4:38 a.m.**
sunrise: **5:39 a.m.**

MOON PHASE

moonrise: **7:30 p.m.**
moonset: **5:26 a.m.**

IN THE SKY

The **full moon** tonight acts like a spotlight. Look around at the desert landscape at how much illumination the moon provides.

Scorpius is visible most of the night. It rises in the southeast around 9:15 p.m. and is at its highest in the south, around 30 degrees, at 1:30 a.m. See the illustration on page 19.

Bright star **Capella** sets in the northwest around 11:05 p.m. Read about Capella on page 14.

Castor and Pollux, the twin stars in the constellation Gemini, set in the northwest around 11:10 p.m. Read how to find Castor and Pollux on page 14.

The bright star **Arcturus** is at its highest, around 75 degrees, at 11:10 p.m.

Just east of **Scorpius**, Sagittarius rises in the southeast around 12:05 a.m. and is at its highest in the south, around 25 degrees, at 3:20 a.m.

Jupiter sets in the west around 1:05 a.m.

Cassiopeia, shaped like a *W*, rises in the northeast around 1:35 a.m. It is in the northeastern sky before dawn. See the illustration on page 11.

Rust-red **Mars** sets in the west around 1:45 a.m.

The **summer triangle**'s three bright stars are **Vega, Altair,** and **Deneb. Vega** is almost directly overhead at 3:35 a.m. **Altair** rises in the east around 10:30 p.m. Altair can be identified by its less bright neighboring stars, one on either side, together called the shaft of Aquila. **Deneb** is in the band of the Milky Way. See the illustration of the summer triangle on page 19.

Saturn rises in the east around 4:15 a.m. and is in the eastern sky before dawn. Unlike bright Jupiter and Venus and ruddy Mars, Saturn is harder to distinguish from the stars. Stars twinkle and shimmer, while the planets, including Saturn, show a steady flat light.

Bright **Venus** rises in the east around 4:45 a.m.

2027

Thursday, May 20

TONIGHT

sunset: **7:44 p.m.**
twilight begins: **8:47 p.m.**

MOON PHASE

IN THE MORNING

twilight begins: **4:37 a.m.**
sunrise: **5:39 a.m.**

moonrise: **8:31 p.m.**
moonset: **6:12 a.m.**

IN THE SKY

The **full moon** tonight acts like a spotlight. Look around at the desert landscape at how much illumination the moon provides.

Scorpius is visible most of the night. It rises in the southeast around 9:10 p.m. and is at its highest in the south, around 30 degrees, at 1:25 a.m. See the illustration on page 19.

The bright star **Arcturus** is at its highest, around 75 degrees, at 11:05 p.m.

Bright star **Capella** sets in the northwest around 11:05 p.m. Read about Capella on page 14.

Castor and Pollux, the twin stars in the constellation Gemini, set in the northwest around 11:05 p.m. Read how to find Castor and Pollux on page 14.

The **full moon** and the bright yellowish star **Antares** appear close together in the evening sky. Look south at about 25 degrees elevation around midnight.

Just east of **Scorpius**, Sagittarius rises in the southeast around midnight and is at its highest in the south, around 25 degrees, at 3:15 a.m.

Jupiter sets in the west around 1 a.m.

Cassiopeia, shaped like a *W*, rises in the northeast around 1:30 a.m. It is in the northeastern sky before dawn. See the illustration on page 11.

Rust-red **Mars** sets in the west around 1:40 a.m.

The **summer triangle**'s three bright stars are **Vega, Altair,** and **Deneb. Vega** is almost directly overhead at 3:30 a.m. **Altair** rises in the east around 10:25 p.m. Altair can be identified by its less bright neighboring stars, one on either side, together called the shaft of Aquila. **Deneb** is in the band of the Milky Way. See the illustration of the summer triangle on page 19.

Saturn rises in the east around 4:10 a.m. and is in the eastern sky before dawn. Unlike bright Jupiter and Venus and ruddy Mars, Saturn is harder to distinguish from the stars. Stars twinkle and shimmer, while the planets, including Saturn, show a steady flat light.

Bright **Venus** rises in the east around 4:45 a.m.

2027

Friday, May 21

TONIGHT

sunset: **7:44 p.m.**
twilight begins: **8:48 p.m.**

IN THE MORNING

twilight begins: **4:36 a.m.**
sunrise: **5:38 a.m.**

MOON PHASE

moonrise: **9:28 p.m.**
moonset: **7:04 a.m.**

IN THE SKY

The **full moon** tonight acts like a spotlight. Look around at the desert landscape at how much illumination the moon provides.

Scorpius is visible most of the night. It rises in the southeast around dusk and is in the eastern sky at dusk and is at its highest in the south, around 30 degrees, at 1:20 a.m. See the illustration on page 19.

Bright star **Capella** sets in the northwest around 11 p.m. Read about Capella on page 14.

Castor and Pollux, the twin stars in the constellation Gemini, set in the northwest around 11 p.m. Read how to find Castor and Pollux on page 14.

The bright star **Arcturus** is at its highest, around 75 degrees, at 11:05 p.m.

Just east of **Scorpius**, Sagittarius rises in the southeast around 11:55 p.m. and is at its highest in the south, around 25 degrees, at 3:10 a.m.

Jupiter sets in the west around 12:55 a.m.

Cassiopeia, shaped like a *W,* rises in the northeast around 1:25 a.m. It is in the northeastern sky before dawn. See the illustration on page 11.

Rust-red **Mars** sets in the west around 1:35 a.m.

The **summer triangle**'s three bright stars are **Vega, Altair,** and **Deneb. Vega** is almost directly overhead at 3:25 a.m. **Altair** rises in the east around 10:20 p.m. Altair can be identified by its less bright neighboring stars, one on either side, together called the shaft of Aquila. **Deneb** is in the band of the Milky Way. See the illustration of the summer triangle on page 19.

Saturn rises in the east around 4:05 a.m. and is in the eastern sky before dawn. Unlike bright Jupiter and Venus and ruddy Mars, Saturn is harder to distinguish from the stars. Stars twinkle and shimmer, while the planets, including Saturn, show a steady flat light.

Bright **Venus** rises in the east around 4:45 a.m.

2027

Saturday, May 22

TONIGHT

sunset: **7:45 p.m.**
twilight begins: **8:48 p.m.**

IN THE MORNING

twilight begins: **4:36 a.m.**
sunrise: **5:38 a.m.**

MOON PHASE

moonrise: **10:18 p.m.**
moonset: **7:04 a.m.**

IN THE SKY

The **full moon** tonight acts like a spotlight. Look around at the desert landscape at how much illumination the moon provides.

Castor and Pollux, the twin stars in the constellation Gemini, set in the northwest around 10:55 p.m. Read how to find Castor and Pollux on page 14.

Bright star **Capella** sets in the northwest around 10:55 p.m. Read about Capella on page 14.

The bright star **Arcturus** is at its highest, around 75 degrees, at 11 p.m.

Just east of **Scorpius,** Sagittarius rises in the southeast around 11:50 p.m. and is at its highest in the south, around 25 degrees, at 3:05 a.m.

Jupiter sets in the west around 12:50 a.m.

Scorpius is visible most of the night and is at its highest in the south, around 30 degrees, at 1:15 a.m. See the illustration on page 19.

Cassiopeia, shaped like a *W,* rises in the northeast around 1:25 a.m. It is in the northeastern sky before dawn. See the illustration on page 11.

Rust-red **Mars** sets in the west around 1:35 a.m.

The **summer triangle**'s three bright stars are **Vega, Altair,** and **Deneb. Vega** is almost directly overhead at 3:25 a.m. **Altair** rises in the east around 10:15 p.m. Altair can be identified by its less bright neighboring stars, one on either side, together called the shaft of Aquila. **Deneb** is in the band of the Milky Way. See the illustration of the summer triangle on page 19.

Saturn rises in the east around 4:05 a.m. and is in the eastern sky before dawn. Unlike bright Jupiter and Venus and ruddy Mars, Saturn is harder to distinguish from the stars. Stars twinkle and shimmer, while the planets, including Saturn, show a steady flat light.

Bright **Venus** rises in the east around 4:45 a.m.

2027

Sunday, May 23

TONIGHT

sunset: **7:46 p.m.**
twilight begins: **8:49 p.m.**

IN THE MORNING

twilight begins: **4:35 a.m.**
sunrise: **5:37 a.m.**

MOON PHASE

moonrise: **11:02 p.m.**
moonset: **7:59 a.m.**

IN THE SKY

Bright star **Capella** sets in the northwest around 10:50 p.m. Read about Capella on page 14.

The bright star **Arcturus** is at its highest, around 75 degrees, at 10:55 p.m.

Castor and Pollux, the twin stars in the constellation Gemini, set in the northwest around 10:55 p.m. Read how to find Castor and Pollux on page 14.

Just east of **Scorpius,** Sagittarius rises in the southeast around 11:50 p.m. and is at its highest in the south, around 25 degrees, at 3:05 a.m.

Jupiter sets in the west around 12:45 a.m.

Scorpius is visible most of the night and is at its highest in the south, around 30 degrees, at 1:15 a.m. See the illustration on page 19.

Cassiopeia, shaped like a *W,* rises in the northeast around 1:20 a.m. It is in the northeastern sky before dawn. See the illustration on page 11.

Rust-red **Mars** sets in the west around 1:30 a.m.

The **summer triangle**'s three bright stars are **Vega, Altair,** and **Deneb. Vega** is almost directly overhead at 3:20 a.m. **Altair** rises in the east around 10:15 p.m. Altair can be identified by its less bright neighboring stars, one on either side, together called the shaft of Aquila. **Deneb** is in the band of the Milky Way. See the illustration of the summer triangle on page 19.

Saturn rises in the east around 4 a.m. and is in the eastern sky before dawn. Unlike bright Jupiter and Venus and ruddy Mars, Saturn is harder to distinguish from the stars. Stars twinkle and shimmer, while the planets, including Saturn, show a steady flat light.

Bright **Venus** rises in the east around 4:45 a.m.

2027

Monday, May 24

TONIGHT

sunset: **7:46 p.m.**
twilight begins: **8:50 p.m.**

IN THE MORNING

twilight begins: **4:34 a.m.**
sunrise: **5:37 a.m.**

MOON PHASE

moonrise: **11:40 p.m.**
moonset: **8:57 a.m.**

IN THE SKY

Castor and Pollux, the twin stars in the constellation Gemini, are in the western sky at dusk and set in the northwest around 10:50 p.m. Read how to find Castor and Pollux on page 14.

Bright star **Capella** sets in the northwest around 10:45 p.m. Read about Capella on page 14.

The bright star **Arcturus** is at its highest, around 75 degrees, at 10:50 p.m.

Before the moon rises, until around 11:20 p.m., use binoculars or a telescope to look for the **moons of Jupiter.** See the guide on page 18.

Explore the **dark sky** after twilight and before the moon rises, around 11:20 p.m.

Just east of **Scorpius,** Sagittarius rises in the southeast around 11:45 p.m. and is at its highest in the south, around 25 degrees, at 3 a.m.

Scorpius is visible most of the night and is at its highest in the south, around 30 degrees, at 1:10 a.m. See the illustration on page 19.

Cassiopeia, shaped like a *W,* rises in the northeast around 1:15 a.m. It is in the northeastern sky before dawn. See the illustration on page 11.

The **summer triangle**'s three bright stars are **Vega, Altair,** and **Deneb. Vega** is almost directly overhead at 3:15 a.m. **Altair** rises in the east around 10:10 p.m. Altair can be identified by its less bright neighboring stars, one on either side, together called the shaft of Aquila. **Deneb** is in the band of the Milky Way. See the illustration of the summer triangle on page 19.

Jupiter sets in the west around 12:40 a.m.

Rust-red **Mars** sets in the west around 1:30 a.m.

Saturn rises in the east around 3:55 a.m. and is in the eastern sky before dawn. Unlike bright Jupiter and Venus and ruddy Mars, Saturn is harder to distinguish from the stars. Stars twinkle and shimmer, while the planets, including Saturn, show a steady flat light.

Bright **Venus** rises in the east around 4:45 a.m.

2027

Tuesday, May 25

TONIGHT

sunset: **7:47 p.m.**
twilight begins: **8:51 p.m.**

IN THE MORNING

twilight begins: **4:34 a.m.**
sunrise: **5:36 a.m.**

MOON PHASE

moonrise: **12:12 a.m.**
moonset: **9:56 a.m.**

IN THE SKY

Castor and Pollux, the twin stars in the constellation Gemini, are in the western sky at dusk and set in the northwest around 10:45 p.m. Read how to find Castor and Pollux on page 14.

The bright star **Arcturus** is at its highest, around 75 degrees, at 10:45 p.m.

Bright star **Capella** sets in the northwest around 10:45 p.m. Read about Capella on page 14.

Just east of **Scorpius**, Sagittarius rises in the southeast around 11:40 p.m. and is at its highest in the south, around 25 degrees, at 2:55 a.m.

Explore the **dark sky** after twilight and before the moon rises, around 11:50 p.m.

Before the moon rises, until around 11:50 p.m., use binoculars or a telescope to look for the **moons of Jupiter.** See the guide on page 18.

Jupiter sets in the west around 12:40 a.m.

Scorpius is visible most of the night and is at its highest in the south, around 30 degrees, at 1:05 a.m. See the illustration on page 19.

Cassiopeia, shaped like a *W,* rises in the northeast around 1:10 a.m. It is in the northeastern sky before dawn. See the illustration on page 11.

Rust-red **Mars** sets in the west around 1:25 a.m.

The **summer triangle**'s three bright stars are **Vega, Altair,** and **Deneb. Vega** is almost directly overhead at 3:10 a.m. **Altair** rises in the east around 10:05 p.m. Altair can be identified by its less bright neighboring stars, one on either side, together called the shaft of Aquila. **Deneb** is in the band of the Milky Way. See the illustration of the summer triangle on page 19.

Saturn rises in the east around 3:50 a.m. and is in the eastern sky before dawn. Unlike bright Jupiter and Venus and ruddy Mars, Saturn is harder to distinguish from the stars. Stars twinkle and shimmer, while the planets, including Saturn, show a steady flat light.

Bright **Venus** rises in the east around 4:45 a.m.

2027

Wednesday, May 26

TONIGHT

sunset: **7:48 p.m.**
twilight begins: **8:52 p.m.**

IN THE MORNING

twilight begins: **4:33 a.m.**
sunrise: **5:36 a.m.**

MOON PHASE

moonrise: **12:41 a.m.**
moonset: **10:54 a.m.**

IN THE SKY

The **summer triangle**'s three bright stars are **Vega, Altair,** and **Deneb. Vega** is almost directly overhead at 3:05 a.m. **Altair** rises in the east around 10 p.m. Altair can be identified by its less bright neighboring stars, one on either side, together called the shaft of Aquila. **Deneb** is in the band of the Milky Way. See the illustration of the summer triangle on page 19.

Bright star **Capella** is in the northwestern sky at dusk and sets in the northwest around 10:40 p.m. Read about Capella on page 14.

Castor and Pollux, the twin stars in the constellation Gemini, are in the western sky at dusk and set in the northwest around 10:40 p.m. Read how to find Castor and Pollux on page 14.

The bright star **Arcturus** is at its highest, around 75 degrees, at 10:45 p.m.

Just east of **Scorpius**, Sagittarius rises in the southeast around 11:35 p.m. and is at its highest in the south, around 25 degrees, at 2:50 a.m.

Explore the **dark sky** after twilight and before the moon rises, around 12:20 a.m.

Jupiter sets in the west around 12:35 a.m.

Scorpius is visible most of the night and is at its highest in the south, around 30 degrees, at 1 a.m. See the illustration on page 19.

Cassiopeia, shaped like a *W,* rises in the northeast around 1:05 a.m. It is in the northeastern sky before dawn. See the illustration on page 11.

Rust-red **Mars** sets in the west around 1:20 a.m.

Saturn rises in the east around 3:50 a.m. and is in the eastern sky before dawn. Unlike bright Jupiter and Venus and ruddy Mars, Saturn is harder to distinguish from the stars. Stars twinkle and shimmer, while the planets, including Saturn, show a steady flat light.

Bright **Venus** rises in the east around 4:45 a.m.

2027

Thursday, May 27

TONIGHT

sunset: **7:49 p.m.**
twilight begins: **8:53 p.m.**

IN THE MORNING

twilight begins: **4:33 a.m.**
sunrise: **5:36 a.m.**

MOON PHASE

moonrise: **1:08 a.m.**
moonset: **11:51 a.m.**

IN THE SKY

Look for **Mercury** in the evening sky, just after sunset in the northwest. It's difficult but possible to see the tiny dot in the glare of the sun. It's easier to find with the naked eye, not using binoculars or a telescope. Never look directly at the sun or point binoculars or a telescope at it.

The **summer triangle**'s three bright stars are **Vega, Altair,** and **Deneb. Vega** is almost directly overhead at 3:05 a.m. **Altair** rises in the east around 9:55 p.m. Altair can be identified by its less bright neighboring stars, one on either side, together called the shaft of Aquila. **Deneb** is in the band of the Milky Way. See the illustration of the summer triangle on page 19.

Bright star **Capella** is in the northwestern sky at dusk and sets in the northwest around 10:35 p.m. Read about Capella on page 14.

Castor and Pollux, the twin stars in the constellation Gemini, are in the western sky at dusk and set in the northwest around 10:35 p.m. Read how to find Castor and Pollux on page 14.

The bright star **Arcturus** is at its highest, around 75 degrees, at 10:40 p.m.

Just east of **Scorpius**, Sagittarius rises in the southeast around 11:30 p.m. and is at its highest in the south, around 25 degrees, at 2:45 a.m.

Jupiter sets in the west around 12:30 a.m.

Explore the **dark sky** after twilight and before the moon rises, around 12:50 a.m.

Scorpius is visible most of the night and is at its highest in the south, around 30 degrees, at 12:55 a.m. See the illustration on page 19.

Cassiopeia, shaped like a *W,* rises in the northeast around 1:05 a.m. It is in the northeastern sky before dawn. See the illustration on page 11.

Rust-red **Mars** sets in the west around 1:20 a.m.

Saturn rises in the east around 3:45 a.m. and is in the eastern sky before dawn. Unlike bright Jupiter and Venus and ruddy Mars, Saturn is harder to distinguish from the stars. Stars twinkle and shimmer, while the planets, including Saturn, show a steady flat light.

Bright **Venus** rises in the east around 4:45 a.m.

2027

Friday, May 28

TONIGHT

sunset: **7:49 p.m.**
twilight begins: **8:53 p.m.**

IN THE MORNING

twilight begins: **4:32 a.m.**
sunrise: **5:35 a.m.**

MOON PHASE

moonrise: **1:33 a.m.**
moonset: **12:47 p.m.**

IN THE SKY

Look for **Mercury** in the evening sky, just after sunset in the northwest. It's difficult but possible to see the tiny dot in the glare of the sun. It's easier to find with the naked eye, not using binoculars or a telescope. Never look directly at the sun or point binoculars or a telescope at it.

The **summer triangle**'s three bright stars are **Vega, Altair,** and **Deneb. Vega** is almost directly overhead at 3 a.m. **Altair** rises in the east around 9:55 p.m. Altair can be identified by its less bright neighboring stars, one on either side, together called the shaft of Aquila. **Deneb** is in the band of the Milky Way. See the illustration of the summer triangle on page 19.

With the dark sky, use binoculars or a telescope to look for the **moons of Jupiter.** See the guide on page 18.

Bright star **Capella** is in the northwestern sky at dusk and sets in the northwest around 10:30 p.m. Read about Capella on page 14.

Castor and Pollux, the twin stars in the constellation Gemini, are in the western sky at dusk and set in the northwest around 10:35 p.m. Read how to find Castor and Pollux on page 14.

The bright star **Arcturus** is at its highest, around 75 degrees, at 10:35 p.m.

Just east of **Scorpius**, Sagittarius rises in the southeast around 11:30 p.m. and is at its highest in the south, around 25 degrees, at 2:45 a.m.

Scorpius is visible most of the night and is at its highest in the south, around 30 degrees, at 12:50 a.m. See the illustration on page 19.

Jupiter sets in the west around 12:25 a.m.

Cassiopeia, shaped like a *W,* rises in the northeast around 1 a.m. It is in the northeastern sky before dawn. See the illustration on page 11.

Rust-red **Mars** sets in the west around 1:15 a.m.

Saturn rises in the east around 3:40 a.m. and is in the eastern sky before dawn. Unlike bright Jupiter and Venus and ruddy Mars, Saturn is harder to distinguish from the stars. Stars twinkle and shimmer, while the planets, including Saturn, show a steady flat light.

Bright **Venus** rises in the east around 4:45 a.m.

2027

Saturday, May 29

TONIGHT

sunset: **7:50 p.m.**
twilight begins: **8:54 p.m.**

IN THE MORNING

twilight begins: **4:32 a.m.**
sunrise: **5:35 a.m.**

MOON PHASE

moonrise: **1:58 a.m.**
moonset: **1:45 p.m.**

IN THE SKY

Look for **Mercury** in the evening sky, just after sunset in the northwest. It's difficult but possible to see the tiny dot in the glare of the sun. It's easier to find with the naked eye, not using binoculars or a telescope. Never look directly at the sun or point binoculars or a telescope at it.

Castor and Pollux, the twin stars in the constellation Gemini, are in the western sky at dusk and set in the northwest around 10:30 p.m. Read how to find Castor and Pollux on page 14.

With the dark sky, use binoculars or a telescope to look for the **moons of Jupiter.** See the guide on page 18.

The **summer triangle**'s three bright stars are **Vega, Altair,** and **Deneb. Vega** is almost directly overhead at 2:55 a.m. **Altair** rises in the east around 9:50 p.m. and is at its highest, around 65 degrees, around dawn. Altair can be identified by its less bright neighboring stars, one on either side, together called the shaft of Aquila. **Deneb** is in the band of the Milky Way. See the illustration of the summer triangle on page 19.

Bright star **Capella** is in the northwestern sky at dusk and sets in the northwest around 10:30 p.m. Read about Capella on page 14.

The bright star **Arcturus** is at its highest, around 75 degrees, at 10:30 p.m.

Just east of **Scorpius**, Sagittarius rises in the southeast around 11:25 p.m. and is at its highest in the south, around 25 degrees, at 2:40 a.m.

Jupiter sets in the west around 12:25 a.m.

Scorpius is visible most of the night and is at its highest in the south, around 30 degrees, at 12:45 a.m. See the illustration on page 19.

Cassiopeia, shaped like a *W,* rises in the northeast around 12:55 a.m. It is in the northeastern sky before dawn. See the illustration on page 11.

Rust-red **Mars** sets in the west around 1:15 a.m.

Saturn rises in the east around 3:40 a.m. and is in the eastern sky before dawn. Unlike bright Jupiter and Venus and ruddy Mars, Saturn is harder to distinguish from the stars. Stars twinkle and shimmer, while the planets, including Saturn, show a steady flat light.

Bright **Venus** rises in the east around 4:45 a.m.

2027

Sunday, May 30

TONIGHT

sunset: **7:50 p.m.**
twilight begins: **8:55 p.m.**

MOON PHASE

IN THE MORNING

twilight begins: **4:31 a.m.**
sunrise: **5:34 a.m.**

moonrise: **2:25 a.m.**
moonset: **2:44 p.m.**

IN THE SKY

Castor and Pollux, the twin stars in the constellation Gemini, are in the western sky at dusk and set in the northwest around 10:25 p.m. Read how to find Castor and Pollux on page 14.

The **summer triangle**'s three bright stars are **Vega, Altair,** and **Deneb. Vega** is almost directly overhead at 2:50 a.m. **Altair** rises in the east around 9:45 p.m. and is at its highest, around 65 degrees, around dawn. Altair can be identified by its less bright neighboring stars, one on either side, together called the shaft of Aquila. **Deneb** is in the band of the Milky Way. See the illustration of the summer triangle on page 19.

With the dark sky, use binoculars or a telescope to look for the **moons of Jupiter.** See the guide on page 18.

Bright star **Capella** is in the northwestern sky at dusk and sets in the northwest around 10:25 p.m. Read about Capella on page 14.

The bright star **Arcturus** is at its highest, around 75 degrees, at 10:25 p.m.

Just east of **Scorpius**, Sagittarius rises in the southeast around 11:20 p.m. and is at its highest in the south, around 25 degrees, at 2:35 a.m.

Jupiter sets in the west around 12:20 a.m.

Scorpius is visible most of the night and is at its highest in the south, around 30 degrees, at 12:40 a.m. See the illustration on page 19.

Cassiopeia, shaped like a *W,* rises in the northeast around 12:50 a.m. It is in the northeastern sky before dawn. See the illustration on page 11.

Rust-red **Mars** sets in the west around 1:10 a.m.

Saturn rises in the east around 3:35 a.m. and is in the eastern sky before dawn. Unlike bright Jupiter and Venus and ruddy Mars, Saturn is harder to distinguish from the stars. Stars twinkle and shimmer, while the planets, including Saturn, show a steady flat light.

Bright **Venus** rises in the east around 4:45 a.m.

2027

Monday, May 31

TONIGHT

sunset: **7:51 p.m.**
twilight begins: **8:56 p.m.**

IN THE MORNING

twilight begins: **4:31 a.m.**
sunrise: **5:34 a.m.**

MOON PHASE

moonrise: **2:55 a.m.**
moonset: **3:46 p.m.**

IN THE SKY

Castor and Pollux, the twin stars in the constellation Gemini, are in the western sky at dusk and set in the northwest around 10:20 p.m. Read how to find Castor and Pollux on page 14.

Bright star **Capella** is in the northwestern sky at dusk and sets in the northwest around 10:20 p.m. Read about Capella on page 14.

The absence of the moon tonight provides excellent **dark-sky** viewing conditions.

With the dark sky, use binoculars or a telescope to look for the **moons of Jupiter.** See the guide on page 18.

The **summer triangle**'s three bright stars are **Vega, Altair,** and **Deneb. Vega** is almost directly overhead at 2:50 a.m. **Altair** rises in the east around 9:40 p.m. and is at its highest, around 65 degrees, around dawn. Altair can be identified by its less bright neighboring stars, one on either side, together called the shaft of Aquila. **Deneb** is in the band of the Milky Way. See the illustration of the summer triangle on page 19.

The bright star **Arcturus** is at its highest, around 75 degrees, at 10:25 p.m.

Just east of **Scorpius**, Sagittarius is visible most of the night. It rises in the southeast around 11:15 p.m. and is at its highest in the south, around 25 degrees, at 2:30 a.m.

Scorpius is visible most of the night and is at its highest in the south, around 30 degrees, at 12:35 a.m. See the illustration on page 19.

Cassiopeia, shaped like a *W,* rises in the northeast around 12:45 a.m. It is in the northeastern sky before dawn. See the illustration on page 11.

Jupiter sets in the west around 12:15 a.m.

Rust-red **Mars** sets in the west around 1:10 a.m.

Saturn rises in the east around 3:30 a.m. and is in the eastern sky before dawn. Unlike bright Jupiter and Venus and ruddy Mars, Saturn is harder to distinguish from the stars. Stars twinkle and shimmer, while the planets, including Saturn, show a steady flat light.

Bright **Venus** rises in the east around 4:45 a.m.

2027

Tuesday, June 1

TONIGHT

sunset: **7:52 p.m.**
twilight begins: **8:56 p.m.**

IN THE MORNING

twilight begins: **4:30 a.m.**
sunrise: **5:34 a.m.**

MOON PHASE

moonrise: **3:30 a.m.**
moonset: **4:51 p.m.**

IN THE SKY

Castor and Pollux, the twin stars in the constellation Gemini, are in the western sky at dusk and set in the northwest around 10:15 p.m. Read how to find Castor and Pollux on page 14.

Bright star **Capella** is in the northwestern sky at dusk and sets in the northwest around 10:15 p.m. Read about Capella on page 14.

The absence of the moon tonight provides excellent **dark-sky** viewing conditions.

With the dark sky, use binoculars or a telescope to look for the **moons of Jupiter.** See the guide on page 18.

The **summer triangle**'s three bright stars are **Vega, Altair,** and **Deneb. Vega** is almost directly overhead at 2:45 a.m. **Altair** rises in the east around 9:40 p.m. and is at its highest, around 65 degrees, around dawn. Altair can be identified by its less bright neighboring stars, one on either side, together called the shaft of Aquila. **Deneb** is in the band of the Milky Way. See the illustration of the summer triangle on page 19.

The bright star **Arcturus** is at its highest, around 75 degrees, at 10:20 p.m.

Just east of **Scorpius**, Sagittarius is visible most of the night. It rises in the southeast around 11:15 p.m. and is at its highest in the south, around 25 degrees, at 2:30 a.m.

Jupiter sets in the west around 12:15 a.m.

Scorpius is visible most of the night and is at its highest in the south, around 30 degrees, at 12:35 a.m. See the illustration on page 19.

Cassiopeia, shaped like a *W,* rises in the northeast around 12:45 a.m. It is in the northeastern sky before dawn. See the illustration on page 11.

Rust-red **Mars** sets in the west around 1:05 a.m.

Before the moon rises, around 3:20 a.m., look for the **Andromeda Galaxy** with binoculars or a telescope. It becomes a good target around 2:55 a.m. See the guide on page 11.

Saturn rises in the east around 3:25 a.m. and is in the eastern sky before dawn. Unlike bright Jupiter and Venus and ruddy Mars, Saturn is harder to distinguish from the stars. Stars twinkle and shimmer, while the planets, including Saturn, show a steady flat light.

Bright **Venus** rises in the east around 4:45 a.m.

2027

Wednesday, June 2

TONIGHT

sunset: **7:52 p.m.**
twilight begins: **8:57 p.m.**

MOON PHASE

IN THE MORNING

twilight begins: **4:30 a.m.**
sunrise: **5:34 a.m.**

moonrise: **4:13 a.m.**
moonset: **6:01 p.m.**

IN THE SKY

Castor and Pollux, the twin stars in the constellation Gemini, are in the western sky at dusk and set in the northwest around 10:15 p.m. Read how to find Castor and Pollux on page 14.

Bright star **Capella** is in the northwestern sky at dusk and sets in the northwest around 10:10 p.m. Read about Capella on page 14.

The absence of the moon tonight provides excellent **dark-sky** viewing conditions.

With the dark sky, use binoculars or a telescope to look for the **moons of Jupiter.** See the guide on page 18.

The **summer triangle**'s three bright stars are **Vega, Altair,** and **Deneb. Vega** is almost directly overhead at 2:40 a.m. **Altair** rises in the east around 9:35 p.m. and is at its highest, around 65 degrees, around dawn. Altair can be identified by its less bright neighboring stars, one on either side, together called the shaft of Aquila. **Deneb** is in the band of the Milky Way. See the illustration of the summer triangle on page 19.

Just east of **Scorpius**, Sagittarius is visible most of the night. It rises in the southeast around 11:10 p.m. and is at its highest in the south, around 25 degrees, at 2:25 a.m.

Scorpius is visible most of the night and is at its highest in the south, around 30 degrees, at 12:30 a.m. See the illustration on page 19.

Cassiopeia, shaped like a *W,* rises in the northeast around 12:40 a.m. It is in the northeastern sky before dawn. See the illustration on page 11.

The bright star **Arcturus** is at its highest, around 75 degrees, at 10:15 p.m.

Jupiter sets in the west around 12:10 a.m.

Rust-red **Mars** sets in the west around 1:05 a.m.

Before the moon rises, around 4:10 a.m., look for the **Andromeda Galaxy** with binoculars or a telescope. It becomes a good target around 2:50 a.m. It is in the northeastern sky before dawn. See the guide on page 11.

Saturn rises in the east around 3:25 a.m. and is in the eastern sky before dawn. Unlike bright Jupiter and Venus and ruddy Mars, Saturn is harder to distinguish from the stars. Stars twinkle and shimmer, while the planets, including Saturn, show a steady flat light.

Bright **Venus** rises in the east around 4:45 a.m.

2027
Thursday, June 3

TONIGHT

sunset: **7:53 p.m.**
twilight begins: **8:58 p.m.**

IN THE MORNING

twilight begins: **4:30 a.m.**
sunrise: **5:33 a.m.**

MOON PHASE

moonrise: **5:04 a.m.**
moonset: **7:12 p.m.**

IN THE SKY

The absence of the moon tonight provides excellent **dark-sky** viewing conditions.

With the dark sky, use binoculars or a telescope to look for the **moons of Jupiter.** See the guide on page 18.

The **summer triangle**'s three bright stars are **Vega, Altair,** and **Deneb. Vega** is almost directly overhead at 2:35 a.m. **Altair** rises in the east around 9:30 p.m. and is at its highest, around 65 degrees, around dawn. Altair can be identified by its less bright neighboring stars, one on either side, together called the shaft of Aquila. **Deneb** is in the band of the Milky Way. See the illustration of the summer triangle on page 19.

Bright star **Capella** is in the northwestern sky at dusk and sets in the northwest around 10:10 p.m. Read about Capella on page 14.

Castor and Pollux, the twin stars in the constellation Gemini, are in the western sky at dusk and set in the northwest around 10:10 p.m. Read how to find Castor and Pollux on page 14.

The bright star **Arcturus** is at its highest, around 75 degrees, at 10:10 p.m.

Just east of **Scorpius,** Sagittarius is visible most of the night. It rises in the southeast around 11:05 p.m. and is at its highest in the south, around 25 degrees, at 2:20 a.m.

Scorpius is visible most of the night and is at its highest in the south, around 30 degrees, at 12:25 a.m. See the illustration on page 19.

Cassiopeia, shaped like a *W,* rises in the northeast around 12:35 a.m. It is in the northeastern sky before dawn. See the illustration on page 11.

Jupiter sets in the west around 12:05 a.m.

Rust-red **Mars** sets in the west around 1 a.m.

Look for the **Andromeda Galaxy** with binoculars or a telescope. It becomes a good target around 2:45 a.m. See the guide on page 11.

Saturn rises in the east around 3:20 a.m. and is in the eastern sky before dawn. Unlike bright Jupiter and Venus and ruddy Mars, Saturn is harder to distinguish from the stars. Stars twinkle and shimmer, while the planets, including Saturn, show a steady flat light.

Bright **Venus** rises in the east around 4:45 a.m.

2027

Friday, June 4

TONIGHT

sunset: **7:53 p.m.**
twilight begins: **8:59 p.m.**

IN THE MORNING

twilight begins: **4:29 a.m.**
sunrise: **5:33 a.m.**

MOON PHASE

moonrise: **6:06 a.m.**
moonset: **8:22 p.m.**

IN THE SKY

The **summer triangle**'s three bright stars are **Vega, Altair,** and **Deneb. Vega** is almost directly overhead at 2:30 a.m. **Altair** is in the eastern sky at dusk and is at its highest, around 65 degrees, around dawn. Altair can be identified by its less bright neighboring stars, one on either side, together called the shaft of Aquila. **Deneb** is in the band of the Milky Way. See the illustration of the summer triangle on page 19.

Castor and Pollux, the twin stars in the constellation Gemini, are in the western sky at dusk and set in the northwest around 10:05 p.m. Read how to find Castor and Pollux on page 14.

The absence of the moon tonight provides excellent **dark-sky** viewing conditions.

After the moon sets, from around 8:40 p.m., use binoculars or a telescope to look for the **moons of Jupiter.** See the guide on page 18.

Bright star **Capella** is in the northwestern sky at dusk and sets in the northwest around 10:05 p.m. Read about Capella on page 14.

The bright star **Arcturus** is at its highest, around 75 degrees, at 10:10 p.m. See page 13 for an easy way to locate the star.

Just east of **Scorpius,** Sagittarius is visible most of the night. It rises in the southeast around 11 p.m. and is at its highest in the south, around 25 degrees, at 2:15 a.m.

Scorpius is visible most of the night and is at its highest in the south, around 30 degrees, at 12:20 a.m. See the illustration on page 19.

Cassiopeia, shaped like a *W,* rises in the northeast around 12:30 a.m. It is in the northeastern sky before dawn. See the illustration on page 11.

Jupiter sets in the west around midnight.

Rust-red **Mars** sets in the west around 12:55 a.m.

Look for the **Andromeda Galaxy** with binoculars or a telescope. It becomes a good target around 2:45 a.m. It is in the northeastern sky before dawn. See the guide on page 11.

Saturn rises in the east around 3:15 a.m. and is in the eastern sky before dawn. Unlike bright Jupiter and Venus and ruddy Mars, Saturn is harder to distinguish from the stars. Stars twinkle and shimmer, while the planets, including Saturn, show a steady flat light.

The **Great Bear (Big Dipper)** is in the northwestern sky before dawn and sets in the northwest around 4:05 a.m. See the illustration on page 10.

Bright star **Capella** rises in the northeast around 4:20 a.m. and is in the northeastern sky before dawn.

Bright **Venus** rises in the east around 4:45 a.m.

2027

Saturday, June 5

MOON PHASE

TONIGHT

sunset: **7:54 p.m.**
twilight begins: **8:59 p.m.**

IN THE MORNING

twilight begins: **4:29 a.m.**
sunrise: **5:33 a.m.**

moonrise: **7:16 a.m.**
moonset: **9:25 p.m.**

IN THE SKY

Castor and Pollux, the twin stars in the constellation Gemini, are in the western sky at dusk and set in the northwest around 10 p.m. Read how to find Castor and Pollux on page 14.

The **summer triangle**'s three bright stars are **Vega, Altair,** and **Deneb. Vega** is almost directly overhead at 2:30 a.m. **Altair** is in the eastern sky at dusk and is at its highest, around 65 degrees, around dawn. Altair can be identified by its less bright neighboring stars, one on either side, together called the shaft of Aquila. **Deneb** is in the band of the Milky Way. See the illustration of the summer triangle on page 19.

After the moon sets, from around 9:45 p.m., use binoculars or a telescope to look for the **moons of Jupiter.** See the guide on page 18.

The absence of the moon tonight provides excellent **dark-sky** viewing conditions.

The bright star **Arcturus** is at its highest, around 75 degrees, at 10:05 p.m. See page 13 for an easy way to locate the star.

Just east of **Scorpius**, Sagittarius is visible most of the night. It rises in the southeast around 10:55 p.m. and is at its highest in the south, around 25 degrees, at 2:10 a.m.

Scorpius is visible most of the night and is at its highest in the south, around 30 degrees, at 12:20 a.m. See the illustration on page 19.

Cassiopeia, shaped like a *W,* rises in the northeast around 12:30 a.m. It is in the northeastern sky before dawn. See the illustration on page 11.

Bright star **Capella** rises in the northeast around 4:15 a.m. and is in the northeastern sky before dawn.

Bright star **Capella** is in the northwestern sky at dusk and sets in the northwest around 10 p.m. Read about Capella on page 14.

Jupiter sets in the west around midnight.

Rust-red **Mars** sets in the west around 12:55 a.m.

Look for the **Andromeda Galaxy** with binoculars or a telescope. It becomes a good target around 2:40 a.m. It is in the northeastern sky before dawn. See the guide on page 11.

Saturn rises in the east around 3:10 a.m. and is in the eastern sky before dawn. Unlike bright Jupiter and Venus and ruddy Mars, Saturn is harder to distinguish from the stars. Stars twinkle and shimmer, while the planets, including Saturn, show a steady flat light.

The **Great Bear (Big Dipper)** is in the northwestern sky before dawn and sets in the northwest around 4:05 a.m. See the illustration on page 10.

Bright **Venus** rises in the northeast around 4:45 a.m.

2027

Sunday, June 6

TONIGHT

sunset: **7:54 p.m.**
twilight begins: **9 p.m.**

MOON PHASE 

IN THE MORNING

twilight begins: **4:29 a.m.**
sunrise: **5:33 a.m.**

moonrise: **7:16 a.m.**
moonset: **10:20 p.m.**

IN THE SKY

Bright star **Capella** is in the northwestern sky at dusk and sets in the northwest around 9:55 p.m. Read about Capella on page 14.

Castor and Pollux, the twin stars in the constellation Gemini, are in the western sky at dusk and set in the northwest around 10 p.m. Read how to find Castor and Pollux on page 14.

The bright star **Arcturus** is at its highest, around 75 degrees, at 10 p.m. See page 13 for an easy way to locate the star.

The absence of the moon tonight provides excellent **dark-sky** viewing conditions. Explore the **dark sky** after the moon sets, from around 10:40 p.m.

After the moon sets, from around 10:40 p.m., use binoculars or a telescope to look for the **moons of Jupiter.** See the guide on page 18.

Just east of **Scorpius**, Sagittarius is visible most of the night. It rises in the southeast around 10:55 p.m. and is at its highest in the south, around 25 degrees, at 2:10 a.m.

The **summer triangle**'s three bright stars are **Vega, Altair,** and **Deneb. Vega** is almost directly overhead at 2:25 a.m. **Altair** is in the eastern sky at dusk and is at its highest, around 65 degrees, around dawn. Altair can be identified by its less bright neighboring stars, one on either side, together called the shaft of Aquila. **Deneb** is in the band of the Milky Way. See the illustration of the summer triangle on page 19.

Jupiter sets in the west around 11:55 p.m.

Scorpius is visible most of the night and is at its highest in the south, around 30 degrees, at 12:15 a.m. See the illustration on page 19.

Cassiopeia, shaped like a *W,* rises in the northeast around 12:25 a.m. It is in the northeastern sky before dawn. See the illustration on page 11.

Rust-red **Mars** sets in the west around 12:50 a.m.

Look for the **Andromeda Galaxy** with binoculars or a telescope. It becomes a good target around 2:35 a.m. It is in the northeastern sky before dawn. See the guide on page 11.

Saturn rises in the east around 3:10 a.m. and is in the eastern sky before dawn. Unlike bright Jupiter and Venus and ruddy Mars, Saturn is harder to distinguish from the stars. Stars twinkle and shimmer, while the planets, including Saturn, show a steady flat light.

The **Great Bear (Big Dipper)** is in the northwestern sky before dawn and sets in the northwest around 4 a.m. See the illustration on page 10.

Bright star **Capella** rises in the northeast around 4:10 a.m. and is in the northeastern sky before dawn.

Bright **Venus** rises in the northeast around 4:45 a.m.

2027
Monday, June 7

TONIGHT

sunset: **7:55 p.m.**
twilight begins: **9 p.m.**

MOON PHASE

IN THE MORNING

twilight begins: **4:28 a.m.**
sunrise: **5:33 a.m.**

moonrise: **8:31 a.m.**
moonset: **11:05 p.m.**

IN THE SKY

The **summer triangle**'s three bright stars are **Vega, Altair,** and **Deneb. Vega** is almost directly overhead at 2:20 a.m. **Altair** is in the eastern sky at dusk and is at its highest, around 65 degrees, around dawn. Altair can be identified by its less bright neighboring stars, one on either side, together called the shaft of Aquila. **Deneb** is in the band of the Milky Way. See the illustration of the summer triangle on page 19.

The bright star **Arcturus** is at its highest, around 75 degrees, around dusk. See page 13 for an easy way to locate the star.

Bright star **Capella** is in the northwestern sky at dusk and sets in the northwest around 9:50 p.m. Read about Capella on page 14.

Castor and Pollux, the twin stars in the constellation Gemini, are in the western sky at dusk and set in the northwest around 9:55 p.m. Read how to find Castor and Pollux on page 14.

Just east of **Scorpius**, Sagittarius is visible most of the night. It rises in the southeast around 10:50 p.m. and is at its highest in the south, around 25 degrees, at 2:05 a.m.

The absence of the moon tonight provides excellent **dark-sky** viewing conditions. Explore the **dark sky** after the moon sets, from around 11:25 p.m.

Jupiter sets in the west around 11:50 p.m.

Scorpius is visible most of the night and is at its highest in the south, around 30 degrees, at 12:10 a.m. See the illustration on page 19.

Cassiopeia, shaped like a *W,* rises in the northeast around 12:20 a.m. It is in the northeastern sky before dawn. See the illustration on page 11.

Rust-red **Mars** sets in the west around 12:45 a.m.

Look for the **Andromeda Galaxy** with binoculars or a telescope. It becomes a good target around 2:30 a.m. It is in the northeastern sky before dawn. See the guide on page 11.

Saturn rises in the east around 3:05 a.m. and is in the eastern sky before dawn. Unlike bright Jupiter and Venus and ruddy Mars, Saturn is harder to distinguish from the stars. Stars twinkle and shimmer, while the planets, including Saturn, show a steady flat light.

The **Great Bear (Big Dipper)** is in the northwestern sky before dawn and sets in the northwest around 3:55 a.m. See the illustration on page 10.

Bright star **Capella** rises in the northeast around 4:05 a.m. and is in the northeastern sky before dawn.

Bright **Venus** rises in the northeast around 4:45 a.m.

2027

Tuesday, June 8

TONIGHT

sunset: **7:55 p.m.**
twilight begins: **9:01 p.m.**

MOON PHASE

IN THE MORNING

twilight begins: **4:28 a.m.**
sunrise: **5:33 a.m.**

moonrise: **9:45 a.m.**
moonset: **11:43 p.m.**

IN THE SKY

Bright star **Capella** is in the northwestern sky at dusk and sets in the northwest around 9:50 p.m. Read about Capella on page 14.

The **summer triangle**'s three bright stars are **Vega, Altair,** and **Deneb. Vega** is almost directly overhead at 2:15 a.m. **Altair** is in the eastern sky at dusk and is at its highest, around 65 degrees, around dawn. Altair can be identified by its less bright neighboring stars, one on either side, together called the shaft of Aquila. **Deneb** is in the band of the Milky Way. See the illustration of the summer triangle on page 19.

Castor and Pollux, the twin stars in the constellation Gemini, are in the western sky at dusk and set in the northwest around 9:50 p.m. Read how to find Castor and Pollux on page 14.

The bright star **Arcturus** is at its highest, around 75 degrees, around dusk. See page 13 for an easy way to locate the star.

The **moon** is prominent in the sky early in the evening. Using a telescope or binoculars, take a look. You should be able to see rough mountains and craters along the boundary between light and dark, in astronomy called the terminator.

The waxing crescent moon and **Jupiter** appear close together in the evening sky. Look west near the horizon around 10:30 p.m.

Just east of **Scorpius**, Sagittarius is visible most of the night. It rises in the southeast around 10:45 p.m. and is at its highest in the south, around 25 degrees, at 2 a.m.

Jupiter sets in the west around 11:50 p.m.

Explore the **dark sky** after the moon sets, from around 12:05 a.m.

Scorpius is visible most of the night and is at its highest in the south, around 30 degrees, at 12:05 a.m. See the illustration on page 19.

Cassiopeia, shaped like a *W,* rises in the northeast around 12:15 a.m. It is in the northeastern sky before dawn. See the illustration on page 11.

Rust-red **Mars** sets in the west around 12:45 a.m.

Look for the **Andromeda Galaxy** with binoculars or a telescope. It becomes a good target around 2:25 a.m. It is in the northeastern sky before dawn. See the guide on page 11.

Saturn rises in the east around 3 a.m. and is in the eastern sky before dawn. Unlike bright Jupiter and Venus and ruddy Mars, Saturn is harder to distinguish from the stars. Stars twinkle and shimmer, while the planets, including Saturn, show a steady flat light.

The **Great Bear (Big Dipper)** is in the northwestern sky before dawn and sets in the northwest around 3:50 a.m. See the illustration on page 10.

Bright **Venus** rises in the northeast around 4:45 a.m.

2027

Wednesday, June 9

MOON PHASE

TONIGHT

sunset: **7:56 p.m.**
twilight begins: **9:02 p.m.**

IN THE MORNING

twilight begins: **4:28 a.m.**
sunrise: **5:33 a.m.**

moonrise: **10:56 a.m.**
moonset: **12:15 a.m.**

IN THE SKY

The **summer triangle**'s three bright stars are **Vega, Altair,** and **Deneb. Vega** is almost directly overhead at 2:10 a.m. **Altair** is in the eastern sky at dusk and is at its highest, around 65 degrees, at 3:25 a.m. Altair can be identified by its less bright neighboring stars, one on either side, together called the shaft of Aquila. **Deneb** is in the band of the Milky Way. See the illustration of the summer triangle on page 19.

Castor and Pollux, the twin stars in the constellation Gemini, are in the western sky at dusk and set in the northwest around 9:45 p.m. Read how to find Castor and Pollux on page 14.

The **moon** is prominent in the sky early in the evening. Using a telescope or binoculars, take a look. You should be able to see rough mountains and craters along the boundary between light and dark, in astronomy called the terminator.

The **crescent moon, Jupiter,** and **Mars** are near each other and near the bright star Regulus in the constellation Leo. Look to the western horizon around 9:30 p.m.

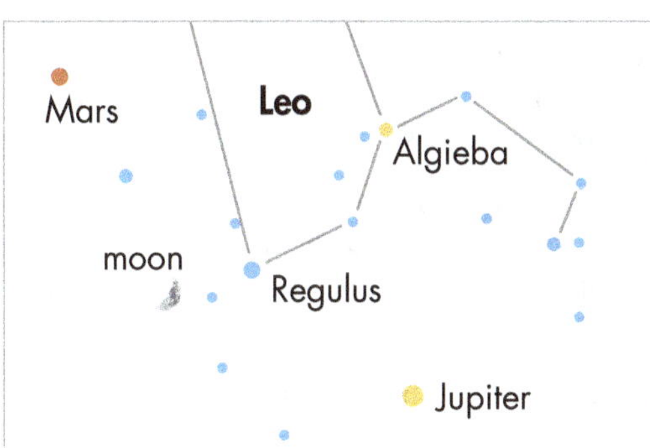

Bright star **Capella** is in the northwestern sky at dusk and sets in the northwest around 9:45 p.m. Read about Capella on page 14.

Just east of **Scorpius**, Sagittarius is visible most of the night. It rises in the southeast around 10:40 p.m. and is at its highest in the south, around 25 degrees, at 1:55 a.m.

Jupiter sets in the west around 11:45 p.m.

Scorpius is visible most of the night and is at its highest in the south, around 30 degrees, at midnight. See the illustration on page 19.

Cassiopeia, shaped like a *W,* rises in the northeast around 12:10 a.m. It is in the northeastern sky before dawn. See the illustration on page 11.

Explore the **dark sky** after the moon sets, from around 12:35 a.m.

Rust-red **Mars** sets in the west around 12:40 a.m.

Saturn rises in the east around 3 a.m. and is in the eastern sky before dawn. Unlike bright Jupiter and Venus and ruddy Mars, Saturn is harder to distinguish from the stars. Stars twinkle and shimmer, while the planets, including Saturn, show a steady flat light.

The **Pleiades** rises in the northeast around 4:15 a.m. and is in the eastern sky before dawn. Read about the cluster on page 18.

Bright **Venus** rises in the northeast around 4:45 a.m.

2027

Thursday, June 10

TONIGHT

sunset: **7:56 p.m.**
twilight begins: **9:02 p.m.**

IN THE MORNING

twilight begins: **4:28 a.m.**
sunrise: **5:33 a.m.**

MOON PHASE

moonrise: **12:03 p.m.**
moonset: **12:45 a.m.**

IN THE SKY

The **summer triangle**'s three bright stars are **Vega, Altair,** and **Deneb. Vega** is almost directly overhead at 2:10 a.m. **Altair** is in the eastern sky at dusk and is at its highest, around 65 degrees, at 3:20 a.m. Altair can be identified by its less bright neighboring stars, one on either side, together called the shaft of Aquila. **Deneb** is in the band of the Milky Way. See the illustration of the summer triangle on page 19.

The bright star **Arcturus** is at its highest, around 75 degrees, around dusk. See page 13 for an easy way to locate the star.

The **moon** is prominent in the sky early in the evening. Using a telescope or binoculars, take a look. You should be able to see rough mountains and craters along the boundary between light and dark, in astronomy called the terminator.

The waxing first-quarter moon and Mars appear close together in the evening sky. Look southwest at about 45 degrees elevation around 9:30 p.m.

Bright star **Capella** is in the northwestern sky at dusk and sets in the northwest around 9:40 p.m. Read about Capella on page 14.

Castor and Pollux, the twin stars in the constellation Gemini, set in the northwest around 9:40 p.m. Read how to find Castor and Pollux on page 14.

Look for the **Andromeda Galaxy** with binoculars or a telescope. It becomes a good target around 2:20 a.m. It is in the northeastern sky before dawn. See the guide on page 11.

Jupiter sets in the west around 11:40 p.m.

Scorpius is visible most of the night and is at its highest in the south, around 30 degrees, at midnight. See the illustration on page 19.

Cassiopeia, shaped like a *W,* rises in the northeast around 12:10 a.m. It is in the northeastern sky before dawn. See the illustration on page 11.

Rust-red Mars sets in the west around 12:40 a.m.

Just east of **Scorpius,** Sagittarius is visible most of the night. It rises in the southeast around 10:35 p.m. and is at its highest in the south, around 25 degrees, at 1:50 a.m.

Explore the **dark sky** after the moon sets, from around 1:05 a.m.

Saturn rises in the east around 2:55 a.m. and is in the eastern sky before dawn. Unlike bright Jupiter and Venus and ruddy Mars, Saturn is harder to distinguish from the stars. Stars twinkle and shimmer, while the planets, including Saturn, show a steady flat light.

The **Pleiades** rises in the northeast around 4:10 a.m. and is in the eastern sky before dawn. Read about the cluster on page 18.

Bright Venus rises in the northeast around 4:45 a.m.

2027

Friday, June 11

TONIGHT

sunset: **7:57 p.m.**
twilight begins: **9:03 p.m.**

IN THE MORNING

twilight begins: **4:28 a.m.**
sunrise: **5:33 a.m.**

MOON PHASE

moonrise: **1:09 p.m.**
moonset: **1:13 a.m.**

IN THE SKY

The **summer triangle**'s three bright stars are **Vega, Altair,** and **Deneb. Vega** is almost directly overhead at 2:05 a.m. **Altair** is in the eastern sky at dusk and is at its highest, around 65 degrees, at 3:20 a.m. Altair can be identified by its less bright neighboring stars, one on either side, together called the shaft of Aquila. **Deneb** is in the band of the Milky Way. See the illustration of the summer triangle on page 19.

The bright star **Arcturus** is at its highest, around 75 degrees, around dusk. See page 13 for an easy way to locate the star.

Bright star **Capella** is in the northwestern sky at dusk and sets in the northwest around 9:35 p.m. Read about Capella on page 14.

The **moon** is prominent in the sky early in the evening. Using a telescope or binoculars, take a look. You should be able to see rough mountains and craters along the boundary between light and dark, in astronomy called the terminator.

Just east of **Scorpius**, Sagittarius is visible most of the night. It rises in the southeast around 10:35 p.m. and is at its highest in the south, around 25 degrees, at 1:50 a.m.

Scorpius is visible most of the night and is at its highest in the south, around 30 degrees, at 11:55 p.m. See the illustration on page 19.

Cassiopeia, shaped like a *W,* rises in the northeast around 12:05 a.m. It is in the northeastern sky before dawn. See the illustration on page 11.

Jupiter sets in the west around 11:40 p.m.

Rust-red **Mars** sets in the west around 12:35 a.m.

Saturn rises in the east around 2:50 a.m. and is in the eastern sky before dawn. Unlike bright Jupiter and Venus and ruddy Mars, Saturn is harder to distinguish from the stars. Stars twinkle and shimmer, while the planets, including Saturn, show a steady flat light.

The **Great Bear (Big Dipper)** is in the northwestern sky before dawn and sets in the northwest around 3:40 a.m. See the illustration on page 10.

Bright star **Capella** rises in the northeast around 3:50 a.m. and is in the northeastern sky before dawn.

The **Pleiades** rises in the northeast around 4:05 a.m. and is in the eastern sky before dawn. Read about the cluster on page 18.

Bright **Venus** rises in the northeast around 4:45 a.m.

Castor and Pollux, the twin stars in the constellation Gemini, set in the northwest around 9:40 p.m. Read how to find Castor and Pollux on page 14.

2027

Saturday, June 12

TONIGHT

sunset: **7:57 p.m.**
twilight begins: **9:03 p.m.**

IN THE MORNING

twilight begins: **4:28 a.m.**
sunrise: **5:33 a.m.**

MOON PHASE

moonrise: **2:12 p.m.**
moonset: **1:42 a.m.**

IN THE SKY

The **summer triangle**'s three bright stars are **Vega, Altair,** and **Deneb. Vega** is almost directly overhead at 2 a.m. **Altair** is in the eastern sky at dusk and is at its highest, around 65 degrees, at 3:15 a.m. Altair can be identified by its less bright neighboring stars, one on either side, together called the shaft of Aquila. **Deneb** is in the band of the Milky Way. See the illustration of the summer triangle on page 19.

The bright star **Arcturus** is at its highest, around 75 degrees, around dusk. See page 13 for an easy way to locate the star.

Bright star **Capella** is in the northwestern sky at dusk and sets in the northwest around 9:35 p.m. Read about Capella on page 14.

The **moon** is prominent in the sky early in the evening. Using a telescope or binoculars, take a look. You should be able to see rough mountains and craters along the boundary between light and dark, in astronomy called the terminator.

Castor and Pollux, the twin stars in the constellation Gemini, set in the northwest around 9:35 p.m. Read how to find Castor and Pollux on page 14.

Just east of **Scorpius**, Sagittarius is visible most of the night. It rises in the southeast around 10:30 p.m. and is at its highest in the south, around 25 degrees, at 1:45 a.m.

Jupiter sets in the west around 11:35 p.m.

Scorpius is visible most of the night and is at its highest in the south, around 30 degrees, at 11:50 p.m. See the illustration on page 19.

Cassiopeia, shaped like a *W,* rises in the northeast around midnight. See the illustration on page 11.

Rust-red **Mars** sets in the west around 12:30 a.m.

Saturn rises in the east around 2:45 a.m. and is in the eastern sky before dawn. Unlike bright Jupiter and Venus and ruddy Mars, Saturn is harder to distinguish from the stars. Stars twinkle and shimmer, while the planets, including Saturn, show a steady flat light.

The **Great Bear (Big Dipper)** is in the northwestern sky before dawn and sets in the northwest around 3:35 a.m. See the illustration on page 10.

Bright star **Capella** rises in the northeast around 3:45 a.m. and is in the northeastern sky before dawn.

The **Pleiades** rises in the northeast around 4 a.m. and is in the eastern sky before dawn. Read about the cluster on page 18.

Bright **Venus** rises in the northeast around 4:45 a.m.

2027

Sunday, June 13

TONIGHT

sunset: **7:58 p.m.**
twilight begins: **9:03 p.m.**

MOON PHASE

IN THE MORNING

twilight begins: **4:28 a.m.**
sunrise: **5:33 a.m.**

moonrise: **3:15 p.m.**
moonset: **2:12 a.m.**

IN THE SKY

The **summer triangle**'s three bright stars are **Vega, Altair,** and **Deneb. Vega** is almost directly overhead at 1:55 a.m. **Altair** is in the eastern sky at dusk and is at its highest, around 65 degrees, at 3:10 a.m. Altair can be identified by its less bright neighboring stars, one on either side, together called the shaft of Aquila. **Deneb** is in the band of the Milky Way. See the illustration of the summer triangle on page 19.

The bright star **Arcturus** is at its highest, around 75 degrees, around dusk. See page 13 for an easy way to locate the star.

Bright star **Capella** is in the northwestern sky at dusk and sets in the northwest around 9:30 p.m. Read about Capella on page 14.

Castor and Pollux, the twin stars in the constellation Gemini, set in the northwest around 9:30 p.m. Read how to find Castor and Pollux on page 14.

Just east of **Scorpius**, Sagittarius is visible most of the night. It rises in the southeast around 10:25 p.m. and is at its highest in the south, around 25 degrees, at 1:40 a.m.

Jupiter sets in the west around 11:30 p.m.

Scorpius is visible most of the night and is at its highest in the south, around 30 degrees, at 11:45 p.m. and sets in the southwest around 4:05 a.m. See the illustration on page 19.

Cassiopeia, shaped like a *W*, rises in the northeast around 11:55 p.m. See the illustration on page 11.

Rust-red **Mars** sets in the west around 12:30 a.m.

Saturn rises in the east around 2:45 a.m. and is in the eastern sky before dawn. Unlike bright Jupiter and Venus and ruddy Mars, Saturn is harder to distinguish from the stars. Stars twinkle and shimmer, while the planets, including Saturn, show a steady flat light.

The **Great Bear (Big Dipper)** is in the northwestern sky before dawn and sets in the northwest around 3:30 a.m. See the illustration on page 10.

Bright star **Capella** rises in the northeast around 3:40 a.m. and is in the northeastern sky before dawn.

The **Pleiades** rises in the northeast around 3:55 a.m. and is in the eastern sky before dawn. Read about the cluster on page 18.

Bright **Venus** rises in the northeast around 4:45 a.m.

2027

Monday, June 14

TONIGHT

sunset: **7:58 p.m.**
twilight begins: **9:04 p.m.**

MOON PHASE

IN THE MORNING

twilight begins: **4:28 a.m.**
sunrise: **5:33 a.m.**

moonrise: **4:19 p.m.**
moonset: **2:46 a.m.**

IN THE SKY

The bright star **Arcturus** is at its highest, around 75 degrees, around dusk. See page 13 for an easy way to locate the star.

The **summer triangle**'s three bright stars are **Vega, Altair,** and **Deneb. Vega** is almost directly overhead at 1:55 a.m. **Altair** is in the eastern sky at dusk and is at its highest, around 65 degrees, at 3:05 a.m. Altair can be identified by its less bright neighboring stars, one on either side, together called the shaft of Aquila. **Deneb** is in the band of the Milky Way. See the illustration of the summer triangle on page 19.

The nearly **full moon** tonight acts like a spotlight. Look around at the desert landscape at how much illumination the moon provides.

Bright star **Capella** sets in the northwest around 9:25 p.m. Read about Capella on page 14.

Castor and Pollux, the twin stars in the constellation Gemini, set in the northwest around 9:25 p.m. Read how to find Castor and Pollux on page 14.

Just east of **Scorpius**, Sagittarius is visible most of the night. It rises in the southeast around 10:20 p.m. and is at its highest in the south, around 25 degrees, at 1:35 a.m.

Jupiter sets in the west around 11:25 p.m.

Scorpius is visible most of the night and is at its highest in the south, around 30 degrees, at 11:40 p.m. and sets in the southwest around 4 a.m. See the illustration on page 19.

Cassiopeia, shaped like a *W,* rises in the northeast around 11:50 p.m. See the illustration on page 11.

Rust-red **Mars** sets in the west around 12:25 a.m.

Saturn rises in the east around 2:40 a.m. and is in the eastern sky before dawn. Unlike bright Jupiter and Venus and ruddy Mars, Saturn is harder to distinguish from the stars. Stars twinkle and shimmer, while the planets, including Saturn, show a steady flat light.

The **Great Bear (Big Dipper)** is in the northwestern sky before dawn and sets in the northwest around 3:25 a.m. See the illustration on page 10.

Bright star **Capella** rises in the northeast around 3:40 a.m. and is in the northeastern sky before dawn.

The **Pleiades** rises in the northeast around 3:55 a.m. and is in the eastern sky before dawn. Read about the cluster on page 18.

Bright **Venus** rises in the northeast around 4:45 a.m.

2027

Tuesday, June 15

TONIGHT

sunset: **7:58 p.m.**
twilight begins: **9:04 p.m.**

IN THE MORNING

twilight begins: **4:28 a.m.**
sunrise: **5:33 a.m.**

MOON PHASE

moonrise: **5:22 p.m.**
moonset: **3:24 a.m.**

IN THE SKY

The **summer triangle**'s three bright stars are **Vega, Altair,** and **Deneb. Vega** is almost directly overhead at 1:50 a.m. **Altair** is in the eastern sky at dusk and is at its highest, around 65 degrees, at 3:05 a.m. Altair can be identified by its less bright neighboring stars, one on either side, together called the shaft of Aquila. **Deneb** is in the band of the Milky Way. See the illustration of the summer triangle on page 19.

The **full moon** tonight acts like a spotlight. Look around at the desert landscape at how much illumination the moon provides.

Bright star **Capella** sets in the northwest around 9:20 p.m. Read about Capella on page 14.

Just east of **Scorpius**, Sagittarius is visible most of the night. It rises in the southeast around 10:20 p.m. and is at its highest in the south, around 25 degrees, at 1:35 a.m.

Jupiter sets in the west around 11:25 p.m.

Scorpius is visible most of the night and is at its highest in the south, around 30 degrees, at 11:40 p.m. and sets in the southwest around 3:55 a.m. See the illustration on page 19.

Cassiopeia, shaped like a *W,* rises in the northeast around 11:50 p.m. See the illustration on page 11.

Rust-red **Mars** sets in the west around 12:25 a.m.

Saturn rises in the east around 2:35 a.m. and is in the eastern sky before dawn. Unlike bright Jupiter and Venus and ruddy Mars, Saturn is harder to distinguish from the stars. Stars twinkle and shimmer, while the planets, including Saturn, show a steady flat light.

The **Great Bear (Big Dipper)** is in the northwestern sky before dawn and sets in the northwest around 3:25 a.m. See the illustration on page 10.

Bright star **Capella** rises in the northeast around 3:35 a.m. and is in the northeastern sky before dawn.

The **Pleiades** rises in the northeast around 3:50 a.m. and is in the eastern sky before dawn. Read about the cluster on page 18.

Bright **Venus** rises in the northeast around 4:45 a.m.

2027

Wednesday, June 16

TONIGHT

sunset: **7:59 p.m.**
twilight begins: **9:05 p.m.**

MOON PHASE

IN THE MORNING

twilight begins: **4:28 a.m.**
sunrise: **5:33 a.m.**

moonrise: **6:23 p.m.**
moonset: **4:08 a.m.**

IN THE SKY

The **summer triangle**'s three bright stars are **Vega, Altair,** and **Deneb. Vega** is almost directly overhead at 1:45 a.m. **Altair** is in the eastern sky at dusk and is at its highest, around 65 degrees, at 3 a.m. Altair can be identified by its less bright neighboring stars, one on either side, together called the shaft of Aquila. **Deneb** is in the band of the Milky Way. See the illustration of the summer triangle on page 19.

The **full moon** tonight acts like a spotlight. Look around at the desert landscape at how much illumination the moon provides.

Bright star **Capella** sets in the northwest around 9:15 p.m. Read about Capella on page 14.

Just east of **Scorpius,** Sagittarius is visible most of the night. It rises in the southeast around 10:15 p.m. and is at its highest in the south, around 25 degrees, at 1:30 a.m.

The **full moon** and the bright yellowish star **Antares** appear close together in the evening sky. Look south at about 30 degrees elevation around 11 p.m.

Jupiter sets in the west around 11:20 p.m.

Scorpius is visible most of the night and is at its highest in the south, around 30 degrees, at 11:35 p.m. and sets in the southwest around 3:55 a.m. Tonight the moon is in **Scorpius,** making it easy to find the constellation. See the illustration on page 19.

Cassiopeia, shaped like a *W,* rises in the northeast around 11:45 p.m. See the illustration on page 11.

Rust-red **Mars** sets in the west around 12:20 a.m.

Saturn rises in the east around 2:30 a.m. and is in the eastern sky before dawn. Unlike bright Jupiter and Venus and ruddy Mars, Saturn is harder to distinguish from the stars. Stars twinkle and shimmer, while the planets, including Saturn, show a steady flat light.

The **Great Bear (Big Dipper)** is in the northwestern sky before dawn and sets in the northwest around 3:20 a.m. See the illustration on page 10.

Bright star **Capella** rises in the northeast around 3:30 a.m. and is in the northeastern sky before dawn.

The **Pleiades** rises in the northeast around 3:45 a.m. and is in the eastern sky before dawn. Read about the cluster on page 18.

Bright **Venus** rises in the northeast around 4:45 a.m.

2027

Thursday, June 17

TONIGHT

sunset: **7:59 p.m.**
twilight begins: **9:05 p.m.**

IN THE MORNING

twilight begins: **4:28 a.m.**
sunrise: **5:33 a.m.**

MOON PHASE

moonrise: **7:21 p.m.**
moonset: **4:57 a.m.**

IN THE SKY

The **summer triangle**'s three bright stars are **Vega, Altair,** and **Deneb. Vega** is almost directly overhead at 1:40 a.m. **Altair** is in the eastern sky at dusk and is at its highest, around 65 degrees, at 2:55 a.m. Altair can be identified by its less bright neighboring stars, one on either side, together called the shaft of Aquila. **Deneb** is in the band of the Milky Way. See the illustration of the summer triangle on page 19.

The **full moon** tonight acts like a spotlight. Look around at the desert landscape at how much illumination the moon provides.

Just east of **Scorpius**, Sagittarius is visible most of the night. It rises in the southeast around 10:10 p.m. and is at its highest in the south, around 25 degrees, at 1:25 a.m.

Jupiter sets in the west around 11:15 p.m.

Scorpius is visible most of the night and is at its highest in the south, around 30 degrees, at 11:30 p.m. and sets in the southwest around 3:50 a.m. See the illustration on page 19.

Cassiopeia, shaped like a *W,* rises in the northeast around 11:40 p.m. See the illustration on page 11.

Rust-red **Mars** sets in the west around 12:20 a.m.

Saturn rises in the east around 2:30 a.m. and is in the eastern sky before dawn. Unlike bright Jupiter and Venus and ruddy Mars, Saturn is harder to distinguish from the stars. Stars twinkle and shimmer, while the planets, including Saturn, show a steady flat light.

The **Great Bear (Big Dipper)** is in the northwestern sky before dawn and sets in the northwest around 3:15 a.m. See the illustration on page 10.

Bright star **Capella** rises in the northeast around 3:25 a.m. and is in the northeastern sky before dawn. Read about Capella on page 14.

The **Pleiades** rises in the northeast around 3:40 a.m. and is in the eastern sky before dawn. Read about the cluster on page 18.

Bright **Venus** rises in the northeast around 4:45 a.m.

2027

Friday, June 18

TONIGHT

sunset: **7:59 p.m.**
twilight begins: **9:05 p.m.**

IN THE MORNING

twilight begins: **4:28 a.m.**
sunrise: **5:33 a.m.**

MOON PHASE

moonrise: **8:13 p.m.**
moonset: **5:52 a.m.**

IN THE SKY

The **full moon** tonight acts like a spotlight. Look around at the desert landscape at how much illumination the moon provides.

Just east of **Scorpius**, Sagittarius is visible most of the night. It rises in the southeast around 10:05 p.m. and is at its highest in the south, around 25 degrees, at 1:20 a.m.

Jupiter sets in the west around 11:15 p.m.

Scorpius is visible most of the night and is at its highest in the south, around 30 degrees, at 11:25 p.m. and sets in the southwest around 3:45 a.m. See the illustration on page 19.

Cassiopeia, shaped like a *W,* rises in the northeast around 11:35 p.m. See the illustration on page 11.

Rust-red Mars sets in the west around 12:15 a.m.

The **summer triangle**'s three bright stars are **Vega, Altair,** and **Deneb. Vega** is almost directly overhead at 1:35 a.m. **Altair** is at its highest, around 65 degrees, at 2:50 a.m. Altair can be identified by its less bright neighboring stars, one on either side, together called the shaft of Aquila. **Deneb** is in the band of the Milky Way. See the illustration of the summer triangle on page 19.

Saturn rises in the east around 2:25 a.m. and is in the eastern sky before dawn. Unlike bright Jupiter and Venus and ruddy Mars, Saturn is harder to distinguish from the stars. Stars twinkle and shimmer, while the planets, including Saturn, show a steady flat light.

The **Great Bear (Big Dipper)** is in the northwestern sky before dawn and sets in the northwest around 3:10 a.m. See the illustration on page 10.

Bright star **Capella** rises in the northeast around 3:25 a.m. It is in the northeastern sky before dawn. Read about Capella on page 14.

The **Pleiades** rises in the northeast around 3:35 a.m. and is in the eastern sky before dawn. Read about the cluster on page 18.

Bright Venus rises in the northeast around 4:45 a.m.

2027
Saturday, June 19

TONIGHT

sunset: **8 p.m.**
twilight begins: **9:06 p.m.**

IN THE MORNING

twilight begins: **4:28 a.m.**
sunrise: **5:34 a.m.**

MOON PHASE

moonrise: **8:59 p.m.**
moonset: **6:49 a.m.**

IN THE SKY

The **full moon** tonight acts like a spotlight. Look around at the desert landscape at how much illumination the moon provides.

Just east of **Scorpius**, Sagittarius is visible most of the night. It rises in the southeast around 10 p.m. and is at its highest in the south, around 25 degrees, at 1:15 a.m.

Jupiter sets in the west around 11:10 p.m.

Scorpius is visible most of the night and is at its highest in the south, around 30 degrees, at 11:25 p.m. and sets in the southwest around 3:40 a.m. See the illustration on page 19.

Cassiopeia, shaped like a *W,* rises in the northeast around 11:30 p.m. See the illustration on page 11.

Rust-red **Mars** sets in the west around 12:15 a.m.

The **summer triangle**'s three bright stars are **Vega, Altair,** and **Deneb. Vega** is almost directly overhead at 1:35 a.m. **Altair** is at its highest, around 65 degrees, at 2:45 a.m. Altair can be identified by its less bright neighboring stars, one on either side, together called the shaft of Aquila. **Deneb** is in the band of the Milky Way. See the illustration of the summer triangle on page 19.

Saturn rises in the east around 2:20 a.m. and is in the eastern sky before dawn. Unlike bright Jupiter and Venus and ruddy Mars, Saturn is harder to distinguish from the stars. Stars twinkle and shimmer, while the planets, including Saturn, show a steady flat light.

The **Great Bear (Big Dipper)** is in the northwestern sky before dawn and sets in the northwest around 3:05 a.m. See the illustration on page 10.

Bright star **Capella** rises in the northeast around 3:20 a.m. It is in the northeastern sky before dawn. Read about Capella on page 14.

The **Pleiades** rises in the northeast around 3:35 a.m. and is in the eastern sky before dawn. Read about the cluster on page 18.

Bright **Venus** rises in the northeast around 4:50 a.m.

2027

Sunday, June 20

TONIGHT

sunset: **8 p.m.**
twilight begins: **9:06 p.m.**

MOON PHASE

IN THE MORNING

twilight begins: **4:29 a.m.**
sunrise: **5:34 a.m.**

moonrise: **9:39 p.m.**
moonset: **7:47 a.m.**

IN THE SKY

The **full moon** tonight acts like a spotlight. Look around at the desert landscape at how much illumination the moon provides.

Just east of **Scorpius**, Sagittarius is visible most of the night. It rises in the southeast around 10 p.m. and is at its highest in the south, around 25 degrees, at 1:15 a.m.

Jupiter sets in the west around 11:05 p.m.

Scorpius is visible most of the night and is at its highest in the south, around 30 degrees, at 11:20 p.m. and sets in the southwest around 3:35 a.m. See the illustration on page 19.

Cassiopeia, shaped like a *W,* rises in the northeast around 11:30 p.m. See the illustration on page 11.

Rust-red Mars sets in the west around 12:10 a.m.

The **summer triangle**'s three bright stars are **Vega, Altair,** and **Deneb. Vega** is almost directly overhead at 1:30 a.m. **Altair** is at its highest, around 65 degrees, at 2:45 a.m. Altair can be identified by its less bright neighboring stars, one on either side, together called the shaft of Aquila. **Deneb** is in the band of the Milky Way. See the illustration of the summer triangle on page 19.

Saturn rises in the east around 2:15 a.m. and is in the eastern sky before dawn. Unlike bright Jupiter and Venus and ruddy Mars, Saturn is harder to distinguish from the stars. Stars twinkle and shimmer, while the planets, including Saturn, show a steady flat light.

The **Great Bear (Big Dipper)** is in the northwestern sky before dawn and sets in the northwest around 3:05 a.m. See the illustration on page 10.

Bright star **Capella** rises in the northeast around 3:15 a.m. It is in the northeastern sky before dawn. Read about Capella on page 14.

The **Pleiades** rises in the northeast around 3:30 a.m. and is in the eastern sky before dawn. Read about the cluster on page 18.

The bright star **Arcturus** is in the western sky before dawn and sets in the northwest around 3:55 a.m. See page 13 for an easy way to locate the star.

Bright Venus rises in the northeast around 4:50 a.m.

2027

Monday, June 21

TONIGHT

sunset: **8 p.m.**
twilight begins: **9:06 p.m.**

IN THE MORNING

twilight begins: **4:29 a.m.**
sunrise: **5:34 a.m.**

MOON PHASE

moonrise: **10:13 p.m.**
moonset: **7:47 a.m.**

IN THE SKY

The summer solstice means tonight is the shortest night of the year.

The nearly **full moon** tonight acts like a spotlight. Look around at the desert landscape at how much illumination the moon provides.

Just east of **Scorpius**, Sagittarius is visible most of the night. It rises in the southeast around 9:55 p.m. and is at its highest in the south, around 25 degrees, at 1:10 a.m.

Jupiter sets in the west around 11:05 p.m.

Scorpius is visible most of the night and is at its highest in the south, around 30 degrees, at 11:15 p.m. and sets in the southwest around 3:35 a.m. See the illustration on page 19.

Cassiopeia, shaped like a *W,* rises in the northeast around 11:25 p.m. See the illustration on page 11.

Rust-red Mars sets in the west around 12:10 a.m.

The bright star **Arcturus** is in the western sky before dawn and sets in the northwest around 3:50 a.m. See page 13 for an easy way to locate the star.

The **summer triangle**'s three bright stars are **Vega, Altair,** and **Deneb. Vega** is almost directly overhead at 1:25 a.m. **Altair** is at its highest, around 65 degrees, at 2:40 a.m. Altair can be identified by its less bright neighboring stars, one on either side, together called the shaft of Aquila. **Deneb** is in the band of the Milky Way. See the illustration of the summer triangle on page 19.

Saturn rises in the east around 2:15 a.m. and is in the eastern sky before dawn. Unlike bright Jupiter and Venus and ruddy Mars, Saturn is harder to distinguish from the stars. Stars twinkle and shimmer, while the planets, including Saturn, show a steady flat light.

The **Great Bear (Big Dipper)** is in the northwestern sky before dawn and sets in the northwest around 3 a.m. See the illustration on page 10.

Bright star **Capella** rises in the northeast around 3:10 a.m. It is in the northeastern sky before dawn. Read about Capella on page 14.

The **Pleiades** rises in the northeast around 3:25 a.m. and is in the eastern sky before dawn. Read about the cluster on page 18.

Bright Venus rises in the northeast around 4:50 a.m.

2027

Tuesday, June 22

TONIGHT

sunset: **8 p.m.**
twilight begins: **9:06 p.m.**

IN THE MORNING

twilight begins: **4:29 a.m.**
sunrise: **5:34 a.m.**

MOON PHASE

moonrise: **10:43 p.m.**
moonset: **8:45 a.m.**

IN THE SKY

Just east of **Scorpius**, Sagittarius is visible most of the night. It rises in the southeast around 9:50 p.m. and is at its highest in the south, around 25 degrees, at 1:05 a.m.

Jupiter sets in the west around 11 p.m.

Scorpius is visible most of the night and is at its highest in the south, around 30 degrees, at 11:10 p.m. and sets in the southwest around 3:30 a.m. See the illustration on page 19.

Cassiopeia, shaped like a *W,* rises in the northeast around 11:20 p.m. See the illustration on page 11.

Rust-red **Mars** sets in the west around 12:05 a.m.

The bright star **Arcturus** is in the western sky before dawn and sets in the northwest around 3:45 a.m.

The **summer triangle**'s three bright stars are **Vega, Altair,** and **Deneb. Vega** is almost directly overhead at 1:20 a.m. **Altair** is at its highest, around 65 degrees, at 2:35 a.m. Altair can be identified by its less bright neighboring stars, one on either side, together called the shaft of Aquila. **Deneb** is in the band of the Milky Way. See the illustration of the summer triangle on page 19.

Saturn rises in the east around 2:10 a.m. and is in the eastern sky before dawn. Unlike bright Jupiter and Venus and ruddy Mars, Saturn is harder to distinguish from the stars. Stars twinkle and shimmer, while the planets, including Saturn, show a steady flat light.

The **Great Bear (Big Dipper)** is in the northwestern sky before dawn and sets in the northwest around 2:55 a.m. See the illustration on page 10.

Bright star **Capella** rises in the northeast around 3:05 a.m. It is in the northeastern sky before dawn. Read about Capella on page 14.

The **Pleiades** rises in the northeast around 3:20 a.m. and is in the eastern sky before dawn. Read about the cluster on page 18.

Bright **Venus** rises in the northeast around 4:50 a.m.

2027

Wednesday, June 23

TONIGHT

sunset: **8 p.m.**
twilight begins: **9:06 p.m.**

IN THE MORNING

twilight begins: **4:29 a.m.**
sunrise: **5:35 a.m.**

MOON PHASE

moonrise: **11:10 p.m.**
moonset: **9:42 a.m.**

IN THE SKY

Just east of **Scorpius**, Sagittarius is visible most of the night. It rises in the southeast around 9:45 p.m. and is at its highest in the south, around 25 degrees, at 1 a.m.

Jupiter sets in the west around 10:55 p.m.

Explore the **dark sky** after twilight and before the moon rises, around 10:50 p.m.

Scorpius is visible most of the night and is at its highest in the south, around 30 degrees, at 11:05 p.m. and sets in the southwest around 3:25 a.m. See the illustration on page 19.

Cassiopeia, shaped like a *W*, rises in the northeast around 11:15 p.m. See the illustration on page 11.

Rust-red **Mars** sets in the west around 12:05 a.m.

The **summer triangle**'s three bright stars are **Vega, Altair,** and **Deneb. Vega** is almost directly overhead at 1:15 a.m. **Altair** is at its highest, around 65 degrees, at 2:30 a.m. Altair can be identified by its less bright neighboring stars, one on either side, together called the shaft of Aquila. **Deneb** is in the band of the Milky Way. See the illustration of the summer triangle on page 19.

Saturn rises in the east around 2:05 a.m. and is in the eastern sky before dawn. Unlike bright Jupiter and Venus and ruddy Mars, Saturn is harder to distinguish from the stars. Stars twinkle and shimmer, while the planets, including Saturn, show a steady flat light.

The **Great Bear (Big Dipper)** is in the northwestern sky before dawn and sets in the northwest around 2:50 a.m. See the illustration on page 10.

Bright star **Capella** rises in the northeast around 3:05 a.m. It is in the northeastern sky before dawn. Read about Capella on page 14.

The **Pleiades** rises in the northeast around 3:20 a.m. and is in the eastern sky before dawn. Read about the cluster on page 18.

The bright star **Arcturus** is in the western sky before dawn and sets in the northwest around 3:45 a.m.

Bright **Venus** rises in the northeast around 4:50 a.m.

2027

Thursday, June 24

TONIGHT

sunset: **8:01 p.m.**
twilight begins: **9:06 p.m.**

IN THE MORNING

twilight begins: **4:30 a.m.**
sunrise: **5:35 a.m.**

MOON PHASE

moonrise: **11:35 p.m.**
moonset: **10:39 a.m.**

IN THE SKY

Just east of **Scorpius**, Sagittarius is visible most of the night. It rises in the southeast around 9:40 p.m. and is at its highest in the south, around 25 degrees, at 12:55 a.m.

Jupiter is in the western sky at dusk and sets in the west around 10:55 p.m. Use binoculars or a telescope to look for the **moons of Jupiter.** See the guide on page 18.

Scorpius is visible most of the night and is at its highest in the south, around 30 degrees, at 11:05 p.m. and sets in the southwest around 3:20 a.m. See the illustration on page 19.

Explore the **dark sky** after twilight and before the moon rises, around 11:15 p.m.

Cassiopeia, shaped like a *W*, rises in the northeast around 11:15 p.m. See the illustration on page 11.

Rust-red **Mars** sets in the west around midnight.

The **summer triangle**'s three bright stars are **Vega, Altair,** and **Deneb. Vega** is almost directly overhead at 1:15 a.m. **Altair** is at its highest, around 65 degrees, at 2:25 a.m. Altair can be identified by its less bright neighboring stars, one on either side, together called the shaft of Aquila. **Deneb** is in the band of the Milky Way. See the illustration of the summer triangle on page 19.

Saturn rises in the east around 2 a.m. and is in the eastern sky before dawn. Unlike bright Jupiter and Venus and ruddy Mars, Saturn is harder to distinguish from the stars. Stars twinkle and shimmer, while the planets, including Saturn, show a steady flat light.

The **Great Bear (Big Dipper)** is in the northwestern sky before dawn and sets in the northwest around 2:50 a.m. See the illustration on page 10.

Bright star **Capella** rises in the northeast around 3 a.m. It is in the northeastern sky before dawn. Read about Capella on page 14.

The **Pleiades** rises in the northeast around 3:15 a.m. and is in the eastern sky before dawn. Read about the cluster on page 18.

The bright star **Arcturus** is in the western sky before dawn and sets in the northwest around 3:40 a.m.

Bright **Venus** rises in the northeast around 4:50 a.m.

2027
Friday, June 25

TONIGHT

sunset: **8:01 p.m.**
twilight begins: **9:07 p.m.**

IN THE MORNING

twilight begins: **4:30 a.m.**
sunrise: **5:35 a.m.**

MOON PHASE

moonrise: **midnight.**
moonset: **11:35 a.m.**

IN THE SKY

Just east of **Scorpius**, Sagittarius is visible most of the night. It rises in the southeast around 9:40 p.m. and is at its highest in the south, around 25 degrees, at 12:50 a.m.

Jupiter is in the western sky at dusk and sets in the west around 10:50 p.m. Use binoculars or a telescope to look for the **moons of Jupiter.** See the guide on page 18.

Scorpius is visible most of the night and is at its highest in the south, around 30 degrees, at 11 p.m. and sets in the southwest around 3:15 a.m. See the illustration on page 19.

Cassiopeia, shaped like a *W,* rises in the northeast around 11:10 p.m. See the illustration on page 11.

Explore the **dark sky** after twilight and before the moon rises, around 11:40 p.m.

Rust-red Mars sets in the west around 11:55 p.m.

The **summer triangle**'s three bright stars are **Vega, Altair,** and **Deneb. Vega** is almost directly overhead at 1:10 a.m. **Altair** is at its highest, around 65 degrees, at 2:25 a.m. Altair can be identified by its less bright neighboring stars, one on either side, together called the shaft of Aquila. **Deneb** is in the band of the Milky Way. See the illustration of the summer triangle on page 19.

Saturn rises in the east around 2 a.m. and is in the eastern sky before dawn. Unlike bright Jupiter and Venus and ruddy Mars, Saturn is harder to distinguish from the stars. Stars twinkle and shimmer, while the planets, including Saturn, show a steady flat light.

The **Great Bear (Big Dipper)** is in the northwestern sky before dawn and sets in the northwest around 2:45 a.m. See the illustration on page 10.

Bright star **Capella** rises in the northeast around 2:55 a.m. It is in the northeastern sky before dawn. Read about Capella on page 14.

The **Pleiades** rises in the northeast around 3:10 a.m. and is in the eastern sky before dawn. Read about the cluster on page 18.

The bright star **Arcturus** is in the western sky before dawn and sets in the northwest around 3:35 a.m.

Bright Venus rises in the northeast around 4:50 a.m.

2027

Saturday, June 26

TONIGHT

sunset: **8:01 p.m.**
twilight begins: **9:07 p.m.**

IN THE MORNING

twilight begins: **4:31 a.m.**
sunrise: **5:36 a.m.**

MOON PHASE

moonrise: **12:25 a.m.**
moonset: **12:32 p.m.**

IN THE SKY

Just east of **Scorpius**, Sagittarius is visible most of the night. It rises in the southeast around 9:35 p.m. and is at its highest in the south, around 25 degrees, at 12:45 a.m.

Jupiter is in the western sky at dusk and sets in the west around 10:45 p.m. Use binoculars or a telescope to look for the **moons of Jupiter.** See the guide on page 18.

Scorpius is visible most of the night and is at its highest in the south, around 30 degrees, at 10:55 p.m. and sets in the southwest around 3:15 a.m. See the illustration on page 19.

Cassiopeia, shaped like a *W*, rises in the northeast around 11:05 p.m. See the illustration on page 11.

Rust-red **Mars** sets in the west around 11:55 p.m.

Explore the **dark sky** after twilight and before the moon rises, around 12:05 a.m.

The **summer triangle**'s three bright stars are **Vega, Altair,** and **Deneb. Vega** is almost directly overhead at 1:05 a.m. **Altair** is at its highest, around 65 degrees, at 2:20 a.m. Altair can be identified by its less bright neighboring stars, one on either side, together called the shaft of Aquila. **Deneb** is in the band of the Milky Way. See the illustration of the summer triangle on page 19.

Saturn rises in the east around 1:55 a.m. and is in the eastern sky before dawn. Unlike bright Jupiter and Venus and ruddy Mars, Saturn is harder to distinguish from the stars. Stars twinkle and shimmer, while the planets, including Saturn, show a steady flat light.

The **Great Bear (Big Dipper)** is in the northwestern sky before dawn and sets in the northwest around 2:40 a.m. See the illustration on page 10.

Bright star **Capella** rises in the northeast around 2:50 a.m. It is in the northeastern sky before dawn. Read about Capella on page 14.

The **Pleiades** rises in the northeast around 3:05 a.m. and is in the eastern sky before dawn. Read about the cluster on page 18.

The bright star **Arcturus** is in the western sky before dawn and sets in the northwest around 3:30 a.m.

Bright **Venus** rises in the northeast around 4:55 a.m.

2027
Sunday, June 27

TONIGHT

sunset: **8:01 p.m.**
twilight begins: **9:07 p.m.**

IN THE MORNING

twilight begins: **4:31 a.m.**
sunrise: **5:36 a.m.**

MOON PHASE

moonrise: **12:53 a.m.**
moonset: **1:31 p.m.**

IN THE SKY

Just east of **Scorpius**, Sagittarius is visible most of the night. It rises in the southeast around 9:30 p.m. and is at its highest in the south, around 25 degrees, at 12:40 a.m.

Jupiter is in the western sky at dusk and sets in the west around 10:40 p.m. With the dark sky, use binoculars or a telescope to look for the **moons of Jupiter.** See the guide on page 18.

Scorpius is at its highest in the south, around 30 degrees, at 10:50 p.m. and sets in the southwest around 3:10 a.m. See the illustration on page 19.

Cassiopeia, shaped like a *W,* rises in the northeast around 11 p.m. See the illustration on page 11.

Rust-red and not very bright, **Mars** sets in the west around 11:50 p.m.

The **summer triangle**'s three bright stars are **Vega, Altair,** and **Deneb. Vega** is almost directly overhead at 1 a.m. **Altair** is at its highest, around 65 degrees, at 2:15 a.m. Altair can be identified by its less bright neighboring stars, one on either side, together called the shaft of Aquila. **Deneb** is in the band of the Milky Way. See the illustration of the summer triangle on page 19.

Saturn rises in the east around 1:50 a.m. and is in the eastern sky before dawn. Unlike bright Jupiter and Venus and ruddy Mars, Saturn is harder to distinguish from the stars. Stars twinkle and shimmer, while the planets, including Saturn, show a steady flat light.

The **Great Bear (Big Dipper)** is in the northwestern sky before dawn and sets in the northwest around 2:35 a.m. See the illustration on page 10.

Bright star **Capella** rises in the northeast around 2:45 a.m. It is in the northeastern sky before dawn. Read about Capella on page 14.

The **Pleiades** rises in the northeast around 3 a.m. and is in the eastern sky before dawn. Read about the cluster on page 18.

The bright star **Arcturus** is in the western sky before dawn and sets in the northwest around 3:25 a.m.

Bright **Venus** rises in the northeast around 4:55 a.m.

2027

Monday, June 28

TONIGHT

sunset: **8:01 p.m.**
twilight begins: **9:07 p.m.**

MOON PHASE

IN THE MORNING

twilight begins: **4:31 a.m.**
sunrise: **5:36 a.m.**

moonrise: **1:25 a.m.**
moonset: **2:33 p.m.**

IN THE SKY

Jupiter is in the western sky at dusk and sets in the west around 10:40 p.m. With the dark sky, use binoculars or a telescope to look for the **moons of Jupiter.** See the guide on page 18.

Scorpius is at its highest in the south, around 30 degrees, at 10:45 p.m. and sets in the southwest around 3:05 a.m. See the illustration on page 19.

Cassiopeia, shaped like a *W,* rises in the northeast around 10:55 p.m. See the illustration on page 11.

Rust-red and not very bright, Mars sets in the west around 11:50 p.m.

Just east of **Scorpius**, Sagittarius is visible most of the night and is at its highest in the south, around 25 degrees, at 12:40 a.m.

The **summer triangle**'s three bright stars are **Vega, Altair,** and **Deneb. Vega** is almost directly overhead at 12:55 a.m. **Altair** is at its highest, around 65 degrees, at 2:10 a.m. Altair can be identified by its less bright neighboring stars, one on either side, together called the shaft of Aquila. **Deneb** is in the band of the Milky Way. See the illustration of the summer triangle on page 19.

Saturn rises in the east around 1:45 a.m. and is in the eastern sky before dawn. Unlike bright Jupiter and Venus and ruddy Mars, Saturn is harder to distinguish from the stars. Stars twinkle and shimmer, while the planets, including Saturn, show a steady flat light.

The **Great Bear (Big Dipper)** is in the northwestern sky before dawn and sets in the northwest around 2:30 a.m. See the illustration on page 10.

Bright star **Capella** rises in the northeast around 2:45 a.m. It is in the northeastern sky before dawn. Read about Capella on page 14.

The **Pleiades** rises in the northeast around 3 a.m. and is in the eastern sky before dawn. Read about the cluster on page 18.

The bright star **Arcturus** is in the western sky before dawn and sets in the northwest around 3:25 a.m.

Bright Venus rises in the northeast around 4:55 a.m.

2027

Tuesday, June 29

TONIGHT

sunset: **8:01 p.m.**
twilight begins: **9:06 p.m.**

MOON PHASE

IN THE MORNING

twilight begins: **4:32 a.m.**
sunrise: **5:37 a.m.**

moonrise: **2:02 a.m.**
moonset: **3:39 p.m.**

IN THE SKY

The absence of the moon tonight provides excellent **dark-sky** viewing conditions.

Jupiter is in the western sky at dusk and sets in the west around 10:35 p.m. With the dark sky, use binoculars or a telescope to look for the **moons of Jupiter.** See the guide on page 18.

Scorpius is at its highest in the south, around 30 degrees, at 10:45 p.m. and sets in the southwest around 3 a.m. See the illustration on page 19.

Cassiopeia, shaped like a *W*, rises in the northeast around 10:55 p.m. See the illustration on page 11.

Rust-red and not very bright, Mars sets in the west around 11:45 p.m.

Just east of **Scorpius**, Sagittarius is visible most of the night and is at its highest in the south, around 25 degrees, at 12:35 a.m.

The **summer triangle**'s three bright stars are **Vega, Altair,** and **Deneb. Vega** is almost directly overhead at 12:50 a.m. **Altair** is at its highest, around 65 degrees, at 2:10 a.m. Altair can be identified by its less bright neighboring stars, one on either side, together called the shaft of Aquila. **Deneb** is in the band of the Milky Way. See the illustration of the summer triangle on page 19.

Saturn rises in the east around 1:45 a.m. and is in the eastern sky before dawn. Unlike bright Jupiter and Venus and ruddy Mars, Saturn is harder to distinguish from the stars. Stars twinkle and shimmer, while the planets, including Saturn, show a steady flat light.

Before the moon rises, until around 1:50 a.m., look for the **Andromeda Galaxy** with binoculars or a telescope. It becomes a good target around 1 a.m. It is in the northeastern sky before dawn. See the guide on page 11.

The **Great Bear (Big Dipper)** sets in the northwest around 2:30 a.m. See the illustration on page 10.

Bright star **Capella** rises in the northeast around 2:40 a.m. It is in the northeastern sky before dawn. Read about Capella on page 14.

The **Pleiades** rises in the northeast around 2:55 a.m. and is in the eastern sky before dawn. Read about the cluster on page 18.

The bright star **Arcturus** is in the western sky before dawn and sets in the northwest around 3:20 a.m.

Bright Venus rises in the northeast around 4:55 a.m.

2027

Wednesday, June 30

TONIGHT

sunset: **8:01 p.m.**
twilight begins: **9:06 p.m.**

MOON PHASE

IN THE MORNING

twilight begins: **4:32 a.m.**
sunrise: **5:37 a.m.**

moonrise: **2:49 a.m.**
moonset: **4:49 p.m.**

IN THE SKY

The absence of the moon tonight provides excellent **dark-sky** viewing conditions.

Jupiter is in the western sky at dusk and sets in the west around 10:30 p.m. With the dark sky, use binoculars or a telescope to look for the **moons of Jupiter.** See the guide on page 18.

Scorpius is at its highest in the south, around 30 degrees, at 10:40 p.m. and sets in the southwest around 3 a.m. See the illustration on page 19.

Cassiopeia, shaped like a *W,* rises in the northeast around 10:50 p.m. See the illustration on page 11.

Rust-red and not very bright, **Mars** sets in the west around 11:45 p.m.

Just east of **Scorpius,** Sagittarius is visible most of the night and is at its highest in the south, around 25 degrees, at 12:30 a.m.

The **summer triangle**'s three bright stars are **Vega, Altair,** and **Deneb. Vega** is almost directly overhead at 12:45 a.m. **Altair** is at its highest, around 65 degrees, at 2:05 a.m. Altair can be identified by its less bright neighboring stars, one on either side, together called the shaft of Aquila. **Deneb** is in the band of the Milky Way. See the illustration of the summer triangle on page 19.

Saturn rises in the east around 1:40 a.m. and is in the eastern sky before dawn. Unlike bright Jupiter and Venus and ruddy Mars, Saturn is harder to distinguish from the stars. Stars twinkle and shimmer, while the planets, including Saturn, show a steady flat light.

The **Great Bear (Big Dipper)** sets in the northwest around 2:25 a.m. See the illustration on page 10.

Bright star **Capella** rises in the northeast around 2:35 a.m. It is in the northeastern sky before dawn. Read about Capella on page 14.

Until the moon rises, around 2:40 a.m., look for the **Andromeda Galaxy** with binoculars or a telescope. It becomes a good target around 12:55 a.m. It is in the northeastern sky before dawn. See the guide on page 11.

The **Pleiades** rises in the northeast around 2:50 a.m. and is in the eastern sky before dawn. Read about the cluster on page 18.

The bright star **Arcturus** is in the western sky before dawn and sets in the northwest around 3:15 a.m.

Bright **Venus** rises in the northeast around 4:55 a.m.

2027

Thursday, July 1

TONIGHT

sunset: **8:01 p.m.**
twilight begins: **9:06 p.m.**

IN THE MORNING

twilight begins: **4:33 a.m.**
sunrise: **5:38 a.m.**

MOON PHASE

moonrise: **3:45 a.m.**
moonset: **5:59 p.m.**

IN THE SKY

The absence of the moon tonight provides excellent **dark-sky** viewing conditions.

Jupiter is in the western sky at dusk and sets in the west around 10:30 p.m. With the dark sky, use binoculars or a telescope to look for the **moons of Jupiter.** See the guide on page 18.

Scorpius is at its highest in the south, around 30 degrees, at 10:35 p.m. and sets in the southwest around 2:55 a.m. See the illustration on page 19.

Cassiopeia, shaped like a *W*, rises in the northeast around 10:45 p.m. See the illustration on page 11.

Rust-red and not very bright, **Mars** sets in the west around 11:40 p.m.

Just east of **Scorpius,** Sagittarius is visible most of the night and is at its highest in the south, around 25 degrees, at 12:25 a.m.

The **summer triangle**'s three bright stars are **Vega, Altair,** and **Deneb. Vega** is almost directly overhead at 12:40 a.m. **Altair** is at its highest, around 65 degrees, at 2 a.m. Altair can be identified by its less bright neighboring stars, one on either side, together called the shaft of Aquila. **Deneb** is in the band of the Milky Way. See the illustration of the summer triangle on page 19.

With the dark sky, look for the **Andromeda Galaxy** with binoculars or a telescope. It becomes a good target around 12:55 a.m. It is in the northeastern sky before dawn. See the guide on page 11.

Saturn rises in the east around 1:35 a.m. and is in the eastern sky before dawn. Unlike bright Jupiter and Venus and ruddy Mars, Saturn is harder to distinguish from the stars. Stars twinkle and shimmer, while the planets, including Saturn, show a steady flat light.

The **Great Bear (Big Dipper)** sets in the northwest around 2:20 a.m. See the illustration on page 10.

Bright star **Capella** rises in the northeast around 2:30 a.m. It is in the northeastern sky before dawn. Read about Capella on page 14.

The **Pleiades** rises in the northeast around 2:45 a.m. and is in the eastern sky before dawn. Read about the cluster on page 18.

The bright star **Arcturus** is in the western sky before dawn and sets in the northwest around 3:10 a.m.

Bright **Venus** rises in the northeast around 5 a.m.

2027

Friday, July 2

TONIGHT

sunset: **8:01 p.m.**
twilight begins: **9:06 p.m.**

MOON PHASE

IN THE MORNING

twilight begins: **4:33 a.m.**
sunrise: **5:38 a.m.**

moonrise: **4:52 a.m.**
moonset: **7:06 p.m.**

IN THE SKY

The absence of the moon tonight provides excellent **dark-sky** viewing conditions.

Jupiter is in the western sky at dusk and sets in the west around 10:25 p.m. With the dark sky, use binoculars or a telescope to look for the **moons of Jupiter.** See the guide on page 18.

Scorpius is at its highest in the south, around 30 degrees, at 10:30 p.m. and sets in the southwest around 2:50 a.m. See the illustration on page 19.

Cassiopeia, shaped like a *W,* rises in the northeast around 10:40 p.m. See the illustration on page 11.

Rust-red and not very bright, **Mars** sets in the west around 11:40 p.m.

Just east of **Scorpius**, Sagittarius is visible most of the night and is at its highest in the south, around 25 degrees, at 12:20 a.m.

The **summer triangle**'s three bright stars are **Vega, Altair,** and **Deneb. Vega** is almost directly overhead at 12:40 a.m. **Altair** is at its highest, around 65 degrees, at 1:55 a.m. Altair can be identified by its less bright neighboring stars, one on either side, together called the shaft of Aquila. **Deneb** is in the band of the Milky Way. See the illustration of the summer triangle on page 19.

With the dark sky, look for the **Andromeda Galaxy** with binoculars or a telescope. It becomes a good target around 12:50 a.m. It is in the northeastern sky before dawn. See the guide on page 11.

Saturn rises in the east around 1:30 a.m. and is in the eastern sky before dawn. Unlike bright Jupiter and Venus and ruddy Mars, Saturn is harder to distinguish from the stars. Stars twinkle and shimmer, while the planets, including Saturn, show a steady flat light.

The **Great Bear (Big Dipper)** sets in the northwest around 2:15 a.m. See the illustration on page 10.

Bright star **Capella** rises in the northeast around 2:30 a.m. It is in the northeastern sky before dawn. Read about Capella on page 14.

The **Pleiades** rises in the northeast around 2:40 a.m. and is in the eastern sky before dawn. Read about the cluster on page 18.

The bright star **Arcturus** is in the western sky before dawn and sets in the northwest around 3:10 a.m.

Bright **Venus** rises in the northeast around 5 a.m.

2027

Saturday, July 3

TONIGHT

sunset: **8:01 p.m.**
twilight begins: **9:06 p.m.**

IN THE MORNING

twilight begins: **4:34 a.m.**
sunrise: **5:38 a.m.**

MOON PHASE

moonrise: **6:06 a.m.**
moonset: **8:06 p.m.**

IN THE SKY

The absence of the moon tonight provides excellent **dark-sky** viewing conditions.

Jupiter is in the western sky at dusk and sets in the west around 10:20 p.m.

Scorpius is at its highest in the south, around 30 degrees, at 10:30 p.m. and sets in the southwest around 2:45 a.m. See the illustration on page 19.

Cassiopeia, shaped like a *W,* rises in the northeast around 10:35 p.m. See the illustration on page 11.

Rust-red and not very bright, Mars sets in the west around 11:35 p.m.

Just east of **Scorpius**, Sagittarius is visible most of the night and is at its highest in the south, around 25 degrees, at 12:20 a.m.

The **summer triangle**'s three bright stars are **Vega, Altair,** and **Deneb. Vega** is almost directly overhead at 12:35 a.m. **Altair** is at its highest, around 65 degrees, at 1:50 a.m. Altair can be identified by its less bright neighboring stars, one on either side, together called the shaft of Aquila. **Deneb** is in the band of the Milky Way. See the illustration of the summer triangle on page 19.

Look for the **Andromeda Galaxy** with binoculars or a telescope. It becomes a good target around 12:45 a.m. It is in the northeastern sky before dawn. See the guide on page 11.

Saturn rises in the east around 1:30 a.m. and is in the eastern sky before dawn. Unlike bright Jupiter and Venus and ruddy Mars, Saturn is harder to distinguish from the stars. Stars twinkle and shimmer, while the planets, including Saturn, show a steady flat light.

The **Great Bear (Big Dipper)** sets in the northwest around 2:10 a.m. See the illustration on page 10.

Bright star **Capella** rises in the northeast around 2:25 a.m. It is in the northeastern sky before dawn. Read about Capella on page 14.

The **Pleiades** rises in the northeast around 2:40 a.m. and is in the eastern sky before dawn. Read about the cluster on page 18.

The bright star **Arcturus** is in the western sky before dawn and sets in the northwest around 3:05 a.m.

Bright Venus rises in the northeast around 5 a.m.

2027

Sunday, July 4

TONIGHT

sunset: **8:01 p.m.**
twilight begins: **9:06 p.m.**

MOON PHASE

IN THE MORNING

twilight begins: **4:35 a.m.**
sunrise: **5:39 a.m.**

moonrise: **7:23 a.m.**
moonset: **8:56 p.m.**

IN THE SKY

The absence of the moon tonight provides excellent **dark-sky** viewing conditions.

Jupiter is in the western sky at dusk and sets in the west around 10:20 p.m.

Scorpius is at its highest in the south, around 30 degrees, at 10:25 p.m. and sets in the southwest around 2:40 a.m. See the illustration on page 19.

Cassiopeia, shaped like a *W*, rises in the northeast around 10:35 p.m. See the illustration on page 11.

Rust-red and not very bright, **Mars** sets in the west around 11:35 p.m.

Just east of **Scorpius**, Sagittarius is visible most of the night and is at its highest in the south, around 25 degrees, at 12:15 a.m. It sets in the southwest around 4:15 a.m.

The **summer triangle**'s three bright stars are **Vega, Altair,** and **Deneb. Vega** is almost directly overhead at 12:30 a.m. **Altair** is at its highest, around 65 degrees, at 1:50 a.m. Altair can be identified by its less bright neighboring stars, one on either side, together called the shaft of Aquila. **Deneb** is in the band of the Milky Way. See the illustration of the summer triangle on page 19.

Look for the **Andromeda Galaxy** with binoculars or a telescope. It becomes a good target around 12:40 a.m. It is in the northeastern sky before dawn. See the guide on page 11.

Saturn rises in the east around 1:25 a.m. and is in the eastern sky before dawn. Unlike bright Jupiter and Venus and ruddy Mars, Saturn is harder to distinguish from the stars. Stars twinkle and shimmer, while the planets, including Saturn, show a steady flat light.

The **Great Bear (Big Dipper)** sets in the northwest around 2:10 a.m. See the illustration on page 10.

Bright star **Capella** rises in the northeast around 2:20 a.m. It is in the northeastern sky before dawn. Read about Capella on page 14.

The **Pleiades** rises in the northeast around 2:35 a.m. and is in the eastern sky before dawn. Read about the cluster on page 18.

The bright star **Arcturus** is in the western sky before dawn and sets in the northwest around 3 a.m.

Bright **Venus** rises in the northeast around 5 a.m.

2027

Monday, July 5

MOON PHASE

TONIGHT

sunset: 8 p.m.
twilight begins: 9:05 p.m.

IN THE MORNING

twilight begins: 4:35 a.m.
sunrise: 5:39 a.m.

moonrise: 7:23 a.m.
moonset: 9:38 p.m.

IN THE SKY

The absence of the moon tonight provides excellent **dark-sky** viewing conditions.

Jupiter is in the western sky at dusk and sets in the west around 10:15 p.m.

Scorpius is at its highest in the south, around 30 degrees, at 10:20 p.m. and sets in the southwest around 2:40 a.m. See the illustration on page 19.

Cassiopeia, shaped like a *W,* rises in the northeast around 10:30 p.m. See the illustration on page 11.

Rust-red and not very bright, **Mars** sets in the west around 11:30 p.m.

Just east of **Scorpius**, Sagittarius is visible most of the night and is at its highest in the south, around 25 degrees, at 12:10 a.m. It sets in the southwest around 4:10 a.m.

The **summer triangle**'s three bright stars are **Vega, Altair,** and **Deneb. Vega** is almost directly overhead at 12:25 a.m. **Altair** is at its highest, around 65 degrees, at 1:45 a.m. Altair can be identified by its less bright neighboring stars, one on either side, together called the shaft of Aquila. **Deneb** is in the band of the Milky Way. See the illustration of the summer triangle on page 19.

Look for the **Andromeda Galaxy** with binoculars or a telescope. It becomes a good target around 12:35 a.m. It is in the northeastern sky before dawn. See the guide on page 11.

Saturn rises in the east around 1:20 a.m. and is in the eastern sky before dawn. Unlike bright Jupiter and Venus and ruddy Mars, Saturn is harder to distinguish from the stars. Stars twinkle and shimmer, while the planets, including Saturn, show a steady flat light.

The **Great Bear (Big Dipper)** sets in the northwest around 2:05 a.m. See the illustration on page 10.

Bright star **Capella** rises in the northeast around 2:15 a.m. It is in the northeastern sky before dawn. Read about Capella on page 14.

The **Pleiades** rises in the northeast around 2:30 a.m. and is in the eastern sky before dawn. Read about the cluster on page 18.

The bright star **Arcturus** is in the western sky before dawn and sets in the northwest around 2:55 a.m.

Bright **Venus** rises in the northeast around 5:05 a.m.

2027

Tuesday, July 6

TONIGHT

sunset: **8 p.m.**
twilight begins: **9:05 p.m.**

IN THE MORNING

twilight begins: **4:36 a.m.**
sunrise: **5:40 a.m.**

MOON PHASE

moonrise: **8:38 a.m.**
moonset: **10:14 p.m.**

IN THE SKY

The waxing sliver-crescent moon and **Jupiter** appear close together in the evening sky. Look west near the horizon around 9 p.m.

Jupiter is in the western sky at dusk and sets in the west around 10:10 p.m.

Scorpius is at its highest in the south, around 30 degrees, at 10:15 p.m. It sets in the southwest around 2:35 a.m. See the illustration on page 19.

Cassiopeia, shaped like a *W,* rises in the northeast around 10:25 p.m. See the illustration on page 11.

Rust-red and not very bright, **Mars** sets in the west around 11:30 p.m.

The absence of the moon tonight provides excellent **dark-sky** viewing conditions. Explore the **dark sky** after the moon sets, from around 10:35 p.m.

Just east of **Scorpius,** Sagittarius is visible most of the night and is at its highest in the south, around 25 degrees, at 12:05 a.m. It sets in the southwest around 4:05 a.m.

The **summer triangle**'s three bright stars are **Vega, Altair,** and **Deneb. Vega** is almost directly overhead at 12:20 a.m. **Altair** is at its highest, around 65 degrees, at 1:40 a.m. Altair can be identified by its less bright neighboring stars, one on either side, together called the shaft of Aquila. **Deneb** is in the band of the Milky Way. See the illustration of the summer triangle on page 19.

Look for the **Andromeda Galaxy** with binoculars or a telescope. It becomes a good target around 12:35 a.m. It is in the northeastern sky before dawn. See the guide on page 11.

Saturn rises in the east around 1:15 a.m. and is in the eastern sky before dawn. Unlike bright Jupiter and Venus and ruddy Mars, Saturn is harder to distinguish from the stars. Stars twinkle and shimmer, while the planets, including Saturn, show a steady flat light.

The **Great Bear (Big Dipper)** sets in the northwest around 2 a.m. See the illustration on page 10.

Bright star **Capella** rises in the northeast around 2:10 a.m. It is in the northeastern sky before dawn. Read about Capella on page 14.

The **Pleiades** rises in the northeast around 2:25 a.m. and is in the eastern sky before dawn. Read about the cluster on page 18.

The bright star **Arcturus** is in the western sky before dawn and sets in the northwest around 2:50 a.m.

Bright **Venus** rises in the northeast around 5:05 a.m.

2027

Wednesday, July 7

TONIGHT

sunset: **8 p.m.**
twilight begins: **9:05 p.m.**

IN THE MORNING

twilight begins: **4:36 a.m.**
sunrise: **5:41 a.m.**

MOON PHASE

moonrise: **9:50 a.m.**
moonset: **10:46 p.m.**

IN THE SKY

The **moon** is prominent in the sky early in the evening. Using a telescope or binoculars, take a look. You should be able to see rough mountains and craters along the boundary between light and dark, in astronomy called the terminator.

Jupiter is in the western sky at dusk and sets in the west around 10:10 p.m.

Scorpius is at its highest in the south, around 30 degrees, at 10:10 p.m. It sets in the southwest around 2:30 a.m. See the illustration on page 19.

Cassiopeia, shaped like a *W,* rises in the northeast around 10:20 p.m. See the illustration on page 11.

Explore the **dark sky** after the moon sets, from around 11:05 p.m.

Rust-red and not very bright, **Mars** sets in the west around 11:25 p.m.

Just east of **Scorpius**, Sagittarius is visible most of the night and is at its highest in the south, around 25 degrees, at midnight. It sets in the southwest around 4 a.m.

The **summer triangle**'s three bright stars are **Vega, Altair,** and **Deneb. Vega** is almost directly overhead at 12:20 a.m. **Altair** is at its highest, around 65 degrees, at 1:35 a.m. Altair can be identified by its less bright neighboring stars, one on either side, together called the shaft of Aquila. **Deneb** is in the band of the Milky Way. See the illustration of the summer triangle on page 19.

Look for the **Andromeda Galaxy** with binoculars or a telescope. It becomes a good target around 12:30 a.m. It is in the northeastern sky before dawn. See the guide on page 11.

Saturn rises in the east around 1:15 a.m. and is in the eastern sky before dawn. Unlike bright Jupiter and Venus and ruddy Mars, Saturn is harder to distinguish from the stars. Stars twinkle and shimmer, while the planets, including Saturn, show a steady flat light.

The **Great Bear (Big Dipper)** sets in the northwest around 1:55 a.m. See the illustration on page 10.

Bright star **Capella** rises in the northeast around 2:10 a.m. It is in the northeastern sky before dawn. Read about Capella on page 14.

The **Pleiades** rises in the northeast around 2:25 a.m. and is in the eastern sky before dawn. Read about the cluster on page 18.

The bright star **Arcturus** is in the western sky before dawn and sets in the northwest around 2:50 a.m.

Bright **Venus** rises in the northeast around 5:05 a.m.

2027

Thursday, July 8

TONIGHT

sunset: **8 p.m.**
twilight begins: **9:04 p.m.**

IN THE MORNING

twilight begins: **4:37 a.m.**
sunrise: **5:41 a.m.**

MOON PHASE

moonrise: **10:58 a.m.**
moonset: **11:15 p.m.**

IN THE SKY

The **moon** is prominent in the sky early in the evening. Using a telescope or binoculars, take a look. You should be able to see rough mountains and craters along the boundary between light and dark, in astronomy called the terminator.

The waxing crescent moon and **Mars** appear close together in the evening sky. Look west near the horizon around 9:15 p.m.

Jupiter is in the western sky at dusk and sets in the west around 10:05 p.m.

Scorpius is at its highest in the south, around 30 degrees, at 10:10 p.m. It sets in the southwest around 2:25 a.m. See the illustration on page 19.

Cassiopeia, shaped like a *W,* rises in the northeast around 10:20 p.m. See the illustration on page 11.

Rust-red and not very bright, **Mars** sets in the west around 11:25 p.m.

Explore the **dark sky** after the moon sets, from around 11:35 p.m.

The **summer triangle**'s three bright stars are **Vega, Altair,** and **Deneb. Vega** is almost directly overhead at 12:15 a.m. **Altair** is at its highest, around 65 degrees, at 1:30 a.m. Altair can be identified by its less bright neighboring stars, one on either side, together called the shaft of Aquila. **Deneb** is in the band of the Milky Way. See the illustration of the summer triangle on page 19.

Look for the **Andromeda Galaxy** with binoculars or a telescope. It becomes a good target around 12:25 a.m. It is in the northeastern sky before dawn. See the guide on page 11.

Saturn rises in the east around 1:05 a.m. Unlike bright Jupiter and Venus and ruddy Mars, Saturn is harder to distinguish from the stars. Stars twinkle and shimmer, while the planets, including Saturn, show a steady flat light.

The **Great Bear (Big Dipper)** sets in the northwest around 1:55 a.m. See the illustration on page 10.

Bright star **Capella** rises in the northeast around 2:05 a.m. It is in the northeastern sky before dawn. Read about Capella on page 14.

The **Pleiades** rises in the northeast around 2:20 a.m. and is in the eastern sky before dawn. Read about the cluster on page 18.

The bright star **Arcturus** is in the western sky before dawn and sets in the northwest around 2:45 a.m.

Just east of **Scorpius,** Sagittarius is visible most of the night and is at its highest in the south, around 25 degrees, at midnight. It sets in the southwest around 3:55 a.m.

Bright **Venus** rises in the northeast around 5:10 a.m.

2027

Friday, July 9

TONIGHT

sunset: **8 p.m.**
twilight begins: **9:04 p.m.**

IN THE MORNING

twilight begins: **4:38 a.m.**
sunrise: **5:42 a.m.**

MOON PHASE

moonrise: **12:04 p.m.**
moonset: **11:44 p.m.**

IN THE SKY

The **moon** is prominent in the sky early in the evening. Using a telescope or binoculars, take a look. You should be able to see rough mountains and craters along the boundary between light and dark, in astronomy called the terminator.

Jupiter is in the western sky at dusk and sets in the west around 10 p.m.

Scorpius is at its highest in the south, around 30 degrees, at 10:05 p.m. It sets in the southwest around 2:20 a.m. See the illustration on page 19.

Cassiopeia, shaped like a *W,* rises in the northeast around 10:15 p.m. See the illustration on page 11.

Just east of **Scorpius**, Sagittarius is visible most of the night and is at its highest in the south, around 25 degrees, at 11:55 p.m. It sets in the southwest around 3:55 a.m.

Rust-red and not very bright, **Mars** sets in the west around 11:20 p.m.

Explore the **dark sky** after the moon sets, from around 12:05 a.m.

The **summer triangle**'s three bright stars are **Vega, Altair,** and **Deneb. Vega** is almost directly overhead at 12:10 a.m. **Altair** is at its highest, around 65 degrees, at 1:30 a.m. Altair can be identified by its less bright neighboring stars, one on either side, together called the shaft of Aquila. **Deneb** is in the band of the Milky Way. See the illustration of the summer triangle on page 19.

After the moon sets, look for the **Andromeda Galaxy** with binoculars or a telescope. It becomes a good target around 12:20 a.m. It is in the northeastern sky before dawn. See the guide on page 11.

Saturn rises in the east around 1 a.m. Unlike bright Jupiter and Venus and ruddy Mars, Saturn is harder to distinguish from the stars. Stars twinkle and shimmer, while the planets, including Saturn, show a steady flat light.

The **Great Bear (Big Dipper)** sets in the northwest around 1:50 a.m. See the illustration on page 10.

Bright star **Capella** rises in the northeast around 2 a.m. It is in the northeastern sky before dawn. Read about Capella on page 14.

The **Pleiades** rises in the northeast around 2:15 a.m. and is in the eastern sky before dawn. Read about the cluster on page 18.

The bright star **Arcturus** is in the western sky before dawn and sets in the northwest around 2:40 a.m.

Bright **Venus** rises in the northeast around 5:10 a.m.

2027

Saturday, July 10

TONIGHT

sunset: **7:59 p.m.**
twilight begins: **9:04 p.m.**

IN THE MORNING

twilight begins: **4:38 a.m.**
sunrise: **5:42 a.m.**

MOON PHASE

moonrise: **1:08 p.m.**
moonset: **12:15 a.m.**

IN THE SKY

The **moon** is prominent in the sky early in the evening. Using a telescope or binoculars, take a look. You should be able to see rough mountains and craters along the boundary between light and dark, in astronomy called the terminator.

The waxing first-quarter moon and Spica appear close together in the evening sky. Look southwest at about 30 degrees elevation around 9:30 p.m.

Jupiter is in the western sky at dusk and sets in the west around 10 p.m.

Cassiopeia, shaped like a *W,* rises in the northeast around 10:10 p.m. See the illustration on page 11.

Rust-red and not very bright, Mars sets in the west around 11:20 p.m.

Just east of **Scorpius**, Sagittarius is visible most of the night and is at its highest in the south, around 25 degrees, at 11:50 p.m. It sets in the southwest around 3:50 a.m.

The **summer triangle**'s three bright stars are **Vega, Altair,** and **Deneb. Vega** is almost directly overhead at 12:05 a.m. **Altair** is at its highest, around 65 degrees, at 1:25 a.m. Altair can be identified by its less bright neighboring stars, one on either side, together called the shaft of Aquila. **Deneb** is in the band of the Milky Way. See the illustration of the summer triangle on page 19.

Explore the **dark sky** after the moon sets, from around 12:35 a.m.

After the moon sets, from around 12:35 a.m., look for the **Andromeda Galaxy** with binoculars or a telescope. It is in the northeastern sky before dawn. See the guide on page 11.

Saturn rises in the east around 1 a.m. Unlike bright Jupiter and Venus and ruddy Mars, Saturn is harder to distinguish from the stars. Stars twinkle and shimmer, while the planets, including Saturn, show a steady flat light.

The **Great Bear (Big Dipper)** sets in the northwest around 1:45 a.m. See the illustration on page 10.

Bright star **Capella** rises in the northeast around 1:55 a.m. It is in the northeastern sky before dawn. Read about Capella on page 14.

The **Pleiades** rises in the northeast around 2:10 a.m. and is in the eastern sky before dawn. Read about the cluster on page 18.

Scorpius is at its highest in the south, around 30 degrees, around dusk and sets in the southwest around 2:20 a.m. See the illustration on page 19.

The bright star **Arcturus** is in the western sky before dawn and sets in the northwest around 2:35 a.m.

Bright Venus rises in the northeast around 5:10 a.m.

2027

Sunday, July 11

TONIGHT

sunset: **7:59 p.m.**
twilight begins: **9:03 p.m.**

MOON PHASE

IN THE MORNING

twilight begins: **4:39 a.m.**
sunrise: **5:43 a.m.**

moonrise: **2:12 p.m.**
moonset: **12:47 a.m.**

IN THE SKY

The **moon** is prominent in the sky early in the evening. Using a telescope or binoculars, take a look. You should be able to see rough mountains and craters along the boundary between light and dark, in astronomy called the terminator.

Jupiter is in the western sky at dusk and sets in the west around 9:55 p.m.

Cassiopeia, shaped like a *W,* rises in the northeast around 10:05 p.m. See the illustration on page 11.

Rust-red and not very bright, **Mars** sets in the west around 11:15 p.m.

Just east of **Scorpius**, Sagittarius is visible most of the night and is at its highest in the south, around 25 degrees, at 11:45 p.m. It sets in the southwest around 3:45 a.m.

The **summer triangle**'s three bright stars are **Vega, Altair,** and **Deneb. Vega** is almost directly overhead at 12:05 a.m. **Altair** is at its highest, around 65 degrees, at 1:20 a.m. Altair can be identified by its less bright neighboring stars, one on either side, together called the shaft of Aquila. **Deneb** is in the band of the Milky Way. See the illustration of the summer triangle on page 19.

Saturn rises in the east around 12:55 a.m. Unlike bright Jupiter and Venus and ruddy Mars, Saturn is harder to distinguish from the stars. Stars twinkle and shimmer, while the planets, including Saturn, show a steady flat light.

After the moon sets, from around 1 a.m., look for the **Andromeda Galaxy** with binoculars or a telescope. It is in the northeastern sky before dawn. See the guide on page 11.

The **Great Bear (Big Dipper)** sets in the northwest around 1:40 a.m. See the illustration on page 10.

Bright star **Capella** rises in the northeast around 1:50 a.m. It is in the northeastern sky before dawn. Read about Capella on page 14.

The **Pleiades** rises in the northeast around 2:05 a.m. and is in the eastern sky before dawn. Read about the cluster on page 18.

Scorpius is at its highest in the south, around 30 degrees, around dusk and sets in the southwest around 2:15 a.m. See the illustration on page 19.

The bright star **Arcturus** is in the western sky before dawn and sets in the northwest around 2:30 a.m.

Bright **Venus** rises in the northeast around 5:15 a.m.

2027

Monday, July 12

TONIGHT

sunset: **7:59 p.m.**
twilight begins: **9:03 p.m.**

IN THE MORNING

twilight begins: **4:40 a.m.**
sunrise: **5:43 a.m.**

MOON PHASE

moonrise: **3:15 p.m.**
moonset: **1:24 a.m.**

IN THE SKY

Jupiter is in the western sky at dusk and sets in the west around 9:50 p.m.

Cassiopeia, shaped like a *W,* rises in the northeast around 10 p.m. See the illustration on page 11.

Rust-red and not very bright, **Mars** sets in the west around 11:15 p.m.

Just east of **Scorpius**, Sagittarius is visible most of the night and is at its highest in the south, around 25 degrees, at 11:40 p.m. It sets in the southwest around 3:40 a.m.

Saturn rises in the east around 12:50 a.m. Unlike bright Jupiter and Venus and ruddy Mars, Saturn is harder to distinguish from the stars. Stars twinkle and shimmer, while the planets, including Saturn, show a steady flat light.

The **summer triangle**'s three bright stars are **Vega, Altair,** and **Deneb. Vega** is almost directly overhead at midnight. **Altair** is at its highest, around 65 degrees, at 1:15 a.m. Altair can be identified by its less bright neighboring stars, one on either side, together called the shaft of Aquila. **Deneb** is in the band of the Milky Way. See the illustration of the summer triangle on page 19.

The **Great Bear (Big Dipper)** sets in the northwest around 1:35 a.m. See the illustration on page 10.

Bright star **Capella** rises in the northeast around 1:50 a.m. It is in the northeastern sky before dawn. Read about Capella on page 14.

The **Pleiades** rises in the northeast around 2:05 a.m. and is in the eastern sky before dawn. Read about the cluster on page 18.

Scorpius is at its highest in the south, around 30 degrees, around dusk and sets in the southwest around 2:10 a.m. See the illustration on page 19.

The bright star **Arcturus** sets in the northwest around 2:30 a.m.

Bright **Venus** rises in the northeast around dawn.

2027
Tuesday, July 13

TONIGHT

sunset: **7:58 p.m.**
twilight begins: **9:02 p.m.**

IN THE MORNING

twilight begins: **4:41 a.m.**
sunrise: **5:44 a.m.**

MOON PHASE

moonrise: **4:17 p.m.**
moonset: **2:06 a.m.**

IN THE SKY

Jupiter is in the western sky at dusk and sets in the west around 9:50 p.m.

Cassiopeia, shaped like a *W,* rises in the northeast around 10 p.m. See the illustration on page 11.

Rust-red and not very bright, **Mars** sets in the west around 11:10 p.m.

Just east of **Scorpius**, Sagittarius is visible most of the night and is at its highest in the south, around 25 degrees, at 11:40 p.m. It sets in the southwest around 3:40 a.m.

The **summer triangle**'s three bright stars are **Vega, Altair,** and **Deneb. Vega** is almost directly overhead at 11:55 p.m. **Altair** is at its highest, around 65 degrees, at 1:15 a.m. Altair can be identified by its less bright neighboring stars, one on either side, together called the shaft of Aquila. **Deneb** is in the band of the Milky Way. See the illustration of the summer triangle on page 19.

Saturn rises in the east around 12:45 a.m. Unlike bright Jupiter and Venus and ruddy Mars, Saturn is harder to distinguish from the stars. Stars twinkle and shimmer, while the planets, including Saturn, show a steady flat light.

The **Great Bear (Big Dipper)** sets in the northwest around 1:35 a.m. See the illustration on page 10.

Bright star **Capella** rises in the northeast around 1:45 a.m. It is in the northeastern sky before dawn. Read about Capella on page 14.

The **Pleiades** rises in the northeast around 2 a.m. and is in the eastern sky before dawn. Read about the cluster on page 18.

Scorpius is at its highest in the south, around 30 degrees, around dusk and sets in the southwest around 2:05 a.m. See the illustration on page 19.

The bright star **Arcturus** sets in the northwest around 2:25 a.m.

Bright **Venus** rises in the northeast around dawn.

Look for **Mercury** in the morning sky, just before sunrise in the northeast. It's difficult but possible to see the tiny dot in the glare of the sun. It's easier to find with the naked eye, not using binoculars or a telescope. Never look directly at the sun or point binoculars or a telescope at it.

2027

Wednesday, July 14

TONIGHT

sunset: **7:58 p.m.**
twilight begins: **9:02 p.m.**

IN THE MORNING

twilight begins: **4:41 a.m.**
sunrise: **5:45 a.m.**

MOON PHASE

moonrise: **5:16 p.m.**
moonset: **2:53 a.m.**

IN THE SKY

The nearly **full moon** tonight acts like a spotlight. Look around at the desert landscape at how much illumination the moon provides.

Jupiter is in the western sky at dusk and sets in the west around 9:45 p.m.

Cassiopeia, shaped like a *W,* rises in the northeast around 9:55 p.m. See the illustration on page 11.

Rust-red and not very bright, **Mars** sets in the west around 11:10 p.m.

Just east of **Scorpius,** Sagittarius is visible most of the night and is at its highest in the south, around 25 degrees, at 11:35 p.m. It sets in the southwest around 3:35 a.m.

The **summer triangle**'s three bright stars are **Vega, Altair,** and **Deneb. Vega** is almost directly overhead at 11:50 p.m. **Altair** is at its highest, around 65 degrees, at 1:10 a.m. Altair can be identified by its less bright neighboring stars, one on either side, together called the shaft of Aquila. **Deneb** is in the band of the Milky Way. See the illustration of the summer triangle on page 19.

Saturn rises in the east around 12:45 a.m. Unlike bright Jupiter and Venus and ruddy Mars, Saturn is harder to distinguish from the stars. Stars twinkle and shimmer, while the planets, including Saturn, show a steady flat light.

The **Great Bear (Big Dipper)** sets in the northwest around 1:30 a.m. See the illustration on page 10.

Bright star **Capella** rises in the northeast around 1:40 a.m. It is in the northeastern sky before dawn. Read about Capella on page 14.

The **Pleiades** rises in the northeast around 1:55 a.m. and is in the eastern sky before dawn. Read about the cluster on page 18.

Scorpius is at its highest in the south, around 30 degrees, around dusk and sets in the southwest around 2 a.m. See the illustration on page 19.

The bright star **Arcturus** sets in the northwest around 2:20 a.m.

Bright **Venus** rises in the northeast around dawn.

Look for **Mercury** in the morning sky, just before sunrise in the northeast. It's difficult but possible to see the tiny dot in the glare of the sun. It's easier to find with the naked eye, not using binoculars or a telescope. Never look directly at the sun or point binoculars or a telescope at it.

2027

Thursday, July 15

TONIGHT

sunset: **7:57 p.m.**
twilight begins: **9:01 p.m.**

IN THE MORNING

twilight begins: **4:42 a.m.**
sunrise: **5:45 a.m.**

MOON PHASE

moonrise: **6:09 p.m.**
moonset: **3:46 a.m.**

IN THE SKY

The **full moon** tonight acts like a spotlight. Look around at the desert landscape at how much illumination the moon provides.

Jupiter is in the western sky at dusk and sets in the west around 9:40 p.m.

Cassiopeia, shaped like a *W,* rises in the northeast around 9:50 p.m. See the illustration on page 11.

Rust-red and not very bright, **Mars** sets in the west around 11:05 p.m.

Just east of **Scorpius**, Sagittarius is visible most of the night and is at its highest in the south, around 25 degrees, at 11:30 p.m. It sets in the southwest around 3:30 a.m.

The **summer triangle**'s three bright stars are **Vega, Altair,** and **Deneb. Vega** is almost directly overhead at 11:45 p.m. **Altair** is at its highest, around 65 degrees, at 1:05 a.m. Altair can be identified by its less bright neighboring stars, one on either side, together called the shaft of Aquila. **Deneb** is in the band of the Milky Way. See the illustration of the summer triangle on page 19.

Saturn rises in the east around 12:40 a.m. Unlike bright Jupiter and Venus and ruddy Mars, Saturn is harder to distinguish from the stars. Stars twinkle and shimmer, while the planets, including Saturn, show a steady flat light.

The **Great Bear (Big Dipper)** sets in the northwest around 1:25 a.m. See the illustration on page 10.

Bright star **Capella** rises in the northeast around 1:35 a.m. It is in the northeastern sky before dawn. Read about Capella on page 14.

The **Pleiades** rises in the northeast around 1:50 a.m. and is in the eastern sky before dawn. Read about the cluster on page 18.

Scorpius is at its highest in the south, around 30 degrees, around dusk and sets in the southwest around 2 a.m. See the illustration on page 19.

The bright star **Arcturus** sets in the northwest around 2:15 a.m.

Bright **Venus** rises in the northeast around dawn.

Look for **Mercury** in the morning sky, just before sunrise in the northeast. It's difficult but possible to see the tiny dot in the glare of the sun. It's easier to find with the naked eye, not using binoculars or a telescope. Never look directly at the sun or point binoculars or a telescope at it.

2027

Friday, July 16

TONIGHT

sunset: **7:57 p.m.**
twilight begins: **9:01 p.m.**

MOON PHASE

IN THE MORNING

twilight begins: **4:43 a.m.**
sunrise: **5:46 a.m.**

moonrise: **6:57 p.m.**
moonset: **4:42 a.m.**

IN THE SKY

The **full moon** tonight acts like a spotlight. Look around at the desert landscape at how much illumination the moon provides.

Jupiter is in the western sky at dusk and sets in the west around 9:40 p.m.

Cassiopeia, shaped like a *W,* rises in the northeast around 9:45 p.m. See the illustration on page 11.

Rust-red and not very bright, Mars sets in the west around 11:05 p.m.

Just east of **Scorpius**, Sagittarius is visible most of the night and is at its highest in the south, around 25 degrees, at 11:25 p.m. It sets in the southwest around 3:25 a.m.

The **summer triangle**'s three bright stars are **Vega, Altair,** and **Deneb. Vega** is almost directly overhead at 11:45 p.m. **Altair** is at its highest, around 65 degrees, at 1 a.m. Altair can be identified by its less bright neighboring stars, one on either side, together called the shaft of Aquila. **Deneb** is in the band of the Milky Way. See the illustration of the summer triangle on page 19.

Saturn rises in the east around 12:35 a.m. Unlike bright Jupiter and Venus and ruddy Mars, Saturn is harder to distinguish from the stars. Stars twinkle and shimmer, while the planets, including Saturn, show a steady flat light.

The **Great Bear (Big Dipper)** sets in the northwest around 1:20 a.m. See the illustration on page 10.

Bright star **Capella** rises in the northeast around 1:35 a.m. It is in the northeastern sky before dawn. Read about Capella on page 14.

The **Pleiades** rises in the northeast around 1:45 a.m. and is in the eastern sky before dawn. Read about the cluster on page 18.

Scorpius is at its highest in the south, around 30 degrees, around dusk and sets in the southwest around 1:55 a.m. See the illustration on page 19.

The bright star **Arcturus** sets in the northwest around 2:15 a.m.

Bright Venus rises in the northeast around dawn.

2027

Saturday, July 17

TONIGHT

sunset: **7:56 p.m.**
twilight begins: **9 p.m.**

IN THE MORNING

twilight begins: **4:44 a.m.**
sunrise: **5:47 a.m.**

MOON PHASE

moonrise: **7:39 p.m.**
moonset: **5:40 a.m.**

IN THE SKY

The **full moon** tonight acts like a spotlight. Look around at the desert landscape at how much illumination the moon provides.

Jupiter is in the western sky at dusk and sets in the west around 9:35 p.m.

Cassiopeia, shaped like a *W,* rises in the northeast around 9:40 p.m. See the illustration on page 11.

Rust-red and not very bright, **Mars** sets in the west around 11 p.m.

Just east of **Scorpius**, Sagittarius is visible most of the night and is at its highest in the south, around 25 degrees, at 11:25 p.m. It sets in the southwest around 3:20 a.m.

The **summer triangle**'s three bright stars are **Vega, Altair,** and **Deneb. Vega** is almost directly overhead at 11:40 p.m. **Altair** is at its highest, around 65 degrees, at 12:55 a.m. Altair can be identified by its less bright neighboring stars, one on either side, together called the shaft of Aquila. **Deneb** is in the band of the Milky Way. See the illustration of the summer triangle on page 19.

Saturn rises in the east around 12:30 a.m. Unlike bright Jupiter and Venus and ruddy Mars, Saturn is harder to distinguish from the stars. Stars twinkle and shimmer, while the planets, including Saturn, show a steady flat light.

The **Great Bear (Big Dipper)** sets in the northwest around 1:15 a.m. See the illustration on page 10.

Bright star **Capella** rises in the northeast around 1:30 a.m. It is in the northeastern sky before dawn. Read about Capella on page 14.

The **Pleiades** rises in the northeast around 1:45 a.m. and is in the eastern sky before dawn. Read about the cluster on page 18.

Scorpius is at its highest in the south, around 30 degrees, around dusk and sets in the southwest around 1:50 a.m. See the illustration on page 19.

The bright star **Arcturus** sets in the northwest around 2:10 a.m.

Bright **Venus** rises in the northeast around dawn.

2027

Sunday, July 18

TONIGHT

sunset: **7:56 p.m.**
twilight begins: **8:59 p.m.**

MOON PHASE

IN THE MORNING

twilight begins: **4:45 a.m.**
sunrise: **5:47 a.m.**

moonrise: **8:14 p.m.**
moonset: **6:39 a.m.**

IN THE SKY

The **full moon** tonight acts like a spotlight. Look around at the desert landscape at how much illumination the moon provides.

Jupiter is in the western sky at dusk and sets in the west around 9:30 p.m.

Cassiopeia, shaped like a *W,* rises in the northeast around 9:40 p.m. It is in the northeastern sky at dusk. See the illustration on page 11.

Rust-red and not very bright, **Mars** sets in the west around 11 p.m.

Just east of **Scorpius,** Sagittarius is visible most of the night and is at its highest in the south, around 25 degrees, at 11:20 p.m. It sets in the southwest around 3:20 a.m.

The **summer triangle**'s three bright stars are **Vega, Altair,** and **Deneb. Vega** is almost directly overhead at 11:35 p.m. **Altair** is at its highest, around 65 degrees, at 12:50 a.m. Altair can be identified by its less bright neighboring stars, one on either side, together called the shaft of Aquila. **Deneb** is in the band of the Milky Way. See the illustration of the summer triangle on page 19.

Saturn rises in the east around 12:30 a.m. Unlike bright Jupiter and Venus and ruddy Mars, Saturn is harder to distinguish from the stars. Stars twinkle and shimmer, while the planets, including Saturn, show a steady flat light.

The **Great Bear (Big Dipper)** sets in the northwest around 1:15 a.m. See the illustration on page 10.

Bright star **Capella** rises in the northeast around 1:25 a.m. It is in the northeastern sky before dawn. Read about Capella on page 14.

The **Pleiades** rises in the northeast around 1:40 a.m. and is in the eastern sky before dawn. Read about the cluster on page 18.

Scorpius is at its highest in the south, around 30 degrees, around dusk and sets in the southwest around 1:45 a.m. See the illustration on page 19.

The bright star **Arcturus** sets in the northwest around 2:05 a.m.

Bright **Venus** rises in the northeast around dawn.

2027

Monday, July 19

TONIGHT

sunset: **7:55 p.m.**
twilight begins: **8:59 p.m.**

IN THE MORNING

twilight begins: **4:46 a.m.**
sunrise: **5:48 a.m.**

MOON PHASE

moonrise: **8:45 p.m.**
moonset: **7:36 a.m.**

IN THE SKY

The **full moon** tonight acts like a spotlight. Look around at the desert landscape at how much illumination the moon provides.

Jupiter sets in the west around 9:30 p.m.

Cassiopeia, shaped like a *W,* rises in the northeast around 9:35 p.m. It is in the northeastern sky at dusk. See the illustration on page 11.

Rust-red and not very bright, Mars sets in the west around 10:55 p.m.

Just east of **Scorpius**, Sagittarius is visible most of the night and is at its highest in the south, around 25 degrees, at 11:15 p.m. It sets in the southwest around 3:15 a.m.

The **summer triangle**'s three bright stars are **Vega, Altair,** and **Deneb. Vega** is almost directly overhead at 11:30 p.m. **Altair** is at its highest, around 65 degrees, at 12:45 a.m. Altair can be identified by its less bright neighboring stars, one on either side, together called the shaft of Aquila. **Deneb** is in the band of the Milky Way. See the illustration of the summer triangle on page 19.

Saturn rises in the east around 12:25 a.m. Unlike bright Jupiter and Venus and ruddy Mars, Saturn is harder to distinguish from the stars. Stars twinkle and shimmer, while the planets, including Saturn, show a steady flat light.

The **Great Bear (Big Dipper)** sets in the northwest around 1:10 a.m. See the illustration on page 10.

Bright star **Capella** rises in the northeast around 1:20 a.m. It is in the northeastern sky before dawn. Read about Capella on page 14.

The **Pleiades** rises in the northeast around 1:35 a.m. and is in the eastern sky before dawn. Read about the cluster on page 18.

Scorpius is at its highest in the south, around 30 degrees, around dusk and sets in the southwest around 1:45 a.m. See the illustration on page 19.

The bright star **Arcturus** sets in the northwest around 2 a.m.

Bright Venus rises in the northeast around dawn.

2027

Tuesday, July 20

TONIGHT

sunset: **7:55 p.m.**
twilight begins: **8:58 p.m.**

IN THE MORNING

twilight begins: **4:46 a.m.**
sunrise: **5:49 a.m.**

MOON PHASE

moonrise: **9:13 p.m.**
moonset: **7:36 a.m.**

IN THE SKY

The **full moon** tonight acts like a spotlight. Look around at the desert landscape at how much illumination the moon provides.

Jupiter sets in the west around 9:25 p.m.

Cassiopeia, shaped like a *W,* rises in the northeast around 9:30 p.m. It is in the northeastern sky at dusk. See the illustration on page 11.

Rust-red and not very bright, **Mars** sets in the west around 10:55 p.m.

Just east of **Scorpius**, Sagittarius is visible most of the night and is at its highest in the south, around 25 degrees, at 11:10 p.m. It sets in the southwest around 3:10 a.m.

The **summer triangle**'s three bright stars are **Vega, Altair,** and **Deneb. Vega** is almost directly overhead at 11:25 p.m. **Altair** is at its highest, around 65 degrees, at 12:40 a.m. Altair can be identified by its less bright neighboring stars, one on either side, together called the shaft of Aquila. **Deneb** is in the band of the Milky Way. See the illustration of the summer triangle on page 19.

Saturn rises in the east around 12:20 a.m. Unlike bright Jupiter and Venus and ruddy Mars, Saturn is harder to distinguish from the stars. Stars twinkle and shimmer, while the planets, including Saturn, show a steady flat light.

The **Great Bear (Big Dipper)** sets in the northwest around 1:05 a.m. See the illustration on page 10.

Bright star **Capella** rises in the northeast around 1:15 a.m. It is in the northeastern sky before dawn. Read about Capella on page 14.

The **Pleiades** rises in the northeast around 1:30 a.m. and is in the eastern sky before dawn. Read about the cluster on page 18.

Scorpius is at its highest in the south, around 30 degrees, around dusk and sets in the southwest around 1:40 a.m. See the illustration on page 19.

The bright star **Arcturus** sets in the northwest around 1:55 a.m.

Bright **Venus** rises in the northeast around dawn.

2027

Wednesday, July 21

MOON PHASE

TONIGHT

sunset: **7:54 p.m.**
twilight begins: **8:57 p.m.**

IN THE MORNING

twilight begins: **4:47 a.m.**
sunrise: **5:49 a.m.**

moonrise: **9:39 p.m.**
moonset: **8:33 a.m.**

IN THE SKY

The nearly **full moon** tonight acts like a spotlight. Look around at the desert landscape at how much illumination the moon provides.

Jupiter sets in the west around 9:20 p.m.

Cassiopeia, shaped like a *W,* rises in the northeast around 9:25 p.m. It is in the northeastern sky at dusk. See the illustration on page 11.

Rust-red and not very bright, **Mars** sets in the west around 10:50 p.m.

Just east of **Scorpius**, Sagittarius is visible most of the night and is at its highest in the south, around 25 degrees, at 11:05 p.m. It sets in the southwest around 3:05 a.m.

The **summer triangle**'s three bright stars are **Vega, Altair,** and **Deneb. Vega** is almost directly overhead at 11:25 p.m. **Altair** is at its highest, around 65 degrees, at 12:35 a.m. Altair can be identified by its less bright neighboring stars, one on either side, together called the shaft of Aquila. **Deneb** is in the band of the Milky Way. See the illustration of the summer triangle on page 19.

Saturn rises in the east around 12:15 a.m. Unlike bright Jupiter and Venus and ruddy Mars, Saturn is harder to distinguish from the stars. Stars twinkle and shimmer, while the planets, including Saturn, show a steady flat light.

The **Great Bear (Big Dipper)** sets in the northwest around 1 a.m. See the illustration on page 10.

Bright star **Capella** rises in the northeast around 1:15 a.m. It is in the northeastern sky before dawn. Read about Capella on page 14.

The **Pleiades** rises in the northeast around 1:30 a.m. and is in the eastern sky before dawn. Read about the cluster on page 18.

Scorpius is at its highest in the south, around 30 degrees, around dusk and sets in the southwest around 1:35 a.m. See the illustration on page 19.

The bright star **Arcturus** sets in the northwest around 1:55 a.m.

Bright **Venus** rises in the northeast around dawn.

2027

Thursday, July 22

TONIGHT

sunset: **7:54 p.m.**
twilight begins: **8:56 p.m.**

IN THE MORNING

twilight begins: **4:48 a.m.**
sunrise: **5:50 a.m.**

MOON PHASE

moonrise: **10:03 p.m.**
moonset: **9:29 a.m.**

IN THE SKY

Jupiter sets in the west around 9:20 p.m.

Cassiopeia, shaped like a *W,* rises in the northeast around 9:25 p.m. It is in the northeastern sky at dusk. See the illustration on page 11.

Rust-red and not very bright, **Mars** sets in the west around 10:50 p.m.

Just east of **Scorpius,** Sagittarius is visible most of the night and is at its highest in the south, around 25 degrees, at 11:05 p.m. It sets in the southwest around 3 a.m.

The **summer triangle**'s three bright stars are **Vega, Altair,** and **Deneb. Vega** is almost directly overhead at 11:20 p.m. **Altair** is at its highest, around 65 degrees, at 12:35 a.m. Altair can be identified by its less bright neighboring stars, one on either side, together called the shaft of Aquila. **Deneb** is in the band of the Milky Way. See the illustration of the summer triangle on page 19.

Saturn rises in the east around 12:15 a.m. Unlike bright Jupiter and Venus and ruddy Mars, Saturn is harder to distinguish from the stars. Stars twinkle and shimmer, while the planets, including Saturn, show a steady flat light.

The **Great Bear (Big Dipper)** sets in the northwest around 12:55 a.m. See the illustration on page 10.

Bright star **Capella** rises in the northeast around 1:05 a.m. Read about Capella on page 14.

The **Pleiades** rises in the northeast around 1:25 a.m. and is in the eastern sky before dawn. Read about the cluster on page 18.

Scorpius sets in the southwest around 1:30 a.m. See the illustration on page 19.

The bright star **Arcturus** sets in the northwest around 1:50 a.m.

Bright **Venus** rises in the northeast around dawn.

2027

Friday, July 23

TONIGHT

sunset: **7:53 p.m.**
twilight begins: **8:55 p.m.**

IN THE MORNING

twilight begins: **4:49 a.m.**
sunrise: **5:51 a.m.**

MOON PHASE

moonrise: **10:28 p.m.**
moonset: **10:25 a.m.**

IN THE SKY

Jupiter sets in the west around 9:15 p.m.

Cassiopeia, shaped like a *W,* rises in the northeast around 9:20 p.m. It is in the northeastern sky at dusk. See the illustration on page 11.

Rust-red and not very bright, **Mars** sets in the west around 10:45 p.m.

Just east of **Scorpius**, Sagittarius is visible most of the night and is at its highest in the south, around 25 degrees, at 11 p.m. It sets in the southwest around 3 a.m.

The **summer triangle**'s three bright stars are **Vega, Altair,** and **Deneb. Vega** is almost directly overhead at 11:15 p.m. **Altair** is at its highest, around 65 degrees, at 12:30 a.m. Altair can be identified by its less bright neighboring stars, one on either side, together called the shaft of Aquila. **Deneb** is in the band of the Milky Way. See the illustration of the summer triangle on page 19.

Saturn rises in the east around 12:10 a.m. Unlike bright Jupiter and Venus and ruddy Mars, Saturn is harder to distinguish from the stars. Stars twinkle and shimmer, while the planets, including Saturn, show a steady flat light.

The **Great Bear (Big Dipper)** sets in the northwest around 12:50 a.m. See the illustration on page 10.

Bright star **Capella** rises in the northeast around 1 a.m. Read about Capella on page 14.

The **Pleiades** rises in the northeast around 1:20 a.m. and is in the eastern sky before dawn. Read about the cluster on page 18.

Scorpius sets in the southwest around 1:25 a.m. See the illustration on page 19.

The bright star **Arcturus** sets in the northwest around 1:45 a.m.

Bright **Venus** rises in the northeast around dawn.

2027

Saturday, July 24

TONIGHT

sunset: **7:52 p.m.**
twilight begins: **8:55 p.m.**

IN THE MORNING

twilight begins: **4:50 a.m.**
sunrise: **5:51 a.m.**

MOON PHASE

moonrise: **10:54 p.m.**
moonset: **11:22 a.m.**

IN THE SKY

Cassiopeia is in the northeastern sky at dusk. See the illustration on page 11.

Jupiter sets in the west around 9:10 p.m.

Explore the **dark sky** after twilight and before the moon rises, around 10:35 p.m.

Rust-red and not very bright, **Mars** is in the western sky at dusk and sets in the west around 10:45 p.m.

Just east of **Scorpius**, Sagittarius is visible most of the night and is at its highest in the south, around 25 degrees, at 10:55 p.m. It sets in the southwest around 2:55 a.m.

The **summer triangle**'s three bright stars are **Vega, Altair,** and **Deneb. Vega** is almost directly overhead at 11:10 p.m. **Altair** is at its highest, around 65 degrees, at 12:25 a.m. Altair can be identified by its less bright neighboring stars, one on either side, together called the shaft of Aquila. **Deneb** is in the band of the Milky Way. See the illustration of the summer triangle on page 19.

Saturn rises in the east around 12:05 a.m. Unlike bright Jupiter and Venus and ruddy Mars, Saturn is harder to distinguish from the stars. Stars twinkle and shimmer, while the planets, including Saturn, show a steady flat light.

The **Great Bear (Big Dipper)** sets in the northwest around 12:45 a.m. See the illustration on page 10.

Bright star **Capella** rises in the northeast around 12:55 a.m. Read about Capella on page 14.

The **Pleiades** rises in the northeast around 1:15 a.m. and is in the eastern sky before dawn. Read about the cluster on page 18.

Scorpius sets in the southwest around 1:25 a.m. See the illustration on page 19.

The bright star **Arcturus** sets in the northwest around 1:40 a.m.

Bright **Venus** rises in the northeast around dawn.

2027

Sunday, July 25

TONIGHT

sunset: **7:52 p.m.**
twilight begins: **8:54 p.m.**

MOON PHASE

IN THE MORNING

twilight begins: **4:51 a.m.**
sunrise: **5:52 a.m.**

moonrise: **11:23 p.m.**
moonset: **12:22 p.m.**

IN THE SKY

Cassiopeia is in the northeastern sky at dusk. See the illustration on page 11.

Jupiter sets in the west around 9:10 p.m.

Rust-red and not very bright, **Mars** is in the western sky at dusk and sets in the west around 10:40 p.m.

Just east of **Scorpius**, Sagittarius is visible most of the night and is at its highest in the south, around 25 degrees, at 10:50 p.m. It sets in the southwest around 2:50 a.m.

Explore the **dark sky** after twilight and before the moon rises, around 11:05 p.m.

The **summer triangle**'s three bright stars are **Vega, Altair,** and **Deneb. Vega** is almost directly overhead at 11:05 p.m. **Altair** is at its highest, around 65 degrees, at 12:20 a.m. Altair can be identified by its less bright neighboring stars, one on either side, together called the shaft of Aquila. **Deneb** is in the band of the Milky Way. See the illustration of the summer triangle on page 19.

Saturn rises in the east around midnight. Unlike bright Jupiter and Venus and ruddy Mars, Saturn is harder to distinguish from the stars. Stars twinkle and shimmer, while the planets, including Saturn, show a steady flat light.

The **Great Bear (Big Dipper)** sets in the northwest around 12:40 a.m. See the illustration on page 10.

Bright star **Capella** rises in the northeast around 12:55 a.m. Read about Capella on page 14.

The **Pleiades** rises in the northeast around 1:10 a.m. and is in the eastern sky before dawn. Read about the cluster on page 18.

Scorpius sets in the southwest around 1:20 a.m. See the illustration on page 19.

The bright star **Arcturus** sets in the northwest around 1:35 a.m.

Bright **Venus** rises in the northeast around dawn.

2027

Monday, July 26

TONIGHT

sunset: **7:51** p.m.
twilight begins: **8:53** p.m.

IN THE MORNING

twilight begins: **4:52** a.m.
sunrise: **5:53** a.m.

MOON PHASE

moonrise: **11:57** p.m.
moonset: **1:25** p.m.

IN THE SKY

Cassiopeia is in the northeastern sky at dusk. See the illustration on page 11.

Jupiter sets in the west around 9:05 p.m.

Rust-red and not very bright, **Mars** is in the western sky at dusk and sets in the west around 10:40 p.m.

Just east of **Scorpius**, Sagittarius is visible most of the night and is at its highest in the south, around 25 degrees, at 10:45 p.m. It sets in the southwest around 2:45 a.m.

The **summer triangle**'s three bright stars are **Vega, Altair,** and **Deneb. Vega** is almost directly overhead at 11:05 p.m. **Altair** is at its highest, around 65 degrees, at 12:20 a.m. Altair can be identified by its less bright neighboring stars, one on either side, together called the shaft of Aquila. **Deneb** is in the band of the Milky Way. See the illustration of the summer triangle on page 19.

Before the moon rises, around 11:45 p.m., look for the **Andromeda Galaxy** with binoculars or a telescope. It becomes a good target around 11:15 p.m. See the guide on page 11.

Saturn rises in the east around midnight. Unlike bright Jupiter and Venus and ruddy Mars, Saturn is harder to distinguish from the stars. Stars twinkle and shimmer, while the planets, including Saturn, show a steady flat light.

The **Great Bear (Big Dipper)** sets in the northwest around 12:40 a.m. See the illustration on page 10.

Bright star **Capella** rises in the northeast around 12:50 a.m. Read about Capella on page 14.

The **Pleiades** rises in the northeast around 1:05 a.m. Read about the cluster on page 18.

Scorpius sets in the southwest around 1:15 a.m. See the illustration on page 19.

The bright star **Arcturus** sets in the northwest around 1:35 a.m.

Bright **Venus** rises in the northeast around dawn.

2027
Tuesday, July 27

TONIGHT

sunset: **7:50 p.m.**
twilight begins: **8:52 p.m.**

IN THE MORNING

twilight begins: **4:52 a.m.**
sunrise: **5:54 a.m.**

MOON PHASE

moonrise: **12:38 a.m.**
moonset: **2:31 p.m.**

IN THE SKY

The **Delta Aquarids meteor shower** happens tonight. A good shower with up to 20 meteors per hour, the best time to watch is after midnight. Tonight the moon will wash out the shower from when it rises, around 12:30 a.m.

Cassiopeia is in the northeastern sky at dusk. See the illustration on page 11.

Rust-red and not very bright, **Mars** is in the western sky at dusk and sets in the west around 10:35 p.m.

Just east of **Scorpius**, Sagittarius is visible most of the night and is at its highest in the south, around 25 degrees, at 10:45 p.m. It sets in the southwest around 2:45 a.m.

The **summer triangle**'s three bright stars are **Vega, Altair,** and **Deneb. Vega** is almost directly overhead at 11 p.m. **Altair** is at its highest, around 65 degrees, at 12:15 a.m. Altair can be identified by its less bright neighboring stars, one on either side, together called the shaft of Aquila. **Deneb** is in the band of the Milky Way. See the illustration of the summer triangle on page 19.

Saturn rises in the east around 11:55 p.m. Unlike bright Jupiter and Venus and ruddy Mars, Saturn is harder to distinguish from the stars. Stars twinkle and shimmer, while the planets, including Saturn, show a steady flat light.

Before the moon rises, around 12:30 a.m., look for the **Andromeda Galaxy** with binoculars or a telescope. It becomes a good target around 11:10 p.m. See the guide on page 11.

The **Great Bear (Big Dipper)** sets in the northwest around 12:35 a.m. See the illustration on page 10.

Bright star **Capella** rises in the northeast around 12:45 a.m. Read about Capella on page 14.

Scorpius sets in the southwest around 1:10 a.m. See the illustration on page 19.

The bright star **Arcturus** sets in the northwest around 1:30 a.m.

Orion rises in the east around 4:30 a.m. and is in the eastern sky before dawn. Look for the bright star **Aldebaran** and the **Pleiades** near Orion. See the illustration on page 17.

Bright **Venus** rises in the northeast around dawn.

2027

Wednesday, July 28

TONIGHT

sunset: **7:50 p.m.**
twilight begins: **8:51 p.m.**

IN THE MORNING

twilight begins: **4:53 a.m.**
sunrise: **5:54 a.m.**

MOON PHASE

moonrise: **1:28 a.m.**
moonset: **3:39 p.m.**

IN THE SKY

The **Delta Aquarids meteor shower** happens tonight. A good shower with up to 20 meteors per hour, the best time to watch is after midnight. Tonight the moon will wash out the shower from when it rises, around 1:20 a.m.

Cassiopeia is in the northeastern sky at dusk. See the illustration on page 11.

Rust-red and not very bright, **Mars** is in the western sky at dusk and sets in the west around 10:35 p.m.

Just east of **Scorpius**, Sagittarius is visible most of the night and is at its highest in the south, around 25 degrees, at 10:40 p.m. It sets in the southwest around 2:40 a.m.

The **summer triangle**'s three bright stars are **Vega, Altair,** and **Deneb. Vega** is almost directly overhead at 10:55 p.m. **Altair** is at its highest, around 65 degrees, at 12:10 a.m. Altair can be identified by its less bright neighboring stars, one on either side, together called the shaft of Aquila. **Deneb** is in the band of the Milky Way. See the illustration of the summer triangle on page 19.

Saturn rises in the east around 11:50 p.m. Unlike bright Jupiter and Venus and ruddy Mars, Saturn is harder to distinguish from the stars. Stars twinkle and shimmer, while the planets, including Saturn, show a steady flat light.

The **Great Bear (Big Dipper)** sets in the northwest around 12:30 a.m. See the illustration on page 10.

Bright star **Capella** rises in the northeast around 12:40 a.m. Read about Capella on page 14.

Scorpius sets in the southwest around 1:05 a.m. See the illustration on page 19.

Before the moon rises, around 1:20 a.m., look for the **Andromeda Galaxy** with binoculars or a telescope. It becomes a good target around 11:05 p.m. See the guide on page 11.

The bright star **Arcturus** sets in the northwest around 1:25 a.m.

Orion rises in the east around 4:25 a.m. and is in the eastern sky before dawn. Look for the bright star **Aldebaran** and the **Pleiades** near Orion. See the illustration on page 17.

Bright **Venus** rises in the northeast around dawn.

2027

Thursday, July 29

TONIGHT

sunset: **7:49 p.m.**
twilight begins: **8:50 p.m.**

IN THE MORNING

twilight begins: **4:54 a.m.**
sunrise: **5:55 a.m.**

MOON PHASE

moonrise: **2:29 a.m.**
moonset: **4:46 p.m.**

IN THE SKY

The **Delta Aquarids meteor shower** happens tonight. A good shower with up to 20 meteors per hour, the best time to watch is after midnight. Tonight has moonless dark skies all night.

Cassiopeia is in the northeastern sky at dusk. See the illustration on page 11.

The absence of the moon tonight provides excellent **dark-sky** viewing conditions.

Rust-red and not very bright, **Mars** is in the western sky at dusk and sets in the west around 10:30 p.m.

Just east of **Scorpius**, Sagittarius is visible most of the night and is at its highest in the south, around 25 degrees, at 10:35 p.m. It sets in the southwest around 2:35 a.m.

The **summer triangle**'s three bright stars are **Vega, Altair,** and **Deneb. Vega** is almost directly overhead at 10:50 p.m. **Altair** is at its highest, around 65 degrees, at 12:05 a.m. Altair can be identified by its less bright neighboring stars, one on either side, together called the shaft of Aquila. **Deneb** is in the band of the Milky Way. See the illustration of the summer triangle on page 19.

Saturn rises in the east around 11:45 p.m. Unlike bright Jupiter and Venus and ruddy Mars, Saturn is harder to distinguish from the stars. Stars twinkle and shimmer, while the planets, including Saturn, show a steady flat light.

The **Great Bear (Big Dipper)** sets in the northwest around 12:25 a.m. See the illustration on page 10.

Bright star **Capella** rises in the northeast around 12:40 a.m. Read about Capella on page 14.

Scorpius sets in the southwest around 1:05 a.m. See the illustration on page 19.

The bright star **Arcturus** sets in the northwest around 1:20 a.m.

Before the moon rises, around 2:15 a.m., look for the **Andromeda Galaxy** with binoculars or a telescope. It becomes a good target around 11 p.m. See the guide on page 11.

Orion rises in the east around 4:20 a.m. and is in the eastern sky before dawn. Look for the bright star **Aldebaran** and the **Pleiades** near Orion. See the illustration on page 17.

2027

Friday, July 30

TONIGHT

sunset: **7:48 p.m.**
twilight begins: **8:49 p.m.**

IN THE MORNING

twilight begins: **4:55 a.m.**
sunrise: **5:56 a.m.**

MOON PHASE

moonrise: **3:39 a.m.**
moonset: **5:48 p.m.**

IN THE SKY

Cassiopeia is in the northeastern sky at dusk. See the illustration on page 11.

The absence of the moon tonight provides excellent **dark-sky** viewing conditions.

Rust-red and not very bright, **Mars** is in the western sky at dusk and sets in the west around 10:30 p.m.

Just east of **Scorpius**, Sagittarius is at its highest in the south, around 25 degrees, at 10:30 p.m. It sets in the southwest around 2:30 a.m.

The **summer triangle**'s three bright stars are **Vega, Altair,** and **Deneb. Vega** is almost directly overhead at 10:50 p.m. **Altair** is at its highest, around 65 degrees, at midnight. Altair can be identified by its less bright neighboring stars, one on either side, together called the shaft of Aquila. **Deneb** is in the band of the Milky Way. See the illustration of the summer triangle on page 19.

Saturn rises in the east around 11:40 p.m. Unlike bright Jupiter and Venus and ruddy Mars, Saturn is harder to distinguish from the stars. Stars twinkle and shimmer, while the planets, including Saturn, show a steady flat light.

Before the moon rises, around 3:30 a.m., look for the **Andromeda Galaxy** with binoculars or a telescope. It becomes a good target around 11 p.m. See the guide on page 11.

The **Great Bear (Big Dipper)** sets in the northwest around 12:20 a.m. See the illustration on page 10.

Bright star **Capella** rises in the northeast around 12:35 a.m. Read about Capella on page 14.

Scorpius sets in the southwest around 1 a.m. See the illustration on page 19.

The bright star **Arcturus** sets in the northwest around 1:20 a.m.

Orion rises in the east around 4:15 a.m. and is in the eastern sky before dawn. Look for the bright star **Aldebaran** and the **Pleiades** near Orion. See the illustration on page 17.

2027

Saturday, July 31

TONIGHT

sunset: 7:47 p.m.
twilight begins: 8:48 p.m.

IN THE MORNING

twilight begins: 4:56 a.m.
sunrise: 5:56 a.m.

MOON PHASE

moonrise: 4:55 a.m.
moonset: 6:43 p.m.

IN THE SKY

Cassiopeia is in the northeastern sky at dusk. See the illustration on page 11.

The absence of the moon tonight provides excellent **dark-sky** viewing conditions.

Rust-red and not very bright, **Mars** is in the western sky at dusk and sets in the west around 10:25 p.m.

Just east of **Scorpius**, Sagittarius is at its highest in the south, around 25 degrees, at 10:30 p.m. It sets in the southwest around 2:25 a.m.

The **summer triangle**'s three bright stars are **Vega, Altair,** and **Deneb. Vega** is almost directly overhead at 10:45 p.m. **Altair** is at its highest, around 65 degrees, at midnight. Altair can be identified by its less bright neighboring stars, one on either side, together called the shaft of Aquila. **Deneb** is in the band of the Milky Way. See the illustration of the summer triangle on page 19.

With the dark sky, look for the **Andromeda Galaxy** with binoculars or a telescope. It becomes a good target around 10:55 p.m. See the guide on page 11.

Saturn rises in the east around 11:40 p.m. Unlike bright Jupiter and Venus and ruddy Mars, Saturn is harder to distinguish from the stars. Stars twinkle and shimmer, while the planets, including Saturn, show a steady flat light.

The **Great Bear (Big Dipper)** sets in the northwest around 12:20 a.m. See the illustration on page 10.

Bright star **Capella** rises in the northeast around 12:30 a.m. Read about Capella on page 14.

Scorpius sets in the southwest around 12:55 a.m. See the illustration on page 19.

The bright star **Arcturus** sets in the northwest around 1:15 a.m.

Orion rises in the east around 4:15 a.m. and is in the eastern sky before dawn. Look for the bright star **Aldebaran** and the **Pleiades** near Orion. See the illustration on page 17.

2027

Sunday, August 1

TONIGHT

sunset: **7:46 p.m.**
twilight begins: **8:47 p.m.**

IN THE MORNING

twilight begins: **4:57 a.m.**
sunrise: **5:57 a.m.**

MOON PHASE

moonrise: **6:12 a.m.**
moonset: **7:29 p.m.**

IN THE SKY

Cassiopeia is in the northeastern sky at dusk. See the illustration on page 11.

The absence of the moon tonight provides excellent **dark-sky** viewing conditions.

Rust-red and not very bright, **Mars** is in the western sky at dusk and sets in the west around 10:25 p.m.

Just east of **Scorpius**, Sagittarius is at its highest in the south, around 25 degrees, at 10:25 p.m. It sets in the southwest around 2:25 a.m.

The **summer triangle**'s three bright stars are **Vega, Altair,** and **Deneb. Vega** is almost directly overhead at 10:40 p.m. **Altair** is at its highest, around 65 degrees, at 11:55 p.m. Altair can be identified by its less bright neighboring stars, one on either side, together called the shaft of Aquila. **Deneb** is in the band of the Milky Way. See the illustration of the summer triangle on page 19.

Look for the **Andromeda Galaxy** with binoculars or a telescope. It becomes a good target around 10:50 p.m. See the guide on page 11.

Saturn rises in the east around 11:35 p.m. Unlike bright Jupiter and Venus and ruddy Mars, Saturn is harder to distinguish from the stars. Stars twinkle and shimmer, while the planets, including Saturn, show a steady flat light.

The **Great Bear (Big Dipper)** sets in the northwest around 12:15 a.m. See the illustration on page 10.

Bright star **Capella** rises in the northeast around 12:25 a.m. Read about Capella on page 14.

Scorpius sets in the southwest around 12:50 a.m. See the illustration on page 19.

The bright star **Arcturus** sets in the northwest around 1:10 a.m.

Orion rises in the east around 4:10 a.m. and is in the eastern sky before dawn. Look for the bright star **Aldebaran** and the **Pleiades** near Orion. See the illustration on page 17.

2027

Monday, August 2

MOON PHASE

TONIGHT

sunset: **7:45 p.m.**
twilight begins: **8:46 p.m.**

IN THE MORNING

twilight begins: **4:58 a.m.**
sunrise: **5:58 a.m.**

moonrise: **7:26 a.m.**
moonset: **8:08 p.m.**

IN THE SKY

Cassiopeia is in the northeastern sky at dusk. See the illustration on page 11.

The absence of the moon tonight provides excellent **dark-sky** viewing conditions.

Rust-red and not very bright, **Mars** is in the western sky at dusk and sets in the west around 10:20 p.m.

Just east of **Scorpius**, Sagittarius is at its highest in the south, around 25 degrees, at 10:20 p.m. It sets in the southwest around 2:20 a.m.

The **summer triangle**'s three bright stars are **Vega, Altair,** and **Deneb. Vega** is almost directly overhead at 10:35 p.m. **Altair** is at its highest, around 65 degrees, at 11:50 p.m. Altair can be identified by its less bright neighboring stars, one on either side, together called the shaft of Aquila. **Deneb** is in the band of the Milky Way. See the illustration of the summer triangle on page 19.

Look for the **Andromeda Galaxy** with binoculars or a telescope. It becomes a good target around 10:45 p.m. See the guide on page 11.

Saturn rises in the east around 11:30 p.m. Unlike bright Jupiter and Venus and ruddy Mars, Saturn is harder to distinguish from the stars. Stars twinkle and shimmer, while the planets, including Saturn, show a steady flat light.

The **Great Bear (Big Dipper)** sets in the northwest around 12:10 a.m. See the illustration on page 10.

Bright star **Capella** rises in the northeast around 12:20 a.m. Read about Capella on page 14.

Scorpius sets in the southwest around 12:50 a.m. See the illustration on page 19.

The bright star **Arcturus** sets in the northwest around 1:05 a.m.

Orion rises in the east around 4:05 a.m. and is in the eastern sky before dawn. Look for the bright star **Aldebaran** and the **Pleiades** near Orion. See the illustration on page 17.

2027

Tuesday, August 3

TONIGHT

sunset: 7:44 p.m.
twilight begins: 8:45 p.m.

IN THE MORNING

twilight begins: 4:59 a.m.
sunrise: 5:59 a.m.

MOON PHASE

moonrise: 7:26 a.m.
moonset: 8:43 p.m.

IN THE SKY

Cassiopeia is in the northeastern sky at dusk. See the illustration on page 11.

The absence of the moon tonight provides excellent **dark-sky** viewing conditions.

Just east of **Scorpius**, Sagittarius is at its highest in the south, around 25 degrees, at 10:15 p.m. It sets in the southwest around 2:15 a.m.

Rust-red and not very bright, **Mars** is in the western sky at dusk and sets in the west around 10:20 p.m.

The **summer triangle**'s three bright stars are **Vega, Altair,** and **Deneb. Vega** is almost directly overhead at 10:30 p.m. **Altair** is at its highest, around 65 degrees, at 11:45 p.m. Altair can be identified by its less bright neighboring stars, one on either side, together called the shaft of Aquila. **Deneb** is in the band of the Milky Way. See the illustration of the summer triangle on page 19.

Look for the **Andromeda Galaxy** with binoculars or a telescope. It becomes a good target around 10:45 p.m. See the guide on page 11.

Saturn rises in the east around 11:25 p.m. Unlike bright Jupiter and Venus and ruddy Mars, Saturn is harder to distinguish from the stars. Stars twinkle and shimmer, while the planets, including Saturn, show a steady flat light.

The **Great Bear (Big Dipper)** sets in the northwest around 12:05 a.m. See the illustration on page 10.

Bright star **Capella** rises in the northeast around 12:20 a.m. Read about Capella on page 14.

Scorpius sets in the southwest around 12:45 a.m. See the illustration on page 19.

The bright star **Arcturus** sets in the northwest around 1 a.m.

Orion rises in the east around 4 a.m. and is in the eastern sky before dawn. Look for the bright star **Aldebaran** and the **Pleiades** near Orion. See the illustration on page 17.

2027
Wednesday, August 4

TONIGHT
sunset: **7:44 p.m.**
twilight begins: **8:44 p.m.**

IN THE MORNING
twilight begins: **5 a.m.**
sunrise: **5:59 a.m.**

MOON PHASE

moonrise: **8:38 a.m.**
moonset: **9:14 p.m.**

IN THE SKY

Cassiopeia is in the northeastern sky at dusk. See the illustration on page 11.

The absence of the moon tonight provides excellent **dark-sky** viewing conditions. Explore the **dark sky** after the moon sets, from around 9:35 p.m.

Just east of **Scorpius**, Sagittarius is at its highest in the south, around 25 degrees, at 10:10 p.m. It sets in the southwest around 2:10 a.m.

Rust-red and not very bright, **Mars** is in the western sky at dusk and sets in the west around 10:15 p.m.

The **summer triangle**'s three bright stars are **Vega, Altair,** and **Deneb. Vega** is almost directly overhead at 10:30 p.m. **Altair** is at its highest, around 65 degrees, at 11:40 p.m. Altair can be identified by its less bright neighboring stars, one on either side, together called the shaft of Aquila. **Deneb** is in the band of the Milky Way. See the illustration of the summer triangle on page 19.

Look for the **Andromeda Galaxy** with binoculars or a telescope. It becomes a good target around 10:40 p.m. See the guide on page 11.

Saturn rises in the east around 11:25 p.m. Unlike bright Jupiter and Venus and ruddy Mars, Saturn is harder to distinguish from the stars. Stars twinkle and shimmer, while the planets, including Saturn, show a steady flat light.

The **Great Bear (Big Dipper)** sets in the northwest around 12:05 a.m. See the illustration on page 10.

Bright star **Capella** rises in the northeast around 12:15 a.m. Read about Capella on page 14.

Scorpius sets in the southwest around 12:40 a.m. See the illustration on page 19.

The bright star **Arcturus** sets in the northwest around 1 a.m.

Orion rises in the east around 3:55 a.m. and is in the eastern sky before dawn. Look for the bright star **Aldebaran** and the **Pleiades** near Orion. See the illustration on page 17.

2027

Thursday, August 5

TONIGHT

sunset: **7:43 p.m.**
twilight begins: **8:43 p.m.**

MOON PHASE

IN THE MORNING

twilight begins: **5:01 a.m.**
sunrise: **6 a.m.**

moonrise: **9:48 a.m.**
moonset: **9:44 p.m.**

IN THE SKY

Cassiopeia is in the northeastern sky at dusk. See the illustration on page 11.

The **moon** is prominent in the sky early in the evening. Using a telescope or binoculars, take a look. You should be able to see rough mountains and craters along the boundary between light and dark, in astronomy called the terminator.

Explore the **dark sky** after the moon sets, from around 10:05 p.m.

Just east of **Scorpius**, Sagittarius is at its highest in the south, around 25 degrees, at 10:10 p.m. It sets in the southwest around 2:05 a.m.

Rust-red and not very bright, **Mars** is in the western sky at dusk and sets in the west around 10:15 p.m.

The **summer triangle**'s three bright stars are **Vega, Altair,** and **Deneb. Vega** is almost directly overhead at 10:25 p.m. **Altair** is at its highest, around 65 degrees, at 11:40 p.m. Altair can be identified by its less bright neighboring stars, one on either side, together called the shaft of Aquila. **Deneb** is in the band of the Milky Way. See the illustration of the summer triangle on page 19.

Look for the **Andromeda Galaxy** with binoculars or a telescope. It becomes a good target around 10:35 p.m. See the guide on page 11.

Saturn rises in the east around 11:20 p.m. Unlike bright Jupiter and Venus and ruddy Mars, Saturn is harder to distinguish from the stars. Stars twinkle and shimmer, while the planets, including Saturn, show a steady flat light.

The **Great Bear (Big Dipper)** sets in the northwest around midnight. See the illustration on page 10.

Bright star **Capella** rises in the northeast around 12:10 a.m. Read about Capella on page 14.

Scorpius sets in the southwest around 12:35 a.m. See the illustration on page 19.

The bright star **Arcturus** sets in the northwest around 12:55 a.m.

Orion rises in the east around 3:55 a.m. and is in the eastern sky before dawn. Look for the bright star **Aldebaran** and the **Pleiades** near Orion. See the illustration on page 17.

2027

Friday, August 6

TONIGHT

sunset: **7:42 p.m.**
twilight begins: **8:42 p.m.**

IN THE MORNING

twilight begins: **5:01 a.m.**
sunrise: **6:01 a.m.**

MOON PHASE

moonrise: **10:55 a.m.**
moonset: **10:15 p.m.**

IN THE SKY

Cassiopeia is in the northeastern sky at dusk. See the illustration on page 11.

The **moon** is prominent in the sky early in the evening. Using a telescope or binoculars, take a look. You should be able to see rough mountains and craters along the boundary between light and dark, in astronomy called the terminator.

The waxing crescent moon and **Mars** appear close together in the evening sky. Look west near the horizon around 9 p.m.

Just east of **Scorpius**, Sagittarius is at its highest in the south, around 25 degrees, at 10:05 p.m. It sets in the southwest around 2:05 a.m.

Rust-red and not very bright, **Mars** is in the western sky at dusk and sets in the west around 10:10 p.m.

The **summer triangle**'s three bright stars are **Vega, Altair,** and **Deneb. Vega** is almost directly overhead at 10:20 p.m. **Altair** is at its highest, around 65 degrees, at 11:35 p.m. Altair can be identified by its less bright neighboring stars, one on either side, together called the shaft of Aquila. **Deneb** is in the band of the Milky Way. See the illustration of the summer triangle on page 19.

Explore the **dark sky** after the moon sets, from around 10:35 p.m.

After the moon sets, from around 10:35 p.m., look for the **Andromeda Galaxy** with binoculars or a telescope. See the guide on page 11.

Saturn rises in the east around 11:15 p.m. Unlike bright Jupiter and Venus and ruddy Mars, Saturn is harder to distinguish from the stars. Stars twinkle and shimmer, while the planets, including Saturn, show a steady flat light.

The **Great Bear (Big Dipper)** sets in the northwest around 11:55 p.m. See the illustration on page 10.

Bright star **Capella** rises in the northeast around 12:05 a.m. Read about Capella on page 14.

Scorpius sets in the southwest around 12:30 a.m. See the illustration on page 19.

The bright star **Arcturus** sets in the northwest around 12:50 a.m.

Orion rises in the east around 3:50 a.m. and is in the eastern sky before dawn. Look for the bright star **Aldebaran** and the **Pleiades** near Orion. See the illustration on page 17.

2027

Saturday, August 7

TONIGHT

sunset: **7:41 p.m.**
twilight begins: **8:40 p.m.**

IN THE MORNING

twilight begins: **5:02 a.m.**
sunrise: **6:02 a.m.**

MOON PHASE

moonrise: **12:01 p.m.**
moonset: **10:47 p.m.**

IN THE SKY

Cassiopeia is in the northeastern sky at dusk. See the illustration on page 11.

The **moon** is prominent in the sky early in the evening. Using a telescope or binoculars, take a look. You should be able to see rough mountains and craters along the boundary between light and dark, in astronomy called the terminator.

Just east of **Scorpius**, Sagittarius is at its highest in the south, around 25 degrees, at 10 p.m. It sets in the southwest around 2 a.m.

Rust-red and not very bright, **Mars** is in the western sky at dusk and sets in the west around 10:10 p.m.

The **summer triangle**'s three bright stars are **Vega, Altair,** and **Deneb. Vega** is almost directly overhead at 10:15 p.m. **Altair** is at its highest, around 65 degrees, at 11:30 p.m. Altair can be identified by its less bright neighboring stars, one on either side, together called the shaft of Aquila. **Deneb** is in the band of the Milky Way. See the illustration of the summer triangle on page 19.

Explore the **dark sky** after the moon sets, from around 11:05 p.m.

After the moon sets, from around 11:05 p.m., look for the **Andromeda Galaxy** with binoculars or a telescope. See the guide on page 11.

Saturn rises in the east around 11:10 p.m. Unlike bright Jupiter and Venus and ruddy Mars, Saturn is harder to distinguish from the stars. Stars twinkle and shimmer, while the planets, including Saturn, show a steady flat light.

The **Great Bear (Big Dipper)** sets in the northwest around 11:50 p.m. See the illustration on page 10.

Bright star **Capella** rises in the northeast around midnight. Read about Capella on page 14.

Scorpius sets in the southwest around 12:25 a.m. See the illustration on page 19.

The bright star **Arcturus** sets in the northwest around 12:40 a.m.

Orion rises in the east around 3:45 a.m. and is in the eastern sky before dawn. Look for the bright star **Aldebaran** and the **Pleiades** near Orion. See the illustration on page 17.

Castor and Pollux, the twin stars in the constellation Gemini, rise in the northeast around 4:40 a.m. and are in the eastern sky before dawn.

2027

Sunday, August 8

MOON PHASE

TONIGHT

sunset: 7:40 p.m.
twilight begins: 8:39 p.m.

IN THE MORNING

twilight begins: 5:03 a.m.
sunrise: 6:02 a.m.

moonrise: 1:06 p.m.
moonset: 11:23 p.m.

IN THE SKY

Cassiopeia is in the northeastern sky at dusk. See the illustration on page 11.

The **moon** is prominent in the sky early in the evening. Using a telescope or binoculars, take a look. You should be able to see rough mountains and craters along the boundary between light and dark, in astronomy called the terminator.

Just east of **Scorpius**, Sagittarius is at its highest in the south, around 25 degrees, at 9:55 p.m. It sets in the southwest around 1:55 a.m.

Rust-red and not very bright, **Mars** is in the western sky at dusk and sets in the west around 10:05 p.m.

The **summer triangle**'s three bright stars are **Vega, Altair,** and **Deneb. Vega** is almost directly overhead at 10:10 p.m. **Altair** is at its highest, around 65 degrees, at 11:25 p.m. Altair can be identified by its less bright neighboring stars, one on either side, together called the shaft of Aquila. **Deneb** is in the band of the Milky Way. See the illustration of the summer triangle on page 19.

Saturn rises in the east around 11:05 p.m. Unlike bright Jupiter and Venus and ruddy Mars, Saturn is harder to distinguish from the stars. Stars twinkle and shimmer, while the planets, including Saturn, show a steady flat light.

The **Great Bear (Big Dipper)** sets in the northwest around 11:45 p.m. See the illustration on page 10.

Explore the **dark sky** after the moon sets, from around 11:45 p.m.

After the moon sets, from around 11:45 p.m., look for the **Andromeda Galaxy** with binoculars or a telescope. See the guide on page 11.

Bright star **Capella** rises in the northeast around midnight. Read about Capella on page 14.

Scorpius sets in the southwest around 12:20 a.m. See the illustration on page 19.

The bright star **Arcturus** sets in the northwest around 12:40 a.m.

Orion rises in the east around 3:40 a.m. and is in the eastern sky before dawn. Look for the bright star **Aldebaran** and the **Pleiades** near Orion. See the illustration on page 17.

Castor and Pollux, the twin stars in the constellation Gemini, rise in the northeast around 4:35 a.m. and are in the eastern sky before dawn.

2027

Monday, August 9

TONIGHT

sunset: **7:39 p.m.**
twilight begins: **8:38 p.m.**

IN THE MORNING

twilight begins: **5:04 a.m.**
sunrise: **6:03 a.m.**

MOON PHASE

moonrise: **2:09 p.m.**
moonset: **12:04 a.m.**

IN THE SKY

The **moon** is prominent in the sky early in the evening. Using a telescope or binoculars, take a look. You should be able to see rough mountains and craters along the boundary between light and dark, in astronomy called the terminator.

Just east of **Scorpius**, Sagittarius is at its highest in the south, around 25 degrees, at 9:50 p.m. It sets in the southwest around 1:50 a.m.

Rust-red and not very bright, **Mars** is in the western sky at dusk and sets in the west around 10:05 p.m.

The **summer triangle**'s three bright stars are **Vega, Altair,** and **Deneb. Vega** is almost directly overhead at 10:10 p.m. **Altair** is at its highest, around 65 degrees, at 11:25 p.m. Altair can be identified by its less bright neighboring stars, one on either side, together called the shaft of Aquila. **Deneb** is in the band of the Milky Way. See the illustration of the summer triangle on page 19.

Saturn rises in the east around 11:05 p.m. Unlike bright Jupiter and Venus and ruddy Mars, Saturn is harder to distinguish from the stars. Stars twinkle and shimmer, while the planets, including Saturn, show a steady flat light.

The **Great Bear (Big Dipper)** sets in the northwest around 11:45 p.m. See the illustration on page 10.

Bright star **Capella** rises in the northeast around 11:55 p.m. Read about Capella on page 14.

Scorpius sets in the southwest around 12:15 a.m. See the illustration on page 19.

The bright star **Arcturus** sets in the northwest around 12:35 a.m.

Orion rises in the east around 3:35 a.m. and is in the eastern sky before dawn. Look for the bright star **Aldebaran** and the **Pleiades** near Orion. See the illustration on page 17.

Castor and Pollux, the twin stars in the constellation Gemini, rise in the northeast around 4:35 a.m. and are in the eastern sky before dawn.

Cassiopeia is in the northeastern sky at dusk. Shaped like an *M,* is at its highest, around 65 degrees, around dawn. See the illustration on page 11.

2027

Tuesday, August 10

MOON PHASE

TONIGHT

sunset: **7:38 p.m.**
twilight begins: **8:37 p.m.**

IN THE MORNING

twilight begins: **5:05 a.m.**
sunrise: **6:04 a.m.**

moonrise: **3:10 p.m.**
moonset: **12:50 a.m.**

IN THE SKY

The waxing gibbous moon and **Antares** appear close together in the evening sky. Look south at about 25 degrees elevation around 9 p.m.

Just east of **Scorpius**, Sagittarius is at its highest in the south, around 25 degrees, at 9:50 p.m. It sets in the southwest around 1:45 a.m.

Rust-red and not very bright, **Mars** is in the western sky at dusk and sets in the west around 10 p.m.

The **summer triangle**'s three bright stars are **Vega, Altair,** and **Deneb. Vega** is almost directly overhead at 10:05 p.m. **Altair** is at its highest, around 65 degrees, at 11:20 p.m. Altair can be identified by its less bright neighboring stars, one on either side, together called the shaft of Aquila. **Deneb** is in the band of the Milky Way. See the illustration of the summer triangle on page 19.

Saturn rises in the east around 11 p.m. Unlike bright Jupiter and Venus and ruddy Mars, Saturn is harder to distinguish from the stars. Stars twinkle and shimmer, while the planets, including Saturn, show a steady flat light.

The **Great Bear (Big Dipper)** sets in the northwest around 11:40 p.m. See the illustration on page 10.

Bright star **Capella** rises in the northeast around 11:50 p.m. Read about Capella on page 14.

Scorpius sets in the southwest around 12:10 a.m. Tonight the moon is in **Scorpius**, making it easy to find the constellation. See the illustration on page 19.

The bright star **Arcturus** sets in the northwest around 12:30 a.m.

Orion rises in the east around 3:35 a.m. and is in the eastern sky before dawn. Look for the bright star **Aldebaran** and the **Pleiades** near Orion. See the illustration on page 17.

Castor and Pollux, the twin stars in the constellation Gemini, rise in the northeast around 4:30 a.m. and are in the eastern sky before dawn.

Cassiopeia is in the northeastern sky at dusk. Shaped like an *M,* is at its highest, around 65 degrees, around dawn. See the illustration on page 11.

2027

Wednesday, August 11

TONIGHT

sunset: **7:37 p.m.**
twilight begins: **8:36 p.m.**

IN THE MORNING

twilight begins: **5:06 a.m.**
sunrise: **6:04 a.m.**

MOON PHASE

moonrise: **4:05 p.m.**
moonset: **1:41 a.m.**

IN THE SKY

The **Perseids meteor shower** happens tonight. A strong shower with up to 60 meteors per hour, the best time to watch is after midnight. Tonight unfortunately moonlight washes out the shower.

Rust-red and not very bright, **Mars** is in the western sky at dusk and sets in the west around 10 p.m.

Just east of **Scorpius**, Sagittarius is at its highest in the south, around 25 degrees, at 9:45 p.m. It sets in the southwest around 1:45 a.m.

The **summer triangle**'s three bright stars are **Vega, Altair,** and **Deneb. Vega** is almost directly overhead at 10 p.m. **Altair** is at its highest, around 65 degrees, at 11:15 p.m. Altair can be identified by its less bright neighboring stars, one on either side, together called the shaft of Aquila. **Deneb** is in the band of the Milky Way. See the illustration of the summer triangle on page 19.

Saturn rises in the east around 10:55 p.m. Unlike bright Jupiter and Venus and ruddy Mars, Saturn is harder to distinguish from the stars. Stars twinkle and shimmer, while the planets, including Saturn, show a steady flat light.

The **Great Bear (Big Dipper)** sets in the northwest around 11:35 p.m. See the illustration on page 10.

Bright star **Capella** rises in the northeast around 11:45 p.m. Read about Capella on page 14.

Scorpius sets in the southwest around 12:10 a.m. See the illustration on page 19.

The bright star **Arcturus** sets in the northwest around 12:25 a.m.

Orion rises in the east around 3:30 a.m. and is in the eastern sky before dawn. Look for the bright star **Aldebaran** and the **Pleiades** near Orion. See the illustration on page 17.

Castor and Pollux, the twin stars in the constellation Gemini, rise in the northeast around 4:25 a.m. and are in the eastern sky before dawn.

Cassiopeia is in the northeastern sky at dusk. Shaped like an *M,* is at its highest, around 65 degrees, around dawn. See the illustration on page 11.

2027

Thursday, August 12

TONIGHT

sunset: **7:36 p.m.**
twilight begins: **8:34 p.m.**

IN THE MORNING

twilight begins: **5:07 a.m.**
sunrise: **6:05 a.m.**

MOON PHASE

moonrise: **4:55 p.m.**
moonset: **2:36 a.m.**

IN THE SKY

The **Perseids meteor shower** happens tonight. A strong shower with up to 60 meteors per hour, the best time to watch is after midnight. Tonight unfortunately moonlight washes out the shower.

Rust-red and not very bright, **Mars** is in the western sky at dusk and sets in the west around 10 p.m.

Cassiopeia is in the northeastern sky at dusk. Shaped like an *M,* is at its highest, around 65 degrees, around dawn. See the illustration on page 11.

Just east of **Scorpius**, Sagittarius is at its highest in the south, around 25 degrees, at 9:40 p.m. It sets in the southwest around 1:40 a.m.

The **summer triangle**'s three bright stars are **Vega, Altair,** and **Deneb. Vega** is almost directly overhead at 9:55 p.m. **Altair** is at its highest, around 65 degrees, at 11:10 p.m. Altair can be identified by its less bright neighboring stars, one on either side, together called the shaft of Aquila. **Deneb** is in the band of the Milky Way. See the illustration of the summer triangle on page 19.

Saturn rises in the east around 10:50 p.m. Unlike bright Jupiter and Venus and ruddy Mars, Saturn is harder to distinguish from the stars. Stars twinkle and shimmer, while the planets, including Saturn, show a steady flat light.

The **Great Bear (Big Dipper)** sets in the northwest around 11:30 p.m. See the illustration on page 10.

Bright star **Capella** rises in the northeast around 11:45 p.m. Read about Capella on page 14.

Scorpius sets in the southwest around 12:05 a.m. See the illustration on page 19.

The bright star **Arcturus** sets in the northwest around 12:25 a.m.

Orion rises in the east around 3:25 a.m. and is in the eastern sky before dawn. Look for the bright star **Aldebaran** and the **Pleiades** near Orion. See the illustration on page 17.

Castor and Pollux, the twin stars in the constellation Gemini, rise in the northeast around 4:20 a.m. and are in the eastern sky before dawn.

2027

Friday, August 13

TONIGHT

sunset: **7:34 p.m.**
twilight begins: **8:33 p.m.**

IN THE MORNING

twilight begins: **5:08 a.m.**
sunrise: **6:06 a.m.**

MOON PHASE

moonrise: **5:39 p.m.**
moonset: **3:34 a.m.**

IN THE SKY

The **Perseids meteor shower** happens tonight. A strong shower with up to 60 meteors per hour, the best time to watch is after midnight. Tonight unfortunately moonlight washes out the shower.

The **full moon** tonight acts like a spotlight. Look around at the desert landscape at how much illumination the moon provides.

Cassiopeia is in the northeastern sky at dusk. Shaped like an *M,* is at its highest, around 65 degrees, around dawn. See the illustration on page 11.

Just east of **Scorpius,** Sagittarius is at its highest in the south, around 25 degrees, at 9:35 p.m. It sets in the southwest around 1:35 a.m.

Rust-red and not very bright, **Mars** is in the western sky at dusk and sets in the west around 9:55 p.m.

The **summer triangle**'s three bright stars are **Vega, Altair,** and **Deneb. Vega** is almost directly overhead at 9:55 p.m. **Altair** is at its highest, around 65 degrees, at 11:05 p.m. Altair can be identified by its less bright neighboring stars, one on either side, together called the shaft of Aquila. **Deneb** is in the band of the Milky Way. See the illustration of the summer triangle on page 19.

Saturn rises in the east around 10:50 p.m. Unlike bright Jupiter and Venus and ruddy Mars, Saturn is harder to distinguish from the stars. Stars twinkle and shimmer, while the planets, including Saturn, show a steady flat light.

The **Great Bear (Big Dipper)** sets in the northwest around 11:25 p.m. See the illustration on page 10.

Bright star **Capella** rises in the northeast around 11:40 p.m. Read about Capella on page 14.

Scorpius sets in the southwest around midnight. See the illustration on page 19.

The bright star **Arcturus** sets in the northwest around 12:20 a.m.

Orion rises in the east around 3:20 a.m. and is in the eastern sky before dawn. Look for the bright star **Aldebaran** and the **Pleiades** near Orion. See the illustration on page 17.

Castor and Pollux, the twin stars in the constellation Gemini, rise in the northeast around 4:15 a.m. and are in the eastern sky before dawn.

Saturday, August 14

TONIGHT

sunset: **7:33 p.m.**
twilight begins: **8:32 p.m.**

IN THE MORNING

twilight begins: **5:09 a.m.**
sunrise: **6:07 a.m.**

MOON PHASE

moonrise: **6:16 p.m.**
moonset: **4:32 a.m.**

IN THE SKY

The **full moon** tonight acts like a spotlight. Look around at the desert landscape at how much illumination the moon provides.

Rust-red and not very bright, **Mars** is in the western sky at dusk and sets in the west around 9:55 p.m.

Cassiopeia is in the northeastern sky at dusk. Shaped like an *M*, is at its highest, around 65 degrees, around dawn. See the illustration on page 11.

Just east of **Scorpius**, Sagittarius is at its highest in the south, around 25 degrees, at 9:35 p.m. It sets in the southwest around 1:30 a.m.

The **summer triangle**'s three bright stars are **Vega, Altair,** and **Deneb. Vega** is almost directly overhead at 9:50 p.m. **Altair** is at its highest, around 65 degrees, at 11:05 p.m. Altair can be identified by its less bright neighboring stars, one on either side, together called the shaft of Aquila. **Deneb** is in the band of the Milky Way. See the illustration of the summer triangle on page 19.

Saturn rises in the east around 10:45 p.m. Unlike bright Jupiter and Venus and ruddy Mars, Saturn is harder to distinguish from the stars. Stars twinkle and shimmer, while the planets, including Saturn, show a steady flat light.

The **Great Bear (Big Dipper)** sets in the northwest around 11:25 p.m. See the illustration on page 10.

Bright star **Capella** rises in the northeast around 11:35 p.m. Read about Capella on page 14.

Scorpius sets in the southwest around 11:55 p.m. See the illustration on page 19.

The bright star **Arcturus** sets in the northwest around 12:15 a.m.

Orion rises in the east around 3:20 a.m. and is in the eastern sky before dawn. Look for the bright star **Aldebaran** and the **Pleiades** near Orion. See the illustration on page 17.

Castor and Pollux, the twin stars in the constellation Gemini, rise in the northeast around 4:15 a.m. and are in the eastern sky before dawn.

The brightest star in the sky, **Sirius** rises in the east around 4:55 a.m. and is in the eastern sky before dawn.

2027

Sunday, August 15

TONIGHT

sunset: **7:32 p.m.**
twilight begins: **8:30 p.m.**

MOON PHASE

IN THE MORNING

twilight begins: **5:10 a.m.**
sunrise: **6:07 a.m.**

moonrise: **6:48 p.m.**
moonset: **5:30 a.m.**

IN THE SKY

The **full moon** tonight acts like a spotlight. Look around at the desert landscape at how much illumination the moon provides.

Rust-red and not very bright, **Mars** is in the western sky at dusk and sets in the west around 9:50 p.m.

Just east of **Scorpius,** Sagittarius is at its highest in the south, around 25 degrees, around dusk and sets in the southwest around 1:30 a.m.

The **summer triangle**'s three bright stars are **Vega, Altair,** and **Deneb. Vega** is almost directly overhead at 9:45 p.m. **Altair** is at its highest, around 65 degrees, at 11 p.m. Altair can be identified by its less bright neighboring stars, one on either side, together called the shaft of Aquila. **Deneb** is in the band of the Milky Way. See the illustration of the summer triangle on page 19.

Saturn rises in the east around 10:40 p.m. Unlike bright Jupiter and Venus and ruddy Mars, Saturn is harder to distinguish from the stars. Stars twinkle and shimmer, while the planets, including Saturn, show a steady flat light.

The **Great Bear (Big Dipper)** sets in the northwest around 11:20 p.m. See the illustration on page 10.

Bright star **Capella** rises in the northeast around 11:30 p.m. Read about Capella on page 14.

Scorpius sets in the southwest around 11:55 p.m. See the illustration on page 19.

The bright star **Arcturus** sets in the northwest around 12:10 a.m.

Orion rises in the east around 3:15 a.m. and is in the eastern sky before dawn. Look for the bright star **Aldebaran** and the **Pleiades** near Orion. See the illustration on page 17.

Castor and Pollux, the twin stars in the constellation Gemini, rise in the northeast around 4:10 a.m. and are in the eastern sky before dawn.

The brightest star in the sky, **Sirius** rises in the east around 4:50 a.m. and is in the eastern sky before dawn.

Cassiopeia, shaped like an *M,* is at its highest, around 65 degrees, around dawn. See the illustration on page 11.

2027

Monday, August 16

TONIGHT

sunset: 7:31 p.m.
twilight begins: 8:29 p.m.

IN THE MORNING

twilight begins: 5:10 a.m.
sunrise: 6:08 a.m.

MOON PHASE

moonrise: 7:17 p.m.
moonset: 6:27 a.m.

IN THE SKY

Rust-red and not very bright, Mars is in the western sky at dusk and sets in the west around 9:50 p.m.

Just east of **Scorpius**, Sagittarius is at its highest in the south, around 25 degrees, around dusk and sets in the southwest around 1:25 a.m.

The **full moon** tonight acts like a spotlight. Look around at the desert landscape at how much illumination the moon provides.

The **summer triangle**'s three bright stars are **Vega, Altair,** and **Deneb. Vega** is almost directly overhead at 9:40 p.m. **Altair** is at its highest, around 65 degrees, at 10:55 p.m. Altair can be identified by its less bright neighboring stars, one on either side, together called the shaft of Aquila. **Deneb** is in the band of the Milky Way. See the illustration of the summer triangle on page 19.

A lunar eclipse tonight begins at 10:24 p.m. and is at its maximum at 12:13 a.m. The eclipse ends at 2:03 a.m. This penumbral or partial eclipse means the moon will darken but won't be completely in the earth's shadow.

10:20 p.m. 12:15 a.m. 2 a.m.

Saturn rises in the east around 10:35 p.m. Unlike bright Jupiter and Venus and ruddy Mars, Saturn is harder to distinguish from the stars. Stars twinkle and shimmer, while the planets, including Saturn, show a steady flat light.

The **Great Bear (Big Dipper)** sets in the northwest around 11:15 p.m. See the illustration on page 10.

Bright star **Capella** rises in the northeast around 11:25 p.m. Read about Capella on page 14.

Scorpius sets in the southwest around 11:50 p.m. See the illustration on page 19.

The bright star **Arcturus** sets in the northwest around 12:05 a.m.

Orion rises in the east around 3:10 a.m. and is in the eastern sky before dawn. Look for the bright star **Aldebaran** and the **Pleiades** near Orion. See the illustration on page 17.

Castor and Pollux, the twin stars in the constellation Gemini, rise in the northeast around 4:05 a.m. and are in the eastern sky before dawn.

The brightest star in the sky, **Sirius** rises in the east around 4:45 a.m. and is in the eastern sky before dawn.

Cassiopeia, shaped like an *M,* is at its highest, around 65 degrees, around dawn. See the illustration on page 11.

2027

Tuesday, August 17

TONIGHT

sunset: **7:30 p.m.**
twilight begins: **8:28 p.m.**

IN THE MORNING

twilight begins: **5:11 a.m.**
sunrise: **6:09 a.m.**

MOON PHASE

moonrise: **7:43 p.m.**
moonset: **7:23 a.m.**

IN THE SKY

Rust-red and not very bright, **Mars** is in the western sky at dusk and sets in the west around 9:45 p.m.

Just east of **Scorpius**, Sagittarius is at its highest in the south, around 25 degrees, around dusk and sets in the southwest around 1:20 a.m.

The **full moon** tonight acts like a spotlight. Look around at the desert landscape at how much illumination the moon provides.

The **summer triangle**'s three bright stars are **Vega, Altair,** and **Deneb. Vega** is almost directly overhead at 9:35 p.m. **Altair** is at its highest, around 65 degrees, at 10:50 p.m. Altair can be identified by its less bright neighboring stars, one on either side, together called the shaft of Aquila. **Deneb** is in the band of the Milky Way. See the illustration of the summer triangle on page 19.

Saturn rises in the east around 10:30 p.m. and is at its highest, around 65 degrees, around dawn. Unlike bright Jupiter and Venus and ruddy Mars, Saturn is harder to distinguish from the stars. Stars twinkle and shimmer, while the planets, including Saturn, show a steady flat light.

The **Great Bear (Big Dipper)** sets in the northwest around 11:10 p.m. See the illustration on page 10.

Bright star **Capella** rises in the northeast around 11:25 p.m. Read about Capella on page 14.

Scorpius sets in the southwest around 11:45 p.m. See the illustration on page 19.

The bright star **Arcturus** sets in the northwest around 12:05 a.m.

Orion rises in the east around 3:05 a.m. and is in the eastern sky before dawn. Look for the bright star **Aldebaran** and the **Pleiades** near Orion. See the illustration on page 17.

Castor and Pollux, the twin stars in the constellation Gemini, rise in the northeast around 4 a.m. and are in the eastern sky before dawn.

The brightest star in the sky, **Sirius** rises in the east around 4:45 a.m. and is in the eastern sky before dawn.

Cassiopeia, shaped like an *M,* is at its highest, around 65 degrees, around dawn. See the illustration on page 11.

2027

Wednesday, August 18

MOON PHASE

TONIGHT

sunset: **7:29 p.m.**
twilight begins: **8:27 p.m.**

IN THE MORNING

twilight begins: **5:12 a.m.**
sunrise: **6:10 a.m.**

moonrise: **8:08 p.m.**
moonset: **7:23 a.m.**

IN THE SKY

Rust-red and not very bright, **Mars** is in the western sky at dusk and sets in the west around 9:45 p.m.

Just east of **Scorpius**, Sagittarius is at its highest in the south, around 25 degrees, around dusk and sets in the southwest around 1:15 a.m.

The **full moon** tonight acts like a spotlight. Look around at the desert landscape at how much illumination the moon provides.

The **summer triangle**'s three bright stars are **Vega, Altair,** and **Deneb. Vega** is almost directly overhead at 9:35 p.m. **Altair** is at its highest, around 65 degrees, at 10:45 p.m. Altair can be identified by its less bright neighboring stars, one on either side, together called the shaft of Aquila. **Deneb** is in the band of the Milky Way. See the illustration of the summer triangle on page 19.

Saturn rises in the east around 10:30 p.m. and is at its highest, around 65 degrees, around dawn. Unlike bright Jupiter and Venus and ruddy Mars, Saturn is harder to distinguish from the stars. Stars twinkle and shimmer, while the planets, including Saturn, show a steady flat light.

The **Great Bear (Big Dipper)** sets in the northwest around 11:10 p.m. See the illustration on page 10.

Bright star **Capella** rises in the northeast around 11:20 p.m. Read about Capella on page 14.

Scorpius sets in the southwest around 11:40 p.m. See the illustration on page 19.

The bright star **Arcturus** sets in the northwest around midnight.

Orion rises in the east around 3 a.m. and is in the eastern sky before dawn. Look for the bright star **Aldebaran** and the **Pleiades** near Orion. See the illustration on page 17.

Castor and Pollux, the twin stars in the constellation Gemini, rise in the northeast around 3:55 a.m. and are in the eastern sky before dawn.

Cassiopeia, shaped like an *M,* is at its highest, around 65 degrees, at 4:10 a.m. See the illustration on page 11.

The brightest star in the sky, **Sirius** rises in the east around 4:40 a.m. and is in the eastern sky before dawn.

2027

Thursday, August 19

TONIGHT

sunset: **7:28 p.m.**
twilight begins: **8:25 p.m.**

MOON PHASE

IN THE MORNING

twilight begins: **5:13 a.m.**
sunrise: **6:10 a.m.**

moonrise: **8:33 p.m.**
moonset: **8:20 a.m.**

IN THE SKY

Rust-red and not very bright, **Mars** is in the western sky at dusk and sets in the west around 9:40 p.m.

Just east of **Scorpius**, Sagittarius is at its highest in the south, around 25 degrees, around dusk and sets in the southwest around 1:10 a.m.

The **full moon** tonight acts like a spotlight. Look around at the desert landscape at how much illumination the moon provides.

The **summer triangle**'s three bright stars are **Vega, Altair,** and **Deneb. Vega** is almost directly overhead at 9:30 p.m. **Altair** is at its highest, around 65 degrees, at 10:45 p.m. Altair can be identified by its less bright neighboring stars, one on either side, together called the shaft of Aquila. **Deneb** is in the band of the Milky Way. See the illustration of the summer triangle on page 19.

Saturn rises in the east around 10:25 p.m. and is at its highest, around 65 degrees, around dawn. Unlike bright Jupiter and Venus and ruddy Mars, Saturn is harder to distinguish from the stars. Stars twinkle and shimmer, while the planets, including Saturn, show a steady flat light.

The **Great Bear (Big Dipper)** sets in the northwest around 11:05 p.m. See the illustration on page 10.

Bright star **Capella** rises in the northeast around 11:15 p.m. Read about Capella on page 14.

Scorpius sets in the southwest around 11:35 p.m. See the illustration on page 19.

The bright star **Arcturus** sets in the northwest around 11:55 p.m.

Orion rises in the east around 3 a.m. and is in the eastern sky before dawn. Look for the bright star **Aldebaran** and the **Pleiades** near Orion. See the illustration on page 17.

Castor and Pollux, the twin stars in the constellation Gemini, rise in the northeast around 3:55 a.m. and are in the eastern sky before dawn.

Cassiopeia, shaped like an *M,* is at its highest, around 65 degrees, at 4:05 a.m. See the illustration on page 11.

The brightest star in the sky, **Sirius** rises in the east around 4:35 a.m. and is in the eastern sky before dawn.

2027

Friday, August 20

TONIGHT

sunset: **7:26 p.m.**
twilight begins: **8:24 p.m.**

IN THE MORNING

twilight begins: **5:14 a.m.**
sunrise: **6:11 a.m.**

MOON PHASE

moonrise: **8:58 p.m.**
moonset: **9:17 a.m.**

IN THE SKY

Rust-red and not very bright, Mars is in the western sky at dusk and sets in the west around 9:40 p.m.

Just east of **Scorpius**, Sagittarius is at its highest in the south, around 25 degrees, around dusk and sets in the southwest around 1:10 a.m.

The **summer triangle**'s three bright stars are **Vega, Altair,** and **Deneb. Vega** is almost directly overhead at 9:25 p.m. **Altair** is at its highest, around 65 degrees, at 10:40 p.m. Altair can be identified by its less bright neighboring stars, one on either side, together called the shaft of Aquila. **Deneb** is in the band of the Milky Way. See the illustration of the summer triangle on page 19.

Saturn rises in the east around 10:20 p.m. and is at its highest, around 65 degrees, around dawn. Unlike bright Jupiter and Venus and ruddy Mars, Saturn is harder to distinguish from the stars. Stars twinkle and shimmer, while the planets, including Saturn, show a steady flat light.

The **Great Bear (Big Dipper)** sets in the northwest around 11 p.m. See the illustration on page 10.

Bright star **Capella** rises in the northeast around 11:10 p.m. Read about Capella on page 14.

Scorpius sets in the southwest around 11:35 p.m. See the illustration on page 19.

The bright star **Arcturus** sets in the northwest around 11:50 p.m.

Orion rises in the east around 2:55 a.m. and is in the eastern sky before dawn. Look for the bright star **Aldebaran** and the **Pleiades** near Orion. See the illustration on page 17.

Castor and Pollux, the twin stars in the constellation Gemini, rise in the northeast around 3:50 a.m. and are in the eastern sky before dawn.

Cassiopeia, shaped like an *M,* is at its highest, around 65 degrees, at 4 a.m. See the illustration on page 11.

The brightest star in the sky, **Sirius** rises in the east around 4:30 a.m. and is in the eastern sky before dawn.

2027

Saturday, August 21

TONIGHT

sunset: **7:25 p.m.**
twilight begins: **8:22 p.m.**

IN THE MORNING

twilight begins: **5:15 a.m.**
sunrise: **6:12 a.m.**

MOON PHASE

moonrise: **9:26 p.m.**
moonset: **10:15 a.m.**

IN THE SKY

Explore the **dark sky** after twilight and before the moon rises, around 9:05 p.m.

Rust-red and not very bright, **Mars** is in the western sky at dusk and sets in the west around 9:35 p.m.

Saturn rises in the east around 10:15 p.m. and is at its highest, around 65 degrees, around dawn. Unlike bright Jupiter and Venus and ruddy Mars, Saturn is harder to distinguish from the stars. Stars twinkle and shimmer, while the planets, including Saturn, show a steady flat light.

The **summer triangle**'s three bright stars are **Vega, Altair,** and **Deneb. Vega** is almost directly overhead around dusk. **Altair** is at its highest, around 65 degrees, at 10:35 p.m. Altair can be identified by its less bright neighboring stars, one on either side, together called the shaft of Aquila. **Deneb** is in the band of the Milky Way. See the illustration of the summer triangle on page 19.

The **Great Bear (Big Dipper)** sets in the northwest around 10:55 p.m. See the illustration on page 10.

Bright star **Capella** rises in the northeast around 11:05 p.m. Read about Capella on page 14.

Scorpius sets in the southwest around 11:30 p.m. See the illustration on page 19.

The bright star **Arcturus** sets in the northwest around 11:45 p.m.

Just east of **Scorpius**, Sagittarius is at its highest in the south, around 25 degrees, around dusk and sets in the southwest around 1:05 a.m.

Orion rises in the east around 2:50 a.m. and is in the eastern sky before dawn. Look for the bright star **Aldebaran** and the **Pleiades** near Orion. See the illustration on page 17.

Castor and Pollux, the twin stars in the constellation Gemini, rise in the northeast around 3:45 a.m. and are in the eastern sky before dawn.

Cassiopeia, shaped like an *M,* is at its highest, around 65 degrees, at 3:55 a.m. See the illustration on page 11.

The brightest star in the sky, **Sirius** rises in the east around 4:25 a.m. and is in the eastern sky before dawn.

2027

Sunday, August 22

TONIGHT

sunset: **7:24 p.m.**

twilight begins: **8:21 p.m.**

IN THE MORNING

twilight begins: **5:16 a.m.**

sunrise: **6:12 a.m.**

MOON PHASE

moonrise: **9:58 p.m.**

moonset: **11:16 a.m.**

IN THE SKY

Rust-red and not very bright, **Mars** is in the western sky at dusk and sets in the west around 9:35 p.m.

Just east of **Scorpius**, Sagittarius is at its highest in the south, around 25 degrees, around dusk and sets in the southwest around 1 a.m.

The **summer triangle**'s three bright stars are **Vega, Altair,** and **Deneb. Vega** is almost directly overhead around dusk. **Altair** is at its highest, around 65 degrees, at 10:30 p.m. Altair can be identified by its less bright neighboring stars, one on either side, together called the shaft of Aquila. **Deneb** is in the band of the Milky Way. See the illustration of the summer triangle on page 19.

Saturn rises in the east around 10:10 p.m. and is at its highest, around 65 degrees, around dawn. Unlike bright Jupiter and Venus and ruddy Mars, Saturn is harder to distinguish from the stars. Stars twinkle and shimmer, while the planets, including Saturn, show a steady flat light.

Bright star **Capella** rises in the northeast around 11:05 p.m. Read about Capella on page 14.

The **Great Bear (Big Dipper)** sets in the northwest around 10:50 p.m. See the illustration on page 10.

Scorpius sets in the southwest around 11:25 p.m. See the illustration on page 19.

The bright star **Arcturus** sets in the northwest around 11:45 p.m.

Orion rises in the east around 2:45 a.m. and is in the eastern sky before dawn. Look for the bright star **Aldebaran** and the **Pleiades** near Orion. See the illustration on page 17.

Castor and Pollux, the twin stars in the constellation Gemini, rise in the northeast around 3:40 a.m. and are in the eastern sky before dawn.

Cassiopeia, shaped like an *M,* is at its highest, around 65 degrees, at 3:50 a.m. See the illustration on page 11.

The brightest star in the sky, **Sirius** rises in the east around 4:25 a.m. and is in the eastern sky before dawn.

2027

Monday, August 23

TONIGHT

sunset: **7:23 p.m.**
twilight begins: **8:20 p.m.**

MOON PHASE

IN THE MORNING

twilight begins: **5:17 a.m.**
sunrise: **6:13 a.m.**

moonrise: **10:35 p.m.**
moonset: **12:20 p.m.**

IN THE SKY

Rust-red and not very bright, **Mars** is in the western sky at dusk and sets in the west around 9:35 p.m.

Just east of **Scorpius**, Sagittarius is at its highest in the south, around 25 degrees, around dusk and sets in the southwest around 12:55 a.m.

Before the moon rises, until around 10:15 p.m., look for the **Andromeda Galaxy** with binoculars or a telescope. It becomes a good target around 9:25 p.m. See the guide on page 11.

The **summer triangle**'s three bright stars are **Vega, Altair,** and **Deneb. Vega** is almost directly overhead around dusk. **Altair** is at its highest, around 65 degrees, at 10:25 p.m. Altair can be identified by its less bright neighboring stars, one on either side, together called the shaft of Aquila. **Deneb** is in the band of the Milky Way. See the illustration of the summer triangle on page 19.

Saturn rises in the east around 10:10 p.m. and is at its highest, around 65 degrees, around dawn. Unlike bright Jupiter and Venus and ruddy Mars, Saturn is harder to distinguish from the stars. Stars twinkle and shimmer, while the planets, including Saturn, show a steady flat light.

Explore the **dark sky** after twilight and before the moon rises, around 10:15 p.m.

The **Great Bear (Big Dipper)** sets in the northwest around 10:50 p.m. See the illustration on page 10.

Bright star **Capella** rises in the northeast around 11 p.m. Read about Capella on page 14.

Scorpius sets in the southwest around 11:20 p.m. See the illustration on page 19.

The bright star **Arcturus** sets in the northwest around 11:40 p.m.

Orion rises in the east around 2:40 a.m. and is in the eastern sky before dawn. Look for the bright star **Aldebaran** and the **Pleiades** near Orion. See the illustration on page 17.

Castor and Pollux, the twin stars in the constellation Gemini, rise in the northeast around 3:40 a.m. and are in the eastern sky before dawn.

Cassiopeia, shaped like an *M,* is at its highest, around 65 degrees, at 3:50 a.m. See the illustration on page 11.

The brightest star in the sky, **Sirius** rises in the east around 4:20 a.m. and is in the eastern sky before dawn.

2027

Tuesday, August 24

TONIGHT

sunset: **7:21 p.m.**
twilight begins: **8:18 p.m.**

IN THE MORNING

twilight begins: **5:17 a.m.**
sunrise: **6:14 a.m.**

MOON PHASE

moonrise: **11:20 p.m.**
moonset: **1:25 p.m.**

IN THE SKY

Rust-red and not very bright, Mars is in the western sky at dusk and sets in the west around 9:30 p.m.

Just east of **Scorpius**, Sagittarius is at its highest in the south, around 25 degrees, around dusk and sets in the southwest around 12:50 a.m.

Before the moon rises, until around 11 p.m., look for the **Andromeda Galaxy** with binoculars or a telescope. It becomes a good target around 9:20 p.m. See the guide on page 11.

Saturn rises in the east around 10:05 p.m. and is at its highest, around 65 degrees, around dawn. Unlike bright Jupiter and Venus and ruddy Mars, Saturn is harder to distinguish from the stars. Stars twinkle and shimmer, while the planets, including Saturn, show a steady flat light.

The **summer triangle**'s three bright stars are **Vega, Altair,** and **Deneb. Vega** is almost directly overhead around dusk. **Altair** is at its highest, around 65 degrees, at 10:25 p.m. and sets in the west around 4:45 a.m. Altair can be identified by its less bright neighboring stars, one on either side, together called the shaft of Aquila. **Deneb** is in the band of the Milky Way. See the illustration of the summer triangle on page 19.

The **Great Bear (Big Dipper)** sets in the northwest around 10:45 p.m. See the illustration on page 10.

Bright star **Capella** rises in the northeast around 10:55 p.m. Read about Capella on page 14.

Explore the **dark sky** after twilight and before the moon rises, around 11 p.m.

Scorpius sets in the southwest around 11:15 p.m. See the illustration on page 19.

The bright star **Arcturus** sets in the northwest around 11:35 p.m.

Orion rises in the east around 2:40 a.m. and is in the eastern sky before dawn. Look for the bright star **Aldebaran** and the **Pleiades** near Orion. See the illustration on page 17.

Castor and Pollux, the twin stars in the constellation Gemini, rise in the northeast around 3:35 a.m. and are in the eastern sky before dawn.

Cassiopeia, shaped like an *M,* is at its highest, around 65 degrees, at 3:45 a.m. See the illustration on page 11.

The brightest star in the sky, **Sirius** rises in the east around 4:15 a.m. and is in the eastern sky before dawn.

2027

Wednesday, August 25

TONIGHT

sunset: **7:20 p.m.**
twilight begins: **8:17 p.m.**

MOON PHASE

IN THE MORNING

twilight begins: **5:18 a.m.**
sunrise: **6:15 a.m.**

moonrise: **12:15 a.m.**
moonset: **2:31 p.m.**

IN THE SKY

Rust-red and not very bright, **Mars** is in the western sky at dusk and sets in the west around 9:30 p.m.

Just east of **Scorpius**, Sagittarius is at its highest in the south, around 25 degrees, around dusk and sets in the southwest around 12:50 a.m.

Before the moon rises, around midnight, look for the **Andromeda Galaxy** with binoculars or a telescope. It becomes a good target around 9:15 p.m. See the guide on page 11.

Saturn rises in the east around 10 p.m. and is at its highest, around 65 degrees, at 4:20 a.m. Unlike bright Jupiter and Venus and ruddy Mars, Saturn is harder to distinguish from the stars. Stars twinkle and shimmer, while the planets, including Saturn, show a steady flat light.

The **summer triangle**'s three bright stars are **Vega, Altair,** and **Deneb. Vega** is almost directly overhead around dusk. **Altair** is at its highest, around 65 degrees, at 10:20 p.m. and sets in the west around 4:40 a.m. Altair can be identified by its less bright neighboring stars, one on either side, together called the shaft of Aquila. **Deneb** is in the band of the Milky Way. See the illustration of the summer triangle on page 19.

The **Great Bear (Big Dipper)** sets in the northwest around 10:40 p.m. See the illustration on page 10.

Bright star **Capella** rises in the northeast around 10:50 p.m. Read about Capella on page 14.

Scorpius sets in the southwest around 11:15 p.m. See the illustration on page 19.

The bright star **Arcturus** sets in the northwest around 11:30 p.m.

Orion rises in the east around 2:35 a.m. and is in the eastern sky before dawn. Look for the bright star **Aldebaran** and the **Pleiades** near Orion. See the illustration on page 17.

Castor and Pollux, the twin stars in the constellation Gemini, rise in the northeast around 3:30 a.m. and are in the eastern sky before dawn.

Cassiopeia, shaped like an *M,* is at its highest, around 65 degrees, at 3:40 a.m. See the illustration on page 11.

The brightest star in the sky, **Sirius** rises in the east around 4:10 a.m. and is in the eastern sky before dawn.

2027
Thursday, August 26

TONIGHT

sunset: **7:19 p.m.**
twilight begins: **8:15 p.m.**

IN THE MORNING

twilight begins: **5:19 a.m.**
sunrise: **6:15 a.m.**

MOON PHASE

moonrise: **1:18 a.m.**
moonset: **3:33 p.m.**

IN THE SKY

Rust-red and not very bright, **Mars** is in the western sky at dusk and sets in the west around 9:25 p.m.

Just east of **Scorpius**, Sagittarius is at its highest in the south, around 25 degrees, around dusk and sets in the southwest around 12:45 a.m.

The **summer triangle**'s three bright stars are **Vega, Altair,** and **Deneb. Vega** is almost directly overhead around dusk. **Altair** is at its highest, around 65 degrees, at 10:15 p.m. and sets in the west around 4:35 a.m. Altair can be identified by its less bright neighboring stars, one on either side, together called the shaft of Aquila. **Deneb** is in the band of the Milky Way. See the illustration of the summer triangle on page 19.

Before the moon rises, around 1:10 a.m., look for the **Andromeda Galaxy** with binoculars or a telescope. It becomes a good target around 9:10 p.m. See the guide on page 11.

Saturn rises in the east around 9:55 p.m. and is at its highest, around 65 degrees, at 4:15 a.m. Unlike bright Jupiter and Venus and ruddy Mars, Saturn is harder to distinguish from the stars. Stars twinkle and shimmer, while the planets, including Saturn, show a steady flat light.

The **Great Bear (Big Dipper)** sets in the northwest around 10:35 p.m. See the illustration on page 10.

Bright star **Capella** rises in the northeast around 10:50 p.m. Read about Capella on page 14.

Scorpius sets in the southwest around 11:10 p.m. See the illustration on page 19.

The bright star **Arcturus** sets in the northwest around 11:30 p.m.

Orion rises in the east around 2:30 a.m. and is in the eastern sky before dawn. Look for the bright star **Aldebaran** and the **Pleiades** near Orion. See the illustration on page 17.

Castor and Pollux, the twin stars in the constellation Gemini, rise in the northeast around 3:25 a.m. and are in the eastern sky before dawn.

Cassiopeia, shaped like an *M,* is at its highest, around 65 degrees, at 3:35 a.m. See the illustration on page 11.

The brightest star in the sky, **Sirius** rises in the east around 4:05 a.m. and is in the eastern sky before dawn.

2027

Friday, August 27

TONIGHT

sunset: **7:18 p.m.**
twilight begins: **8:14 p.m.**

MOON PHASE

IN THE MORNING

twilight begins: **5:20 a.m.**
sunrise: **6:16 a.m.**

moonrise: **2:29 a.m.**
moonset: **4:30 p.m.**

IN THE SKY

Rust-red and not very bright, Mars is in the western sky at dusk and sets in the west around 9:25 p.m.

Just east of **Scorpius**, Sagittarius is at its highest in the south, around 25 degrees, around dusk and sets in the southwest around 12:40 a.m.

The **summer triangle**'s three bright stars are **Vega, Altair,** and **Deneb. Vega** is almost directly overhead around dusk. **Altair** is at its highest, around 65 degrees, at 10:10 p.m. and sets in the west around 4:30 a.m. Altair can be identified by its less bright neighboring stars, one on either side, together called the shaft of Aquila. **Deneb** is in the band of the Milky Way. See the illustration of the summer triangle on page 19.

The absence of the moon tonight provides excellent **dark-sky** viewing conditions.

Before the moon rises, around 2:20 a.m., look for the **Andromeda Galaxy** with binoculars or a telescope. It becomes a good target around 9:10 p.m. See the guide on page 11.

Saturn rises in the east around 9:50 p.m. and is at its highest, around 65 degrees, at 4:10 a.m. Unlike bright Jupiter and Venus and ruddy Mars, Saturn is harder to distinguish from the stars. Stars twinkle and shimmer, while the planets, including Saturn, show a steady flat light.

The **Great Bear (Big Dipper)** sets in the northwest around 10:30 p.m. See the illustration on page 10.

Bright star **Capella** rises in the northeast around 10:45 p.m. Read about Capella on page 14.

Scorpius sets in the southwest around 11:05 p.m. See the illustration on page 19.

The bright star **Arcturus** sets in the northwest around 11:25 p.m.

Orion rises in the east around 2:25 a.m. and is in the eastern sky before dawn. Look for the bright star **Aldebaran** and the **Pleiades** near Orion. See the illustration on page 17.

Castor and Pollux, the twin stars in the constellation Gemini, rise in the northeast around 3:20 a.m. and are in the eastern sky before dawn.

Cassiopeia, shaped like an *M,* is at its highest, around 65 degrees, at 3:35 a.m. See the illustration on page 11.

The brightest star in the sky, **Sirius** rises in the east around 4:05 a.m. and is in the eastern sky before dawn.

2027

Saturday, August 28

TONIGHT

sunset: **7:16 p.m.**
twilight begins: **8:13 p.m.**

IN THE MORNING

twilight begins: **5:21 a.m.**
sunrise: **6:17 a.m.**

MOON PHASE

moonrise: **3:44 a.m.**
moonset: **5:19 p.m.**

IN THE SKY

Rust-red and not very bright, **Mars** is in the western sky at dusk and sets in the west around 9:20 p.m.

Just east of **Scorpius**, Sagittarius is at its highest in the south, around 25 degrees, around dusk and sets in the southwest around 12:35 a.m.

The **summer triangle**'s three bright stars are **Vega, Altair,** and **Deneb. Vega** is almost directly overhead around dusk. **Altair** is at its highest, around 65 degrees, at 10:10 p.m. and sets in the west around 4:30 a.m. Altair can be identified by its less bright neighboring stars, one on either side, together called the shaft of Aquila. **Deneb** is in the band of the Milky Way. See the illustration of the summer triangle on page 19.

The absence of the moon tonight provides excellent **dark-sky** viewing conditions.

Before the moon rises, around 3:30 a.m., look for the **Andromeda Galaxy** with binoculars or a telescope. It becomes a good target around 9:05 p.m. See the guide on page 11.

Saturn rises in the east around 9:50 p.m. and is at its highest, around 65 degrees, at 4:05 a.m. Unlike bright Jupiter and Venus and ruddy Mars, Saturn is harder to distinguish from the stars. Stars twinkle and shimmer, while the planets, including Saturn, show a steady flat light.

The **Great Bear (Big Dipper)** sets in the northwest around 10:30 p.m. See the illustration on page 10.

Bright star **Capella** rises in the northeast around 10:40 p.m. Read about Capella on page 14.

Scorpius sets in the southwest around 11 p.m. See the illustration on page 19.

The bright star **Arcturus** sets in the northwest around 11:20 p.m.

Orion rises in the east around 2:25 a.m. and is in the eastern sky before dawn. Look for the bright star **Aldebaran** and the **Pleiades** near Orion. See the illustration on page 17.

Castor and Pollux, the twin stars in the constellation Gemini, rise in the northeast around 3:20 a.m. and are in the eastern sky before dawn.

Cassiopeia, shaped like an *M,* is at its highest, around 65 degrees, at 3:30 a.m. See the illustration on page 11.

The brightest star in the sky, **Sirius** rises in the east around 4 a.m. and is in the eastern sky before dawn.

2027

Sunday, August 29

TONIGHT

sunset: **7:15 p.m.**
twilight begins: **8:11 p.m.**

MOON PHASE

IN THE MORNING

twilight begins: **5:22 a.m.**
sunrise: **6:17 a.m.**

moonrise: **4:59 a.m.**
moonset: **6:01 p.m.**

IN THE SKY

Rust-red and not very bright, **Mars** is in the western sky at dusk and sets in the west around 9:20 p.m.

Just east of **Scorpius,** Sagittarius is at its highest in the south, around 25 degrees, around dusk and sets in the southwest around 12:35 a.m.

The **summer triangle**'s three bright stars are **Vega, Altair,** and **Deneb. Vega** is almost directly overhead around dusk. **Altair** is at its highest, around 65 degrees, at 10:05 p.m. and sets in the west around 4:25 a.m. Altair can be identified by its less bright neighboring stars, one on either side, together called the shaft of Aquila. **Deneb** is in the band of the Milky Way. See the illustration of the summer triangle on page 19.

The absence of the moon tonight provides excellent **dark-sky** viewing conditions.

With the dark sky, look for the **Andromeda Galaxy** with binoculars or a telescope. It becomes a good target around 9 p.m. See the guide on page 11.

Saturn rises in the east around 9:45 p.m. and is at its highest, around 65 degrees, at 4 a.m. Unlike bright Jupiter and Venus and ruddy Mars, Saturn is harder to distinguish from the stars. Stars twinkle and shimmer, while the planets, including Saturn, show a steady flat light.

The **Great Bear (Big Dipper)** sets in the northwest around 10:25 p.m. See the illustration on page 10.

Bright star **Capella** rises in the northeast around 10:35 p.m. Read about Capella on page 14.

Scorpius sets in the southwest around 11 p.m. See the illustration on page 19.

The bright star **Arcturus** sets in the northwest around 11:15 p.m.

Orion rises in the east around 2:20 a.m. and is in the eastern sky before dawn. Look for the bright star **Aldebaran** and the **Pleiades** near Orion. See the illustration on page 17.

Castor and Pollux, the twin stars in the constellation Gemini, rise in the northeast around 3:15 a.m. and are in the eastern sky before dawn.

Cassiopeia, shaped like an *M,* is at its highest, around 65 degrees, at 3:25 a.m. See the illustration on page 11.

The brightest star in the sky, **Sirius** rises in the east around 3:55 a.m. and is in the eastern sky before dawn.

2027
Monday, August 30

TONIGHT

sunset: **7:14 p.m.**
twilight begins: **8:10 p.m.**

IN THE MORNING

twilight begins: **5:22 a.m.**
sunrise: **6:18 a.m.**

MOON PHASE

moonrise: **6:13 a.m.**
moonset: **6:37 p.m.**

IN THE SKY

Rust-red and not very bright, **Mars** is in the western sky at dusk and sets in the west around 9:15 p.m.

The **summer triangle**'s three bright stars are **Vega, Altair,** and **Deneb. Vega** is almost directly overhead around dusk. **Altair** is at its highest, around 65 degrees, at 10 p.m. and sets in the west around 4:20 a.m. Altair can be identified by its less bright neighboring stars, one on either side, together called the shaft of Aquila. **Deneb** is in the band of the Milky Way. See the illustration of the summer triangle on page 19.

The absence of the moon tonight provides excellent **dark-sky** viewing conditions.

Look for the **Andromeda Galaxy** with binoculars or a telescope. It becomes a good target around 8:55 p.m. See the guide on page 11.

Saturn rises in the east around 9:40 p.m. and is at its highest, around 65 degrees, at 4 a.m. Unlike bright Jupiter and Venus and ruddy Mars, Saturn is harder to distinguish from the stars. Stars twinkle and shimmer, while the planets, including Saturn, show a steady flat light.

The **Great Bear (Big Dipper)** sets in the northwest around 10:20 p.m. See the illustration on page 10.

Bright star **Capella** rises in the northeast around 10:30 p.m. Read about Capella on page 14.

Scorpius sets in the southwest around 10:55 p.m. See the illustration on page 19.

The bright star **Arcturus** sets in the northwest around 11:10 p.m.

Just east of **Scorpius,** Sagittarius sets in the southwest around 12:30 a.m.

Orion rises in the east around 2:15 a.m. and is in the eastern sky before dawn. Look for the bright star **Aldebaran** and the **Pleiades** near Orion. See the illustration on page 17.

Castor and Pollux, the twin stars in the constellation Gemini, rise in the northeast around 3:10 a.m. and are in the eastern sky before dawn.

Cassiopeia, shaped like an *M,* is at its highest, around 65 degrees, at 3:20 a.m. See the illustration on page 11.

The brightest star in the sky, **Sirius** rises in the east around 3:50 a.m. and is in the eastern sky before dawn.

2027

Tuesday, August 31

TONIGHT

sunset: **7:12 p.m.**
twilight begins: **8:08 p.m.**

IN THE MORNING

twilight begins: **5:23 a.m.**
sunrise: **6:19 a.m.**

MOON PHASE

moonrise: **7:24 a.m.**
moonset: **7:10 p.m.**

IN THE SKY

Rust-red and not very bright, Mars is in the western sky at dusk and sets in the west around 9:15 p.m.

Saturn rises in the east around 9:35 p.m. and is at its highest, around 65 degrees, at 3:55 a.m. Unlike bright Jupiter and Venus and ruddy Mars, Saturn is harder to distinguish from the stars. Stars twinkle and shimmer, while the planets, including Saturn, show a steady flat light.

The **summer triangle**'s three bright stars are **Vega, Altair,** and **Deneb. Vega** is almost directly overhead around dusk and sets in the northwest around 4:50 a.m. **Altair** is at its highest, around 65 degrees, at 9:55 p.m. and sets in the west around 4:15 a.m. Altair can be identified by its less bright neighboring stars, one on either side, together called the shaft of Aquila. **Deneb** is in the band of the Milky Way. See the illustration of the summer triangle on page 19.

The absence of the moon tonight provides excellent **dark-sky** viewing conditions.

Look for the **Andromeda Galaxy** with binoculars or a telescope. It becomes a good target around 8:55 p.m. See the guide on page 11.

The **Great Bear (Big Dipper)** sets in the northwest around 10:15 p.m. See the illustration on page 10.

Bright star **Capella** rises in the northeast around 10:30 p.m. Read about Capella on page 14.

Scorpius sets in the southwest around 10:50 p.m. See the illustration on page 19.

The bright star **Arcturus** sets in the northwest around 11:10 p.m.

Just east of **Scorpius**, Sagittarius sets in the southwest around 12:25 a.m.

Orion rises in the east around 2:10 a.m. and is in the eastern sky before dawn. Look for the bright star **Aldebaran** and the **Pleiades** near Orion. See the illustration on page 17.

Castor and Pollux, the twin stars in the constellation Gemini, rise in the northeast around 3:05 a.m. and are in the eastern sky before dawn.

Cassiopeia, shaped like an *M,* is at its highest, around 65 degrees, at 3:15 a.m. See the illustration on page 11.

The brightest star in the sky, **Sirius** rises in the east around 3:50 a.m. and is in the eastern sky before dawn.

Pronunciation Guide

Standard American pronunciation of the names of stars, constellations, moons, and meteor showers described in this book:

Name	IPA	Sounds like
Adhafera	ædəˈfɪərə	add a FEAR uh
Ain	ˈeɪn	eye-n
Albireo	ælˈbɪrioʊ	al BEER ee oh
Aldebaran	ælˈdɛbərən	al DEB uh ren
Algieba	ælˈdʒiːbə	al JEE buh
Algol	ˈælgɒl	AL gall
Almach	ˈælmæk	AL mack
Alnilam	ælˈnaɪlæm	al NYE lam
Alnitak	ælˈnaɪtæk	al NYE tack
Alpheratz	ælˈfɪəræts	al FEE rats
Altair	ˈæltaɪər	al TIE uhr
Andromeda	ænˈdrɒmɪdə	an DROM ih duh
Antares	ænˈtɛəriːz	an TARE eez
Aquarids	ˈakwɑːɹˌɪdz	a KWER ids

Arcturus	ɑːrk'tʊərəs	ark TOUR es
Bellatrix	'bɛlətrɪks	BELL uh tricks
Betelgeuse	'bɛtəldʒuːz	BAY tuhl jooz
Callisto	kə'lɪstoʊ	kuh LIST oh
Camelopardalis	kəˌmɛlə'pɑːrdəlɪs	keh MEL uh PAR deh lis
Capella	kə'pɛlə	keh PELL uh
Cassiopeia	ˌkæsioʊ'piːə	KAH see uh PEE uh
Castor	'kæstər	CASS ter
Cepheus	'siːfiəs	SEE fee us
Chertan	k'ɜːtɔːn	CURR tawn
Cursa	'kɜːrsə	CURR suh
Cygnus	'sɪgnəs	SIG nus
Deneb	'dɛnɛb	DEN ebb
Denebola	də'nɛbələ	duh NEBB uh luh
Draco	'dreɪkoʊ	DRAY co
Draconids	drə'koʊnɪdz	druh CO nids
Elnath	ɛl'næθ	el NATH
Europa	jʊ'roʊpə	yer ROPE uh
Ganymede	'gænɪmiːd	GAN ih meed
Geminids	'dʒɛmənədz	JEM un nids
Hyades	'haɪ.ədiːz	HI uh deez
Io	'aɪ.oʊ	EYE oh
Leonids	'liːənɪdz	LEE uh nids
Lyra	'laɪrə	LIE ruh

Lyrae	ˈlaɪriː	LIE ree
Lyrids	lˈɹɪdz	LEER ids
Meissa	ˈmaɪsə	MY suh
Mintaka	ˈmɪntəkə	MIN tuh kuh
Mirach	ˈmaɪræk	MY rack
Mirfak	ˈmɪərfæk	MERE fack
Orionids	ɒɹˈaɪənɪdz	oh RYAN ids
Perseids	ˈpɜːrsi.ədz	PURR see ids
Perseus	ˈpɜːrsiəs,	PURR see us
Pleiades	ˈpliː.ədiːz	PLEE uh deez
Polaris	pəˈlɛərɪs	puh LAIR iss
Pollux	ˈpɒləks	PAUL uks
Procyon	ˈproʊsi.ɒn	PRO see on
Quadrantids	kwɒˈdræntədz	kwa DRAN tids
Rasalas	ˈræsəlæs	RASS uh lass
Regulus	ˈrɛɡˌjʊlʊs	REGG you luss
Rigel	ˈraɪdʒəl	RYE jell
Sadr	ˈsædər	SAY dur
Saiph	ˈseɪf	safe
Scorpius	ˈskɔːrpiəs	SCORE pee us
Shaula	ˈʃɔːlə	SHAW luh
Sirius	ˈsɪriəs	SEAR ee us
Spica	ˈspaɪkə	SPIKE uh
Taurids	tˈɔːɹɪdz	TORE ids

Ursa	ˈɜːrsə	UHR suh
Ursids	ˈɜːsɪdz	UHR sids
Vega	ˈveɪgə	VAY guh
Zosma	ˈzɒzmə	ZOSS muh

www.ingramcontent.com/pod-product-compliance
Lightning Source LLC
LaVergne TN
LVHW080013210926
848481LV00011B/3016